THEORY
OF
VIBRATION
WITH
APPLICATIONS

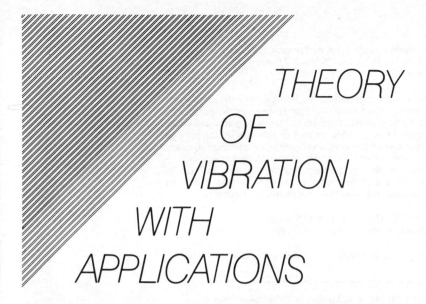

THEORY OF VIBRATION WITH APPLICATIONS

SECOND EDITION

WILLIAM T. THOMSON, *Professor Emeritus*

Department of Mechanical and Environmental Engineering
University of California
Santa Barbara, California

London
GEORGE ALLEN & UNWIN
Sydney

First published in the USA by Prentice-Hall, Inc., 1981
First published by George Allen & Unwin 1983
Reprinted 1984

GEORGE ALLEN & UNWIN LTD
40 Museum Street, London WC1A 1LU

© 1981 by Prentice-Hall, Inc., Englewood Cliffs, NJ 07632

British Library Cataloguing in Publication Data

Thomson, William Tyrell
  Theory of vibration with applications. – 2nd ed.
  1. Vibration
  I. Title
  531'.32   QC136

  ISBN 0–04–620012–6

Printed in Great Britain by
Butler & Tanner Ltd, Frome and London

CONTENTS

PREFACE

1 OSCILLATORY MOTION   1

  1.1   Harmonic Motion   *2*
  1.2   Periodic Motion   *5*
  1.3   Vibration Terminology   *8*

2 FREE VIBRATION   13

  2.1   Equation of Motion—Natural
        Frequency   *13*
  2.2   Energy Method   *18*
  2.3   Viscously Damped Free Vibration   *25*
  2.4   Logarithmic Decrement   *30*
  2.5   Coulomb Damping   *34*

3 HARMONICALLY EXCITED VIBRATION   48

  3.1   Forced Harmonic Vibration   *48*
  3.2   Rotating Unbalance   *52*

3.3   Balancing of Rotors   *55*
3.4   Whirling of Rotating Shafts   *58*
3.5   Support Motion   *62*
3.6   Vibration Isolation   *64*
3.7   Energy Dissipated by Damping   *68*
3.8   Equivalent Viscous Damping   *72*
3.9   Structural Damping   *74*
3.10   Sharpness of Resonance   *76*
3.11   Response to Periodic Forces   *77*
3.12   Vibration Measuring Instruments   *78*

4   TRANSIENT VIBRATION   92

4.1   Impulse Excitation   *92*
4.2   Arbitrary Excitation   *94*
4.3   Laplace Transform Formulation   *100*
4.4   Response Spectrum   *103*
4.5   Finite Difference Numerical Computation   *110*
4.6   Runge-Kutta Method   *119*

5   TWO DEGREES OF FREEDOM   132

5.1   Normal Mode Vibration   *132*
5.2   Coordinate Coupling   *139*
5.3   Forced Harmonic Vibration   *143*
5.4   Digital Computation   *145*
5.5   Vibration Absorber
5.6   Centrifugal Pendulum Vibration Absorber   *151*
5.7   Vibration Damper   *153*
5.8   Gyroscopic Effect on Rotating
       Shafts   *158*

6   PROPERTIES OF VIBRATING SYSTEMS   174

6.1   Flexibility and Stiffness Matrix   *174*
6.2   Reciprocity Theorem   *182*
6.3   Eigenvalues and Eigenvectors   *183*
6.4   Equations Based on Flexibility   *187*

6.5   Orthogonal Properties of Eigenvectors   *188*
6.6   Repeated Roots   *190*
6.7   Modal Matrix P *192*
6.8   Modal Damping in Forced Vibration   *196*
6.9   Normal Mode Summation   *197*

7   **NORMAL MODE VIBRATION OF CONTINUOUS SYSTEMS**   **209**

7.1   Vibrating String   *210*
7.2   Longitudinal Vibration of Rods   *212*
7.3   Torsional Vibration of Rods   *215*
7.4   Euler Equation for the Beam   *218*
7.5   Effect of Rotary Inertia and Shear Deformation   *221*
7.6   Vibration of Membranes   *223*
7.7   Digital Computation   *225*

8   **LAGRANGE'S EQUATION**   **238**

8.1   Generalized Coordinates   *238*
8.2   Virtual Work   *244*
8.3   Kinetic Energy, Potential Energy, and Generalized Force   *247*
8.4   Lagrange's Equations   *252*
8.5   Vibration of Framed Structures   *256*
8.6   Consistent Mass   *258*

9   **APPROXIMATE NUMERICAL METHODS**   **268**

9.1   Rayleigh Method   *268*
9.2   Dunkerley's Equation   *276*
9.3   Rayleigh-Ritz Method   *281*
9.4   Method of Matrix Iteration   *285*
9.5   Calculation of Higher Modes   *287*

10   CALCULATION PROCEDURES FOR LUMPED PARAMETER SYSTEMS   296

10.1   Holzer Method   *296*
10.2   Digital Computer Program for the Torsional System   *300*
10.3   Holzer's Procedure for the Linear Spring-Mass System   *302*
10.4   Myklestad's Method for Beams   *304*
10.5   Myklestad's Method for Rotating Beams   *310*
10.6   Coupled Flexure-Torsion Vibration   *311*
10.7   Transfer Matrices   *313*
10.8   Systems with Damping   *316*
10.9   Geared System   *319*
10.10 Branched System   *320*
10.11 Transfer Matrices for Beams   *323*
10.12 Transfer Matrix for Repeated Structures   *325*
10.13 Difference Equation   *328*

11   MODE SUMMATION PROCEDURES FOR CONTINUOUS SYSTEMS   340

11.1   Mode Summation Method   *340*
11.2   Beam Orthogonality Including Rotary Inertia and Shear Deformation   *345*
11.3   Normal Modes of Constrained Structures   *347*
11.4   Mode Acceleration Method   *353*
11.5   Component Mode Synthesis   *355*

12   NONLINEAR VIBRATION   367

12.1   Phase Plane   *368*
12.2   Conservative System   *369*
12.3   Stability of Equilibrium   *372*
12.4   Method of Isoclines   *375*
12.5   Delta Method   *377*

12.6   Perturbation Method   *380*
12.7   Method of Iteration   *383*
12.8   Self-Excited Oscillations   *388*
12.9   Analog Computer Circuits for Non-
        linear Systems   *390*
12.10 Runge-Kutta Method   *391*

13 RANDOM VIBRATION   400

13.1   Random Phenomena   *400*
13.2   Time Averaging and Expected Value   *402*
13.3   Probability Distribution   *404*
13.4   Correlation   *410*
13.5   Power Spectrum and Power Spectral
        Density   *414*
13.6   Fourier Transforms   *420*
13.7   Frequency Response Function   *427*

APPENDIXES

A   SPECIFICATIONS OF VIBRATION BOUNDS   444

**B   INTRODUCTION TO
     LAPLACE TRANSFORMATION**   447

C   DETERMINANT AND MATRICES   453

    I   Determinant   *453*
    II   Matrices   *455*
    III Rules of Matrix Operations   *457*
    IV   Determination of Eigenvectors   *461*
    V   Cholesky's Method of Solution   *463*

D   NORMAL MODES OF UNIFORM BEAMS   465

ANSWERS TO SELECTED PROBLEMS   475

INDEX   487

31 OCT 84

012.95.   — 8
302.95.   — 8
322.90.   CASH

PREFACE

This book is an updating revision of the former texts, Mechanical Vibration 1948, Second Edition 1953, Vibration Theory and Applications 1965, and Theory of Vibration with Applications 1972. In keeping with continuing advances in modern technology, a number of changes have been made in the subject matter, mode of presentation and emphasis. Outdated material has been deleted and techniques found to be useful have been emphasized. Attempt is made to extend the coverage over broad areas important to the field, yet retain the simplicity of the earlier editions.

In the first five chapters, which deal with single degree of freedom systems and with two degrees of freedom systems, the simple physical approach of the previous edition has been adhered to and, hopefully, improved upon. Several different digital computer methods such as the Euler Method, Modified Euler Method, Average Acceleration Method, etc., were tried out, and from the consideration of accuracy of results and simplicity of use, the Central Difference and the Runge-Kutta Methods were chosen for the text. Some simple examples are presented for the encouragement of their use. The important concept of mode summation is first introduced in these chapters.

In Chapter 6, the concepts of the two degrees of freedom systems are generalized to those of multiple degrees of freedom systems. The emphasis in this chapter is theory focusing on the general properties of all vibrating systems. The use of matrix algebra greatly facilitates the clear presentation of these theories and is extensively used. Modal techniques are greatly expanded in this chapter with the aid of matrices.

Continuous systems are next introduced in Chapter 7. The interrelationship of normal modes to the physical boundary conditions is stressed. Structural members broken down into elementary components of rods, beams, etc., are used for this purpose. The chapter provides a basis for the understanding of normal coordinates in general.

Lagrange's equation, introduced in Chapter 8, strengthens again the understanding of dynamical systems presented earlier and broadens one's view for further extensions. For example, the important concepts of the mode summation procedure are a natural consequence of the Lagrangian generalized coordinates which are extensively discussed with many examples. Application of Lagrange's method is illustrated with a number of problems including framed structures.

Chapters 9 and 10 face the problem of obtaining numerical results for the natural frequencies and mode shapes of multi-degree of freedom systems. Approximate numerical methods for the estimation of the fundamental and few higher modes are presented in Chapter 9. Digital computer methods for lumped parameter systems follow in Chapter 10. These are the algebraic methods of Holzer, Mykelstad, and Transfer Matrices, which are ideally suited for the high speed digital computer. For the repeated structures, analytical results are available from the method of difference equations.

Chapter 11 deals more extensively with the mode summation procedures for the continuous and constrained systems. A matrix method for the synthesis of systems in terms of non-orthogonal functions is illustrated.

In Chapter 12 the treatment of nonlinear systems is introduced with emphasis on the phase plane method. When the nonlinearities are small, the methods of perturbation or iteration offer an analytic approach. Results of machine computations for a nonlinear system illustrate what can be done.

Chapter 13 treats dynamical systems excited by random forces or displacements. Such problems must be examined from a statistical point of view. In many cases, the probability density of the random excitation is normally distributed, which facilitates the computation. The point of view taken here is that, given a random record, an autocorrelation can be easily determined from which the spectral density and mean square response can be calculated. Under simplifying assumptions, a number of problems on random vibrations can be solved, however, the digital computer and the newly developed Fourier Spectrum Analyzer are essential tools for extensive analysis. Many examples are given throughout the text. Various mathematical techniques are illustrated through applications in these examples.

WILLIAM T. THOMSON

Santa Barbara, California

THE
SI SYSTEM
OF UNITS

THE SI SYSTEM OF UNITS

The English system of units that has dominated the United States from historical times, is now being replaced by the SI system of units. Major industries throughout the United States either have already made or, are in the process of making, the transition and engineering students and teachers must deal with the new SI units as well as the present English system. We present here a short discussion of the SI units as they apply to the vibration field and outline a simple procedure to convert from one set of units to the other.

The basic units of the SI system are:

Units	Name	Symbol
Length	Meter	m
Mass	Kilogram	kg
Time	Second	s

The following quantities pertinent to the vibration field are derived from these basic units.

force	Newton	$N \ (= kg \ m/s^2)$
stress	Pascal	$Pa \ (= N/m^2)$
work	Joule	$J \ (= N \ m)$
power	Watt	$W \ (= J/s)$
frequency	Hertz	$Hz \ (= 1/s)$
moment of a force		$N \ m \ (= kg \ m^2/s^2)$
acceleration		m/s^2

xiii

velocity	m/s
angular velocity	1/s
moment of inertia (area)	m^4 ($mm^4 \times 10^{-8}$)
moment of inertia (mass)	$kg\ m^2$ ($kg\ cm^2 \times 10^{-4}$)

Because the meter is a large unit of length, it will be more convenient to express it as the number of millimeters multiplied by 10^{-3}. Vibration instruments, such as accelerometers, are in general calibrated in terms of $g = 9.81$ m/s^2, and hence expressed in non-dimensional units. It is advisable to use non-dimensional presentation whenever possible.

In the English system the weight of an object is generally specified. In the SI system, it is more common to specify the mass, a quantity of matter which remains unchanged with location.

In working with the SI system, it is advisable to think directly in SI units. This will require some time but the following round numbers will help to develop a feeling of confidence in the use of SI units.

The Newton is a smaller unit of force than the pound. One pound of force is equal to 4.4482 Newtons or approximately four and a half times the value for the pound. (An apple weighs approximately $\frac{1}{4}$ lb or approximately one Newton).

One inch is 2.54 cm or .0254 meter. Thus the acceleration of gravity which is 386 in/s^2 in the English system becomes $386 \times .0254 = 9.81$ m/s^2, or approximately 10 m/s^2.

TABLE OF APPROXIMATE EQUIVALENTS

1 lb	≈	4.5 N
accel. of gravity g	≈	10 m/s^2
mass of 1 slug	≈	15 kg
1 ft	≈	1/3 m

SI Conversion. A simple procedure to convert from one set of units to another follows: Write the desired SI units equal to the English units, and put in canceling unit factors. For example, suppose one wished to convert torque in the English units into the SI units, we proceed as follows;

EXAMPLE 1

$$[\text{Torque SI}] = [\text{Torque English}] \times [\text{multiplying factors}]$$

$$[N\ m] = [lb\ in]\left(\frac{N}{lb}\right)\left(\frac{m}{in}\right)$$

$$= [lb\ in](4.448)(.0254)$$

$$= [lb\ in](0.1129)$$

EXAMPLE 2

$$[\text{Moment of inertia SI}] = [\text{Moment of inertia, Eng}] \times [\text{mult. factors}]$$

$$\left[\begin{array}{c} \text{kg m}^2 \\ = \text{N m s}^2 \end{array}\right] = [\cancel{lb}\ \cancel{in}\ s^2]\left(\frac{\text{N}}{\cancel{lb}} \cdot \frac{\text{m}}{\cancel{in}}\right)$$

$$= [\text{lb in sec}^2](4.448 \times .0254)$$

$$= [\text{lb. in. sec}^2](0.1129)$$

EXAMPLE 3

Modulus of Elasticity:

$$[\text{E N/m}^2] = \left[\text{E}\frac{\cancel{lb}}{\cancel{in}^2}\right]\left(\frac{\text{N}}{\cancel{lb}}\right)\left(\frac{\cancel{in}}{\text{m}}\right)^2$$

$$= \left[\text{E}\frac{\text{lb}}{\text{in}^2}\right](4.448)\left(\frac{1}{.0254}\right)^2$$

$$= \left[\text{E}\frac{\text{lb}}{\text{in}^2}\right](6894.7)$$

$$\text{E}_{\text{steel}}\text{N/m}^2 = (29 \times 10^6\ \text{lb/in}^2)(6894.7) = \underline{\underline{200 \times 10^9\ \text{N/m}^2}}$$

EXAMPLE 4

Spring stiffness:

$$[\text{k} \sim \text{N/m}] = [\text{k lb/in}] \times (175.13)$$

Mass:

$$[\text{m} \sim \text{kg}] = [\text{m lb sec}^2/\text{in}] \times (175.13)$$

CONVERSION FACTORS*
U.S.-BRITISH UNITS TO SI UNITS

To Convert From	To	Multiply By
(Acceleration)		
foot/second^2 (ft/s^2)	metre/second^2 (m/s^2)	3.048×10^{-1}*
inch/second^2 (in./s^2)	metre/second^2 (m/s^2)	2.54×10^{-2}*
(Area)		
foot^2 (ft^2)	metre^2 (m^2)	9.2903×10^{-2}
inch^2 (in.^{2})	metre^2 (m^2)	6.4516×10^{-4}*
yard^2 (yd^2)	metre^2 (m^2)	8.3613×10^{-1}

(Density)

pound mass/inch^3 (lbm/in.^{3})	kilogram/metre^3 (kg/m^3)	2.7680×10^4
pound mass/foot^3 (lbm/ft^3)	kilogram/metre^3 (kg/m^3)	1.6018×10

(Energy, Work)

British thermal unit (BTU)	joule (J)	1.0551×10^3
foot-pound force (ft · lbf)	joule (J)	1.3558
kilowatt-hour (kw · h)	joule (J)	$3.60 \times 10^{6*}$

(Force)

kip (1000 lbf)	newton (N)	4.4482×10^3
pound force (lbf)	newton (N)	4.4482
ounce force	newton (N)	2.7801×10^{-1}

(Length)

foot (ft)	metre(m)	$3.048 \times 10^{-1*}$
inch (in.)	metre (m)	$2.54 \times 10^{-2*}$
mile (mi), (U.S. statute)	metre (m)	1.6093×10^3
mile (mi), (international nautical)	metre (m)	$1.852 \times 10^{3*}$
yard (yd)	metre (m)	$9.144 \times 10^{-1*}$

(Mass)

pound · mass (lbm)	kilogram (kg)	4.5359×10^{-1}
slug (lbf · s^2/ft)	kilogram (kg)	1.4594×10
ton (2000 lbm)	kilogram (kg)	9.0718×10^2

(Power)

foot-pound/minute (ft · lbf/min)	watt (W)	2.2597×10^{-2}
horsepower (550 ft · lbf/s)	watt (W)	7.4570×10^2

(Pressure, stress)

atmosphere (std) (14.7 lbf/in.^{2})	newton/metre^2 (N/m^2 or Pa)	1.0133×10^5
pound/foot^2 (lbf/ft^2)	newton/metre^2 (N/m^2 or Pa)	4.7880×10
pound/inch^2 (lbf/in.^{2} or psi)	newton/metre^2 (N/m^2 or Pa)	6.8948×10^3

(Velocity)

foot/minute (ft/min)	metre/second (m/s)	$5.08 \times 10^{-3*}$
foot/second (ft/s)	metre/second (m/s)	$3.048 \times 10^{-1*}$
knot (nautical mi/h)	metre/second (m/s)	5.1444×10^{-1}
mile/hour (mi/h)	metre/second (m/s)	$4.4704 \times 10^{-1*}$
mile/hour (mi/h)	kilometre/hour (km/h)	1.6093
mile/second (mi/s)	kilometre/second (km/s)	1.6093

(Volume)

foot^3 (ft^3)	metre^3 (m^3)	2.8317×10^{-2}
inch^3 (in.^{3})	metre^3 (m^3)	1.6387×10^{-5}

*Exact value

Source: Meriam, J. L., Dynamics, 2nd Ed. (SI Version). New York: John Wiley and Sons, Inc., 1975.

1 OSCILLATORY MOTION

The study of vibration is concerned with the oscillatory motions of bodies and the forces associated with them. All bodies possessing mass and elasticity are capable of vibration. Thus most engineering machines and structures experience vibration to some degree, and their design generally requires consideration of their oscillatory behavior.

Oscillatory systems can be broadly characterized as *linear* or *nonlinear*. For linear systems the principle of superposition holds, and the mathematical techniques available for their treatment are well-developed. In contrast, techniques for the analysis of nonlinear systems are less well known, and difficult to apply. However, some knowledge of nonlinear systems is desirable, since all systems tend to become nonlinear with increasing amplitude of oscillation.

There are two general classes of vibrations—free and forced. *Free vibration* takes place when a system oscillates under the action of forces inherent in the system itself, and when external impressed forces are absent. The system under free vibration will vibrate at one or more of its *natural frequencies*, which are properties of the dynamical system established by its mass and stiffness distribution.

Vibration that takes place under the excitation of external forces is called *forced vibration*. When the excitation is oscillatory, the system is forced to vibrate at the excitation frequency. If the frequency of excitation coincides with one of the natural frequencies of the system, a condition of *resonance* is encountered, and dangerously large oscillations may result.

The failure of major structures, such as bridges, buildings, or airplane wings, is an awesome possibility under resonance. Thus, the calculation of the natural frequencies is of major importance in the study of vibrations.

Vibrating systems are all subject to *damping* to some degree because energy is dissipated by friction and other resistances. If the damping is small, it has very little influence on the natural frequencies of the system, and hence the calculations for the natural frequencies are generally made on the basis of no damping. On the other hand, damping is of great importance in limiting the amplitude of oscillation at resonance.

The number of independent coordinates required to describe the motion of a system is called the *degrees of freedom* of the system. Thus a free particle undergoing general motion in space will have three degrees of freedom, while a rigid body will have six degrees of freedom, i.e., three components of position and three angles defining its orientation. Furthermore, a continuous elastic body will require an infinite number of coordinates (three for each point on the body) to describe its motion; hence its degrees of freedom must be infinite. However, in many cases, parts of such bodies may be assumed to be rigid, and the system may be considered to be dynamically equivalent to one having finite degrees of freedom. In fact, a surprisingly large number of vibration problems can be treated with sufficient accuracy by reducing the system to one having a single degree of freedom.

1.1   HARMONIC MOTION

Oscillatory motion may repeat itself regularly, as in the balance wheel of a watch, or display considerable irregularity, as in earthquakes. When the motion is repeated in equal intervals of time τ, it is called *periodic motion*. The repetition time τ is called the *period* of the oscillation, and its reciprocal, $f = 1/\tau$, is called the *frequency*. If the motion is designated by the time function $x(t)$, then any periodic motion must satisfy the relationship $x(t) = x(t + \tau)$.

The simplest form of periodic motion is *harmonic motion*. It can be demonstrated by a mass suspended from a light spring, as shown in Fig. 1.1-1. If the mass is displaced from its rest position and released, it will oscillate up and down. By placing a light source on the oscillating mass, its motion can be recorded on a light-sensitive film strip which is made to move past it at constant speed.

The motion recorded on the film strip can be expressed by the equation

$$x = A \sin 2\pi \frac{t}{\tau} \qquad (1.1\text{-}1)$$

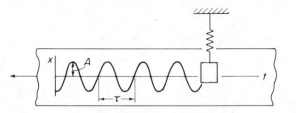

Figure 1.1-1.   Recording of harmonic motion.

where A is the amplitude of oscillation, measured from the equilibrium position of the mass, and τ is the period. The motion is repeated when $t = \tau$.

Harmonic motion is often represented as the projection on a straight line of a point that is moving on a circle at constant speed, as shown in Fig. 1.1-2. With the angular speed of the line op designated by ω, the displacement x can be written as

$$x = A \sin \omega t \tag{1.1-2}$$

The quantity ω is generally measured in radians per second, and is referred to as the *circular frequency*. Since the motion repeats itself in 2π radians, we have the relationship

$$\omega = \frac{2\pi}{\tau} = 2\pi f \tag{1.1-3}$$

where τ and f are the period and frequency of the harmonic motion, usually measured in seconds and cycles per second respectively.

The velocity and acceleration of harmonic motion can be simply determined by differentiation of Eq. (1.1-2). Using the dot notation for the derivative, we obtain

$$\dot{x} = \omega A \cos \omega t = \omega A \sin\left(\omega t + \frac{\pi}{2}\right) \tag{1.1-4}$$

$$\ddot{x} = -\omega^2 A \sin \omega t = \omega^2 A \sin(\omega t + \pi) \tag{1.1-5}$$

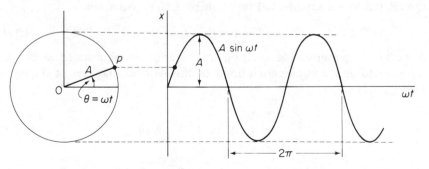

Figure 1.1-2.   Harmonic motion as projection of a point moving on a circle.

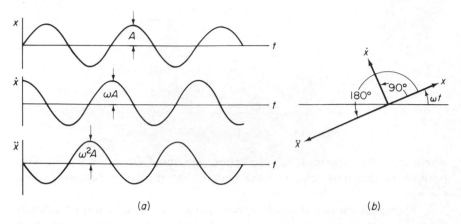

Figure 1.1-3. In harmonic motion, the velocity and acceleration lead the displacement by $\pi/2$ and π.

Thus the velocity and acceleration are also harmonic with the same frequency of oscillation, but lead the displacement by $\pi/2$ and π radians respectively. Figure 1.1-3 shows both the time variation and the vector phase relationship between the displacement, velocity, and acceleration in harmonic motion.

Examination of Eqs. (1.1-2) and (1.1-5) reveals that

$$\ddot{x} = -\omega^2 x \qquad (1.1\text{-}6)$$

so that in harmonic motion the acceleration is proportional to the displacement and is directed towards the origin. Since Newton's second law of motion states that the acceleration is proportional to the force, harmonic motion can be expected for systems with linear springs with force varying as kx.

Exponential Form. The trigonometric functions of sine and cosine are related to the exponential function by Euler's equation

$$e^{i\theta} = \cos \theta + i \sin \theta \qquad (1.1\text{-}7)$$

A vector of amplitude A rotating at constant angular speed ω can be represented as a complex quantity z in the Argand diagram as shown in Fig. 1.1-4.

$$\begin{aligned} z &= A e^{i\omega t} \\ &= A \cos \omega t + iA \sin \omega t \qquad (1.1\text{-}8) \\ &= x + iy \end{aligned}$$

The quantity z is referred to as the *complex sinusoid* with x and y as the

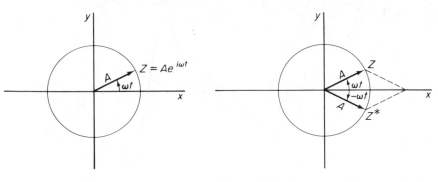

Figure 1.1-4.   **Figure 1.1-5.**   Vector z and its conjugate z^*.

real and imaginary components. The quantity $z = Ae^{i\omega t}$ also satisfies the differential equation (1.1-6) for harmonic motion.

Figure 1.1-5 shows z and its conjugate $z^* = Ae^{-i\omega t}$ which is rotating in the negative direction with angular speed $-\omega$. It is evident from this diagram that the real component x is expressible in terms of z and z^* by the equation

$$x = \tfrac{1}{2}(z + z^*) = A \cos \omega t = ReAe^{i\omega t} \qquad (1.1\text{-}9)$$

where Re stands for the real part of the quantity z. We will find that the exponential form of the harmonic motion often offers mathematical advantages.

Some of the rules of exponential operations between $z_1 = A_1 e^{i\theta_1}$ and $z_2 = A_2 e^{i\theta_2}$ are:

Multiplication $\qquad\qquad z_1 z_2 = A_1 A_2 e^{i(\theta_1 + \theta_2)}$

Division $\qquad\qquad \dfrac{z_1}{z_2} = \left(\dfrac{A_1}{A_2}\right) e^{i(\theta_1 - \theta_2)}$ $\qquad (1.1\text{-}10)$

Powers $\qquad\qquad z^n = A^n e^{in\theta}$

$\qquad\qquad\qquad\qquad z^{1/n} = A^{1/n} e^{i\theta/n}$

1.2 PERIODIC MOTION

It is quite common for vibrations of several different frequencies to exist simultaneously. For example, the vibration of a violin string is composed of the fundamental frequency f and all its harmonics $2f$, $3f$, etc. Another example is the free vibration of a multidegree-of-freedom system, to which the vibrations at each natural frequency contribute. Such vibrations result in a complex waveform which is repeated periodically as shown in Fig. 1.2-1.

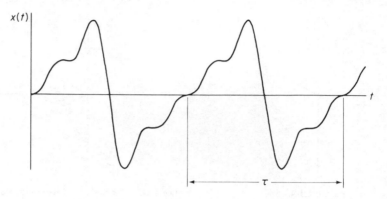

Figure 1.2-1. Periodic motion of period τ.

The French mathematician J. Fourier (1768–1830) showed that any periodic motion can be represented by a series of sines and cosines which are harmonically related. If $x(t)$ is a periodic function of the period τ, it is represented by the Fourier series

$$x(t) = \frac{a_0}{2} + a_1 \cos \omega_1 t + a_2 \cos \omega_2 t + \cdots$$
$$+ b_1 \sin \omega_1 t + b_2 \sin \omega_2 t + \cdots \qquad (1.2\text{-}1)$$

where

$$\omega_1 = \frac{2\pi}{\tau}$$
$$\omega_n = n\omega_1$$

To determine the coefficients a_n and b_n, we multiply both sides of Eq. (1.2-1) by $\cos \omega_n t$ or $\sin \omega_n t$ and integrate each term over the period τ. Recognizing the following relations,

$$\int_{-\tau/2}^{\tau/2} \cos \omega_n t \, \cos \omega_m t \, dt = \begin{cases} 0 & \text{if } m \neq n \\ \tau/2 & \text{if } m = n \end{cases}$$

$$\int_{-\tau/2}^{\tau/2} \sin \omega_n t \, \sin \omega_m t \, dt = \begin{cases} 0 & \text{if } m \neq n \\ \tau/2 & \text{if } m = n \end{cases} \qquad (1.2\text{-}2)$$

$$\int_{-\tau/2}^{\tau/2} \cos \omega_n t \, \sin \omega_m t \, dt = \begin{cases} 0 & \text{if } m \neq n \\ 0 & \text{if } m = n \end{cases}$$

all terms except one on the right side of the equation will be zero, and we obtain the result

$$a_n = \frac{2}{\tau} \int_{-\tau/2}^{\tau/2} x(t) \cos \omega_n t \, dt$$

$$(1.2\text{-}3)$$

$$b_n = \frac{2}{\tau} \int_{-\tau/2}^{\tau/2} x(t) \sin \omega_n t \, dt$$

The Fourier series can also be represented in terms of the exponential function. Substituting

$$\cos \omega_n t = \tfrac{1}{2}(e^{i\omega_n t} + e^{-i\omega_n t})$$

$$\sin \omega_n t = -\tfrac{i}{2}(e^{i\omega_n t} - e^{-i\omega_n t})$$

in Eq. (1.2-1), we obtain

$$x(t) = \frac{a_0}{2} + \sum_{n=1}^{\infty} \left[\tfrac{1}{2}(a_n - ib_n)e^{i\omega_n t} + \tfrac{1}{2}(a_n + ib_n)e^{-i\omega_n t} \right]$$

$$= \frac{a_0}{2} + \sum_{n=1}^{\infty} \left[c_n e^{i\omega_n t} + c_n^* e^{-i\omega_n t} \right] \qquad (1.2\text{-}4)$$

$$= \sum_{n=-\infty}^{\infty} c_n e^{i\omega_n t}$$

where

$$c_0 = \tfrac{1}{2}a_0$$
$$c_n = \tfrac{1}{2}(a_n - ib_n) \qquad (1.2\text{-}5)$$

Substituting for a_n and b_n from Eq. (1.2-3), we find c_n to be

$$c_n = \frac{1}{\tau} \int_{-\tau/2}^{\tau/2} x(t)(\cos \omega_n t - i \sin \omega_n t)\, dt$$

$$= \frac{1}{\tau} \int_{-\tau/2}^{\tau/2} x(t)e^{-i\omega_n t}\, dt \qquad (1.2\text{-}6)$$

Some computational effort can be minimized when the function $x(t)$ is recognizable in terms of the even and odd functions

$$x(t) = E(t) + O(t) \qquad (1.2\text{-}7)$$

An even function $E(t)$ is symmetric about the origin so that $E(t) = E(-t)$, i.e., $\cos \omega t = \cos(-\omega t)$. An odd function satisfies the relationship $O(t) = -O(-t)$, i.e., $\sin \omega t = -\sin(-\omega t)$. The following integrals are then help-ful:

$$\int_{-\tau/2}^{\tau/2} E(t) \sin \omega_n t\, dt = 0$$

$$\int_{-\tau/2}^{\tau/2} O(t) \cos \omega_n t\, dt = 0 \qquad (1.2\text{-}8)$$

When the coefficients of the Fourier series are plotted against frequency ω_n, the result is a series of discrete lines called the *Fourier spectrum*. Generally plotted are the absolute value $|2c_n| = \sqrt{a_n^2 + b_n^2}$ and the phase $\phi_n = \tan^{-1} b_n/a_n$, an example of which is shown in Fig. 1.2-2.

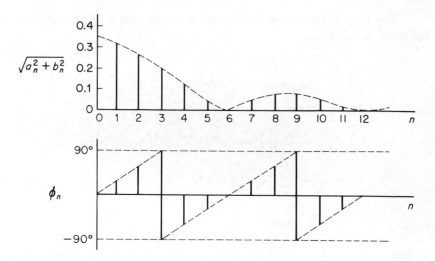

Figure 1.2-2. Fourier spectrum for pulses shown in Prob. 1–16, $k = 1/3$.

With the aid of the digital computer, harmonic analysis today is efficiently carried out. A computer algorithm known as the *Fast Fourier Transform*[*] (FFT) is commonly used to minimize the computation time.

1.3   VIBRATION TERMINOLOGY

Certain terminologies used in vibration need to be represented here. The simplest of these are the *peak value* and the *average value*.

The peak value will generally indicate the maximum stress which the vibrating part is undergoing. It also places a limitation on the "rattle space" requirement.

The average value indicates a steady or static value somewhat like the DC level of an electrical current. It can be found by the time integral

$$\bar{x} = \lim_{T \to \infty} \frac{1}{T} \int_0^T x(t)\, dt \tag{1.3-1}$$

For example, the average value for a complete cycle of a sine wave, $A \sin t$, is zero; whereas its average value for a half-cycle is

$$\bar{x} = \frac{A}{\pi} \int_0^\pi \sin t\, dt = \frac{2A}{\pi} = 0.637\, A$$

It is evident that this is also the average value of the rectified sine wave shown in Fig. 1.3-1.

---

[*]See J. S. Bendat & A. G. Piersol, "Random Data" (New York: John Wiley & Sons, 1971), p. 305–306.

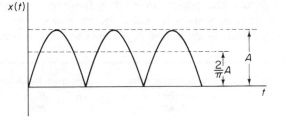

Figure 1.3-1. Average value of a rectified sine wave.

The square of the displacement generally is associated with the energy of the vibration for which the mean square value is a measure. The *mean square value* of a time function $x(t)$ is found from the average of the squared values, integrated over some time interval T:

$$\overline{x^2} = \lim_{T\to\infty} \frac{1}{T} \int_0^T x^2(t)\, dt \qquad (1.3\text{-}2)$$

For example, if $x(t) = A \sin \omega t$, its mean square value is

$$\overline{x^2} = \lim_{T\to\infty} \frac{A^2}{T} \int_0^T \frac{1}{2}(1 - \cos 2\omega t)\, dt = \frac{1}{2} A^2$$

The *root mean square* (rms) value is the square root of the mean square value. From the previous example, the rms of the sine wave of amplitude A is $A/\sqrt{2} = 0.707\ A$. Vibrations are commonly measured by rms meters.

Decibel: The decibel is a unit of measurement that is frequently used in vibration measurements. It is defined in terms of a power ratio.

$$\text{Db} = 10 \log_{10}\!\left(\frac{P_1}{P_2}\right)$$

$$= 10 \log_{10}\!\left(\frac{x_1}{x_2}\right)^2 \qquad (1.3\text{-}3)$$

The second equation results from the fact that power is proportional to the square of the amplitude or voltage. The decibel is often expressed in terms of the first power of amplitude or voltage as

$$\text{Db} = 20 \log_{10}\!\left(\frac{x_1}{x_2}\right) \qquad (1.3\text{-}4)$$

Thus an amplifier with a voltage gain of 5 has a decibel gain of

$$20 \log_{10}(5) = +14$$

Because the decibel is a logarithmic unit, it compresses or expands the scale.

Octave: When the upper limit of a frequency range is twice its lower limit, the frequency span is said to be an *octave*. For example, each of the frequency bands given below represents an octave band.

Band	Frequency range (Hz)	Frequency Bandwidth
1	10–20	10
2	20–40	20
3	40–80	40
4	200–400	200

PROBLEMS

1-1 A harmonic motion has an amplitude of 0.20 cm and a period of 0.15 sec. Determine the maximum velocity and acceleration.

1-2 An accelerometer indicates that a structure is vibrating harmonically at 82 cps with a maximum acceleration of 50 g. Determine the amplitude of vibration.

1-3 A harmonic motion has a frequency of 10 cps and its maximum velocity is 4.57 m/sec. Determine its amplitude, its period, and its maximum acceleration.

1-4 Find the sum of two harmonic motions of equal amplitude but of slightly different frequencies. Discuss the beating phenomena that result from this sum.

1-5 Express the complex vector $4 + 3i$ in the exponential form $Ae^{i\theta}$.

1-6 Add two complex vectors $(2 + 3i)$ and $(4 - i)$ expressing the result as $A \angle \theta$.

1-7 Show that the multiplication of a vector $z = Ae^{i\omega t}$ by i rotates it by 90°.

1-8 Determine the sum of two vectors $5e^{i\pi/6}$ and $4e^{i\pi/3}$ and find the angle between the resultant and the first vector.

1-9 Determine the Fourier series for the rectangular wave shown in Fig. P1-9.

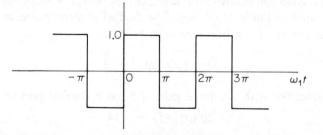

Figure P1-9.

1-10   If the origin of the square wave of Prob. 1-9 is shifted to the right by $\pi/2$, determine the Fourier series.

1-11   Determine the Fourier series for the triangular wave shown in Fig. P1-11.

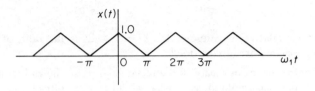

Figure P1-11.

1-12   Determine the Fourier series for the saw tooth curve shown in Fig. P1-12. Express the result of Prob. 1-12 in the exponential form of Eq. (1.2-4).

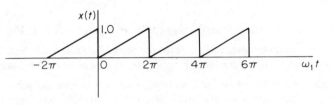

Figure P1-12.

1-13   Determine the rms value of a wave consisting of the positive portions of a sine wave.

1-14   Determine the mean square value of the saw tooth wave of Prob. 1-12. Do this two ways, from the squared curve and from the Fourier series.

1-15   Plot the frequency spectrum for the triangular wave of Prob. 1-11.

1-16   Determine the Fourier series of a series of rectangular pulses shown in Fig. P1-16. Plot c_n and ϕ_n vs. n when $k = \frac{2}{3}$.

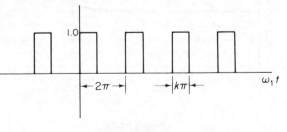

Figure P1-16.

1-17   Write the equation for the displacement s of the piston in the crank-piston mechanism shown in Fig. P1-17, and determine the harmonic components

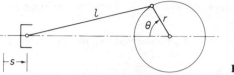

Figure P1-17.

and their relative magnitudes. If $r/l = \frac{1}{3}$, what is the ratio of the second harmonic compared to the first?

1-18   Determine the mean square of the rectangular pulse shown in Fig. P1-18 for $k = 0.10$. If the amplitude is A, what would an rms voltmeter read?

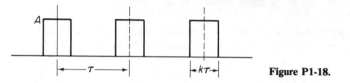

Figure P1-18.

1-19   Determine the mean square value of the triangular wave of Fig. P1-11.

1-20   An rms voltmeter specifies an accuracy of ± 0.5 Db. If a vibration of 2.5 mm rms is measured, determine the millimeter accuracy as read by the voltmeter.

1-21   Amplification factors on a voltmeter used to measure the vibration output from an accelerometer are given as 10, 50, and 100. What are the decibel steps?

1-22   The calibration curve of a piezoelectric accelerometer is shown in Fig. P1-22 where the ordinate is in decibels. If the peak is 32 Db, what is the ratio of the resonance response to that at some low frequency, say 1000 cps?

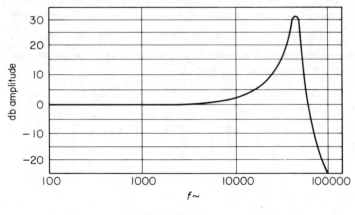

Figure P1-22.

1-23   Using coordinate paper similar to that of Appendix A, outline the bounds for the following vibration specifications. Max. acceleration = 2g, max. displacement = 0.08 inch, min. and max. frequencies: 1 Hz and 200 Hz.

2
FREE VIBRATION

All systems possessing mass and elasticity are capable of free vibration, or vibration which takes place in the absence of external excitation. Of primary interest for such systems is its natural frequency of vibration. Our object here is to learn to write its equation of motion and evaluate its natural frequency which is mainly a function of the mass and stiffness of the system.

Damping in moderate amounts has little influence on the natural frequency and may be neglected in its calculation. The system can then be considered to be conservative and the principle of conservation of energy offers another approach to the calculation of the natural frequency. The effect of damping is mainly evident in the diminishing of the vibration amplitude with time. Although there are many models of damping, only those which lead to simple analytic procedures are considered in this chapter.

2.1 EQUATION OF MOTION—NATURAL FREQUENCY

The simplest oscillatory system consists of a mass and a spring as shown in Fig. 2.1-1. The spring supporting the mass is assumed to be negligible in mass with a stiffness k Newtons per meter of deflection. The system possesses one degree of freedom since its motion is described by a single coordinate x.

When placed into motion, oscillation will take place at the natural frequency f_n, which is a property of the system. We now examine some of the basic concepts associated with the free vibration of systems with one degree of freedom.

Newton's second law is the first basis for examining the motion of the system. As shown in Fig. 2.1-1 the deformation of the spring in the static equilibrium position is Δ, and the spring force $k\Delta$ is equal to the gravitational force w acting on the mass m:

$$k\Delta = w = mg \qquad (2.1\text{-}1)$$

Measuring the displacement x from the static equilibrium position, the forces acting on m are $k(\Delta + x)$ and w. With x chosen to be positive in the downward direction, all quantities—force, velocity, and acceleration—are also positive in the downward direction.

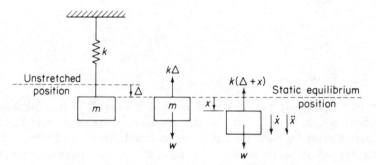

Figure 2.1-1.   Spring-mass system and free-body diagram.

We now apply Newton's second law of motion to the mass m

$$m\ddot{x} = \Sigma F = w - k(\Delta + x)$$

and since $k\Delta = w$, we obtain

$$m\ddot{x} = -kx \qquad (2.1\text{-}2)$$

It is evident that the choice of the static equilibrium position as reference for x has eliminated w, the force due to gravity, and the static spring force $k\Delta$ from the equation of motion, and the resultant force on m is simply the spring force due to the displacement x.

Defining the circular frequency ω_n by the equation

$$\omega_n^2 = \frac{k}{m} \qquad (2.1\text{-}3)$$

Eq. (2.1-2) may be written as

$$\ddot{x} + \omega_n^2 x = 0 \qquad (2.1\text{-}4)$$

and we conclude by comparison with Eq. (1.1-6) that the motion is harmonic. Equation (2.1-4), a homogeneous second-order linear differential equation, has the following general solution:

$$x = A \sin \omega_n t + B \cos \omega_n t \qquad (2.1\text{-}5)$$

where A and B are the two necessary constants. These constants are evaluated from initial conditions $x(0)$ and $\dot{x}(0)$, and Eq. (2.1-5) can be shown to reduce to

$$x = \frac{\dot{x}(0)}{\omega_n} \sin \omega_n t + x(0) \cos \omega_n t \qquad (2.1\text{-}6)$$

The natural period of the oscillation is established from $\omega_n \tau = 2\pi$, or

$$\tau = 2\pi \sqrt{\frac{m}{k}} \qquad (2.1\text{-}7)$$

and the natural frequency is

$$f_n = \frac{1}{\tau} = \frac{1}{2\pi} \sqrt{\frac{k}{m}} \qquad (2.1\text{-}8)$$

These quantities may be expressed in terms of the statical deflection Δ by observing Eq. (2.1-1), $k\Delta = mg$. Thus Eq. (2.1-8) may be expressed in terms of the statical deflection Δ as

$$f_n = \frac{1}{2\pi} \sqrt{\frac{g}{\Delta}} \qquad (2.1\text{-}9)$$

and the natural frequency of a single degree of freedom system is uniquely determined by the statical deflection Δ.

The units used in the above equation must be consistent. For example, if g is given in inches/sec^2, then Δ must be in inches. Using $g = 9.81$ m/s^2, Δ must be in meters. However, it is more convenient to use Δ in millimeters, $\Delta_m = \Delta_{mm} \times 10^{-3}$, in which case Eq. (2.1-9) becomes

$$f_n = \frac{1}{2\pi} \sqrt{\frac{9.81}{\Delta_{mm} \times 10^{-3}}} = \frac{15.76}{\sqrt{\Delta_{mm}}} \qquad (2.1\text{-}10)$$

A logarithmic plot of Eq. (2.1-10) is shown in Fig. 2.1-2.

EXAMPLE 2.1-1

A $\frac{1}{4}$-kg mass is suspended by a spring having a stiffness of 0.1533 N/mm. Determine its natural frequency in cycles per second. Determine its statical deflection and check the natural frequency from Fig. 2.1-2.

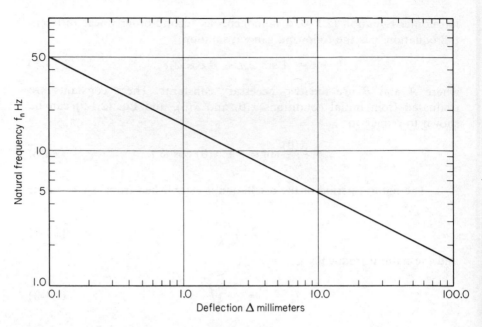

Figure 2.1-2.

Solution: The stiffness is

$$k = 153.3 \text{ N/m}$$

Substituting into Eq. (2.1-8), the natural frequency is

$$f = \frac{1}{2\pi}\sqrt{\frac{k}{m}} = \frac{1}{2\pi}\sqrt{\frac{153.3}{0.25}} = 3.941 \text{ Hz}$$

The statical deflection of the spring suspending the $\frac{1}{4}$-kg mass is obtained from the relationship $mg = k\Delta$

$$\Delta = \frac{mg}{k_{N/mm}} = \frac{0.25 \times 9.81}{0.1533} = 16.0 \text{ mm}$$

From Fig. 2.1-2 the natural frequency is approximately 3.9 Hz.

EXAMPLE 2.1-2

Determine the natural frequency of the mass M on the end of a cantilever beam of negligible mass shown in Fig. 2.1-3.

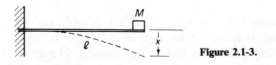

Figure 2.1-3.

Solution: The deflection of the cantilever beam under a concentrated force P is

$$x = \frac{Pl^3}{3EI} = \frac{P}{k}$$

where EI is the flexural rigidity. Thus the stiffness of the beam is $k = 3EI/l^3$ and the natural frequency of the system becomes

$$f_n = \frac{1}{2\pi}\sqrt{\frac{3EI}{Ml^3}}$$

EXAMPLE 2.1-3

An automobile wheel and tire are suspended by a steel rod 0.50 cm in diameter and 2 m long as shown in Fig. 2.1-4. When the wheel is given an angular displacement and released, it makes 10 oscillations in 30.2 s. Determine the polar moment of inertia of the wheel and tire.

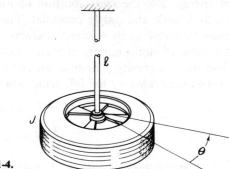

Figure 2.1-4.

Solution: The rotational equation of motion corresponding to Newton's equation is

$$J\ddot{\theta} = -K\theta$$

where J is the rotational mass moment of inertia, K is the rotational stiffness, and θ is the angle of rotation in radians. Thus the natural frequency of oscillation is equal to

$$\omega_n = 2\pi\frac{10}{30.2} = 2.081 \text{ rad/s}$$

The torsional stiffness of the rod is given by the equation $K = GI_p/l$

where $I_p = \pi d^4/32$ = polar moment of inertia of the circular cross-sectional area of the rod, l = length, and $G = 80 \times 10^9$ N/m^2 = shear modulus of steel.

$$I_p = \frac{\pi}{32}(0.5 \times 10^{-2})^4 = 0.006136 \times 10^{-8} \text{ m}^4$$

$$K = \frac{80 \times 10^9 \times 0.006136 \times 10^{-8}}{2} = 2.455 \text{ Nm/rad}$$

Substituting into the natural frequency equation, the polar moment of inertia of the wheel and tire is

$$J = \frac{K}{\omega_n^2} = \frac{2.455}{2.081^2} = 0.567 \text{ kg m}^2$$

2.2 ENERGY METHOD

In a conservative system the total energy is constant, and the differential equation of motion can also be established by the principle of conservation of energy. For the free vibration of an undamped system, the energy is partly kinetic and partly potential. The kinetic energy T is stored in the mass by virtue of its velocity, whereas the potential energy U is stored in the form of strain energy in elastic deformation or work done in a force field such as gravity. The total energy being constant, its rate of change is zero as illustrated by the following equations

$$T + U = \text{constant} \tag{2.2-1}$$

$$\frac{d}{dt}(T + U) = 0 \tag{2.2-2}$$

If our interest is only in the natural frequency of the system, it can be determined by the following considerations. From the principle of conservation of energy we can write

$$T_1 + U_1 = T_2 + U_2 \tag{2.2-3}$$

where $_1$ and $_2$ represent two instances of time. Let $_1$ be the time when the mass is passing through its static equilibrium position and choose $U_1 = 0$ as reference for the potential energy. Let $_2$ be the time corresponding to the maximum displacement of the mass. At this position, the velocity of the mass is zero, and hence $T_2 = 0$. We then have

$$T_1 + 0 = 0 + U_2 \tag{2.2-4}$$

However, if the system is undergoing harmonic motion, then T_1 and U_2 are maximum values, and hence

$$T_{max} = U_{max} \qquad (2.2\text{-}5)$$

The above equation leads directly to the natural frequency.

EXAMPLE 2.2-1

Using the spring-mass system as an example, show that the loss of potential energy of the mass due to displacement from the static equilibrium position will always be cancelled by the work done by the equilibrium force of the spring.

Solution: We will choose the static equilibrium position to be the reference of zero potential energy. Due to displacement x from this reference the increase in the potential energy of the spring which is equal to the area under the force-displacement diagram of Fig. 2.2-1 is $mgx + \frac{1}{2}kx^2$. The loss in potential energy of m due to displacement x is $-mgx$, so that the net change in the potential energy is $\frac{1}{2}kx^2$.

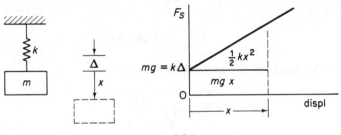

Figure 2.2-1.

Adding to the kinetic energy of m, we have

$$\frac{1}{2}m\dot{x}^2 + \frac{1}{2}kx^2 = \text{constant}$$

and differentiating

$$(m\ddot{x} + kx)\dot{x} = 0$$

or

$$m\ddot{x} + kx = 0$$

EXAMPLE 2.2-2

Determine the natural frequency of the system shown in Fig. 2.2-2.

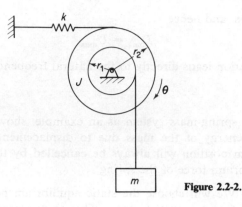

Figure 2.2-2.

Solution: Assume that the system is vibrating harmonically with amplitude θ from its static equilibrium position. The maximum kinetic energy is

$$T_{max} = \frac{1}{2}J\dot{\theta}^2 + \frac{1}{2}m(r_1\dot{\theta})^2$$

The maximum potential energy is the energy stored in the spring which is

$$U_{max} = \frac{1}{2}k(r_2\theta)^2$$

Equating the two, the natural frequency is

$$\omega_n = \sqrt{\frac{kr_2^2}{J + mr_1^2}}$$

The student should verify that the loss of potential energy of m due to position $r_1\theta$ is cancelled by the work done by the equilibrium force of the spring in the position $\theta = 0$.

EXAMPLE 2.2-3

A cylinder of weight w and radius r rolls without slipping on a cylindrical surface of radius R, as shown in Fig. 2.2-3. Determine its differential equation of motion for small oscillations about the lowest point. For no slipping, we have $r\phi = R\theta$.

Solution: In determining the kinetic energy of the cylinder, it must be noted that both translation and rotation take place. The translational velocity of the center of the cylinder is $(R - r)\dot{\theta}$, whereas the rotational velocity is $(\dot{\phi} - \dot{\theta}) = (R/r - 1)\dot{\theta}$, since $\dot{\phi} = (R/r)\dot{\theta}$ for no

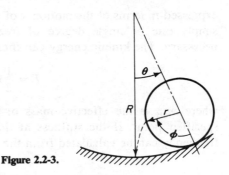

Figure 2.2-3.

slipping. The kinetic energy may now be written as

$$T = \frac{1}{2}\frac{w}{g}[(R-r)\dot{\theta}]^2 + \frac{1}{2}\frac{w}{g}\frac{r^2}{2}\left[\left(\frac{R}{r}-1\right)\dot{\theta}\right]^2$$

$$= \frac{3}{4}\frac{w}{g}(R-r)^2\dot{\theta}^2$$

where $(w/g)(r^2/2)$ is the moment of inertia of the cylinder about its mass center.

The potential energy referred to its lowest position is

$$U = w(R-r)(1-\cos\theta)$$

which is equal to the negative of the work done by the gravity force in lifting the cylinder through the vertical height $(R-r)(1-\cos\theta)$.

Substituting into Eq. (2.2-2)

$$\left[\frac{3}{2}\frac{w}{g}(R-r)^2\ddot{\theta} + w(R-r)\sin\theta\right]\dot{\theta} = 0$$

and letting $\sin\theta = \theta$ for small angles, we obtain the familiar equation for harmonic motion

$$\ddot{\theta} + \frac{2g}{3(R-r)}\theta = 0$$

By inspection, the circular frequency of oscillation is

$$\omega_n = \sqrt{\frac{2g}{3(R-r)}}$$

Rayleigh Method. The energy method can be used for multimass systems or for distributed mass systems, provided the motion of every point in the system is known. In systems in which masses are joined by rigid links, levers, or gears the motion of the various masses can be expressed in terms of the motion $\dot{x}$ of some specific point and the system is simply one of single degree of freedom since only one coordinate is necessary. The kinetic energy can then be written as

$$T = \frac{1}{2} m_{\text{eff}} \dot{x}^2 \qquad (2.2\text{-}6)$$

where m_{eff} is the effective mass or an equivalent lumped mass at the specified point. If the stiffness at that point is also known, the natural frequency can be calculated from the simple equation

$$\omega_n = \sqrt{\frac{k}{m_{\text{eff}}}} \qquad (2.2\text{-}7)$$

In distributed mass systems such as springs and beams, a knowledge of the distribution of the vibration amplitude becomes necessary before the kinetic energy can be calculated. Rayleigh* showed that with a reasonable assumption for the shape of the vibration amplitude, it is possible to take into account previously ignored masses and arrive at a better estimate for the fundamental frequency. The following examples illustrate the use of both of these methods.

EXAMPLE 2.2-4

Determine the effect of the mass of the spring on the natural frequency of the system shown in Fig. 2.2-4.

Solution: With $\dot{x}$ equal to the velocity of the lumped mass m, we will assume the velocity of a spring element located a distance y from the fixed end to vary linearly with y as follows

$$\dot{x}\frac{y}{l}$$

The kinetic energy of the spring may then be integrated to

$$T_{\text{add}} = \frac{1}{2} \int_0^l \left(\dot{x}\frac{y}{l}\right)^2 \frac{m_s}{l}\, dy = \frac{1}{2}\frac{m_s}{3}\dot{x}^2$$

and the effective mass is found to be one-third the mass of the spring. Adding this to the lumped mass, the revised natural frequency is

$$\omega_n = \sqrt{\frac{k}{m + \frac{1}{3}m_s}}$$

*John W. Strutt, Lord Rayleigh, *The Theory of Sound*, Vol. 1, 2nd rev. ed. (New York: Dover Publications, 1937), pp. 109–110.

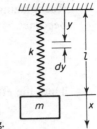

Figure 2.2-4.   Effective mass of spring.

EXAMPLE 2.2-5

Oscillatory systems are often composed of levers, gears, and other linkages that seemingly complicate the analysis. The engine valve system of Fig. 2.2-5 is an example of such a system. The reduction of such a system to a simpler equivalent system is generally desirable.

The rocker arm of moment of inertia J, the valve of mass m_v, and the spring of mass m_s can be reduced to a single mass at A by writing the kinetic energy equation as follows

$$T = \frac{1}{2}J\dot{\theta}^2 + \frac{1}{2}m_v(b\dot{\theta})^2 + \frac{1}{2}\left(\frac{m_s}{3}\right)(b\dot{\theta})^2 = \frac{1}{2}\left(J + m_v b^2 + \frac{1}{3}m_s b^2\right)\dot{\theta}^2$$

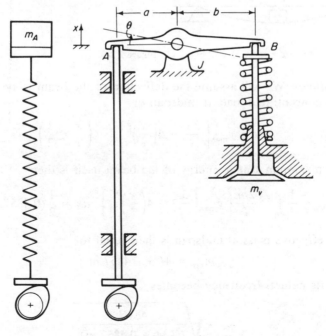

Figure 2.2-5.   Engine valve system.

Recognizing that the velocity at A is $\dot{x} = a\dot{\theta}$, the above equation becomes

$$T = \frac{1}{2}\left(\frac{J + m_v b^2 + \frac{1}{3}m_s b^2}{a^2}\right)\dot{x}^2$$

Thus the effective mass at A is

$$m_A = \left(\frac{J + m_v b^2 + \frac{1}{3}m_s b^2}{a^2}\right)$$

If the push rod is now reduced to a spring and an additional mass at the end A, the entire system is reduced to a single spring and a mass as shown in Fig. 2.2-5.

EXAMPLE 2.2-6

A simply supported beam of total mass m has a concentrated mass M at midspan. Determine the effective mass of the system at midspan and find its fundamental frequency. The deflection under the load due to a concentrated force P applied at midspan is $Pl^3/48EI$. (See Fig. 2.2-6 and table of stiffness at end of chapter.)

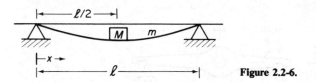

Figure 2.2-6.

Solution: We will assume the deflection of the beam to be that due to a concentrated load at midspan or

$$y = y_{max}\left[\frac{3x}{l} - 4\left(\frac{x}{l}\right)^3\right] \qquad \left(\frac{x}{l} < \frac{1}{2}\right)$$

The maximum kinetic energy of the beam itself is then

$$T_{max} = \frac{1}{2}\int_0^{l/2}\frac{2m}{l}\left\{\dot{y}_{max}\left[\frac{3x}{l} - 4\left(\frac{x}{l}\right)^3\right]\right\}^2 dx = \frac{1}{2}(0.4857 \text{ m})\dot{y}_{max}^2$$

The effective mass at midspan is then equal to

$$m_{eff} = M + 0.4857 \text{ m}$$

and its natural frequency becomes

$$\omega_n = \sqrt{\frac{48EI}{l^3(M + 0.4857 \text{ m})}}$$

2.3 VISCOUSLY DAMPED FREE VIBRATION

When a linear system of one degree of freedom is excited, its response will depend on the type of excitation and the damping which is present. The equation of motion will in general be of the form

$$m\ddot{x} + F_d + kx = F(t) \qquad (2.3-1)$$

where $F(t)$ is the excitation and F_d the damping force. Although the actual description of the damping force is difficult, ideal damping models can be assumed that will often result in satisfactory prediction of the response. Of these models, the viscous damping force, proportional to the velocity, leads to the simplest mathematical treatment.

Viscous damping force is expressed by the equation

$$F_d = c\dot{x} \qquad (2.3-2)$$

where c is a constant of proportionality. Symbolically it is designated by a dashpot as shown in Fig. 2.3-1. From the free-body diagram the equation of motion is seen to be

$$m\ddot{x} + c\dot{x} + kx = F(t) \qquad (2.3-3)$$

The solution of the above equation has two parts. If $F(t) = 0$, we have the homogeneous differential equation whose solution corresponds physically to that of *free-damped vibration*. With $F(t) \neq 0$, we obtain the particular solution that is due to the excitation irrespective of the homogeneous solution. We will first examine the homogeneous equation that will give us some understanding of the role of damping.

With the homogeneous equation

$$m\ddot{x} + c\dot{x} + kx = 0 \qquad (2.3-4)$$

the traditional approach is to assume a solution of the form

$$x = e^{st} \qquad (2.3-5)$$

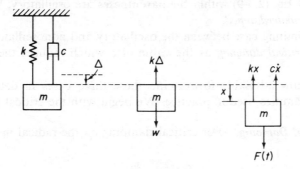

Figure 2.3-1.

25

where s is a constant. Upon substitution into the differential equation, we obtain

$$(ms^2 + cs + k)e^{st} = 0$$

which is satisfied for all values of t when

$$s^2 + \frac{c}{m}s + \frac{k}{m} = 0 \qquad (2.3\text{-}6)$$

Equation (2.3-6), which is known as the *characteristic equation*, has two roots

$$s_{1,2} = -\frac{c}{2m} \pm \sqrt{\left(\frac{c}{2m}\right)^2 - \frac{k}{m}} \qquad (2.3\text{-}7)$$

Hence, the general solution is given by the equation

$$x = Ae^{s_1 t} + Be^{s_2 t} \qquad (2.3\text{-}8)$$

where A and B are constants to be evaluated from the initial conditions $x(0)$ and $\dot{x}(0)$.

Equation (2.3-7) substituted into (2.3-8) gives

$$x = e^{-(c/2m)t}\left(Ae^{\sqrt{(c/2m)^2 - k/m}\,t} + Be^{-\sqrt{(c/2m)^2 - k/m}\,t}\right) \qquad (2.3\text{-}9)$$

The first term $e^{-(c/2m)t}$ is simply an exponentially decaying function of time. The behavior of the terms in the parentheses, however, depends on whether the numerical value within the radical is positive, zero, or negative.

When the damping term $(c/2m)^2$ is larger than k/m, the exponents in the above equation are real numbers and no oscillations are possible. We refer to this case as *overdamped*.

When the damping term $(c/2m)^2$ is less than k/m, the exponent becomes an imaginary number, $\pm i\sqrt{k/m - (c/2m)^2}\,t$. Since

$$e^{\pm i\sqrt{k/m - (c/2m)^2}\,t} = \cos\sqrt{\frac{k}{m} - \left(\frac{c}{2m}\right)^2}\,t \pm i\sin\sqrt{\frac{k}{m} - \left(\frac{c}{2m}\right)^2}\,t$$

the terms of Eq. (2.3-9) within the parentheses are oscillatory. We refer to this case as *underdamped*.

As a limiting case between the oscillatory and nonoscillatory motion, we define *critical damping* as the value of c which reduces the radical to zero.

It is now advisable to examine these three cases in detail, and in terms of quantities used in practice. We begin with the critical damping.

Critical Damping.   For critical damping c_c, the radical in Eq. (2.3-9) is zero.

$$\left(\frac{c_c}{2m}\right)^2 = \frac{k}{m} = \omega_n^2$$

or

$$c_c = 2\sqrt{km} = 2m\omega_n \qquad (2.3\text{-}10)$$

It is convenient to express the value of any damping in terms of the critical damping by the nondimensional ratio

$$\zeta = \frac{c}{c_c} \qquad (2.3\text{-}11)$$

which is called the *damping ratio*. We now express the roots of Eq. (2.3-7) in terms of ζ by noting that

$$\frac{c}{2m} = \zeta \frac{c_c}{2m} = \zeta \omega_n$$

Equation (2.3-7) then becomes

$$s_{1,2} = \left(-\zeta \pm \sqrt{\zeta^2 - 1} \right)\omega_n \qquad (2.3\text{-}12)$$

and the three cases of damping previously discussed now depend on whether ζ is greater than, less than, or equal to unity.

Figure 2.3-2 shows Eq. (2.3-12) plotted in a complex plane with ζ along the horizontal axis. If $\zeta = 0$, Eq. (2.3-12) reduces to $s_{1,2}/\omega_n = \pm i$ so that the roots on the imaginary axis correspond to the undamped case. For $0 \le \zeta \le 1$, Eq. (2.3-12) can be rewritten as

$$\frac{s_{1,2}}{\omega_n} = -\zeta \pm i\sqrt{1 - \zeta^2} \; .$$

The roots s_1 and s_2 are then conjugate complex points on a circular arc

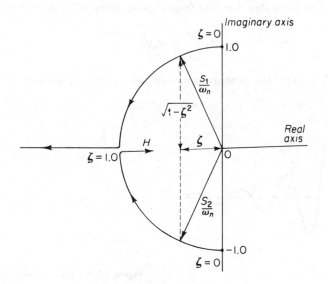

Figure 2.3-2.

converging at the point $s_{1,2}/\omega_n = -1.0$. As ζ increases beyond unity, the roots separate along the horizontal axis and remain real numbers. With this diagram in mind, we are now ready to examine the solution given by Eq. (2.3-9).

Oscillatory Motion.   [$\zeta < 1.0$ (Underdamped Case).] Substituting Eq. (2.3-12) into (2.3-8), the general solution becomes

$$x = e^{-\zeta\omega_n t}\left(Ae^{i\sqrt{1-\zeta^2}\,\omega_n t} + Be^{-i\sqrt{1-\zeta^2}\,\omega_n t}\right) \qquad (2.3\text{-}13)$$

The above equation can also be written in either of the following two forms

$$x = Xe^{-\zeta\omega_n t}\sin\left(\sqrt{1-\zeta^2}\,\omega_n t + \phi\right) \qquad (2.3\text{-}14)$$

$$= e^{-\zeta\omega_n t}\left(C_1\sin\sqrt{1-\zeta^2}\,\omega_n t + C_2\cos\sqrt{1-\zeta^2}\,\omega_n t\right) \qquad (2.3\text{-}15)$$

where the arbitrary constants X, ϕ, or C_1, C_2 are determined from initial conditions. With initial conditions $x(0)$ and $\dot{x}(0)$, Eq. (2.3-15) can be shown to reduce to

$$x = e^{-\zeta\omega_n t}\left(\frac{\dot{x}(0) + \zeta\omega_n x(0)}{\omega_n\sqrt{1-\zeta^2}}\sin\sqrt{1-\zeta^2}\,\omega_n t + x(0)\cos\sqrt{1-\zeta^2}\,\omega_n t\right)$$

$$(2.3\text{-}16)$$

The equation indicates that the *frequency of damped oscillation* is equal to

$$\omega_d = \frac{2\pi}{\tau_d} = \omega_n\sqrt{1-\zeta^2} \qquad (2.3\text{-}17)$$

Figure 2.3-3 shows the general nature of the oscillatory motion.

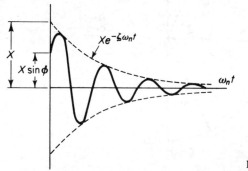

Figure 2.3-3.   Damped oscillation $\zeta < 1.0$.

Nonoscillatory Motion.   [$\zeta > 1.0$ (Overdamped Case).] As ζ exceeds unity, the two roots remain on the real axis of Fig. 2.3-2 and separate, one increasing and the other decreasing. The general solution then becomes

$$x = Ae^{(-\zeta + \sqrt{\zeta^2 - 1})\omega_n t} + Be^{(-\zeta - \sqrt{\zeta^2 - 1})\omega_n t} \qquad (2.3\text{-}18)$$

where

$$A = \frac{\dot{x}(0) + (\zeta + \sqrt{\zeta^2 - 1})\omega_n x(0)}{2\omega_n \sqrt{\zeta^2 - 1}}$$

and

$$B = \frac{-\dot{x}(0) - (\zeta - \sqrt{\zeta^2 - 1})\omega_n x(0)}{2\omega_n \sqrt{\zeta^2 - 1}}$$

The motion is an exponentially decreasing function of time as shown in Fig. 2.3-4, and is referred to as *aperiodic*.

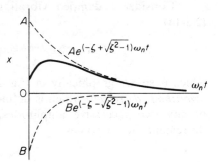

Figure 2.3-4. Aperiodic motion $\zeta > 1.0$.

Critically Damped Motion.   [$\zeta = 1.0$] For $\zeta = 1$, we obtain a double root $s_1 = s_2 = -\omega_n$, and the two terms of Eq. (2.3-8) combine to form a single term

$$x = (A + B)e^{-\omega_n t} = Ce^{-\omega_n t}$$

which is lacking in the number of constants required to satisfy the two initial conditions. The solution for the initial conditions $x(0)$ and $\dot{x}(0)$ can be found from Eq. (2.3-16) by letting $\zeta \to 1$

$$x = e^{-\omega_n t}\{[\dot{x}(0) + \omega_n x(0)]t + x(0)\} \qquad (2.3\text{-}19)$$

Figure 2.3-5 shows three types of response with initial displacement $x(0)$. The moving parts of many electrical meters and instruments are critically damped to avoid overshoot and oscillation.

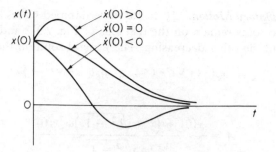

Figure 2.3-5. Critically damped motion $\zeta = 1.0$.

2.4 LOGARITHMIC DECREMENT

A convenient way to determine the amount of damping present in a system is to measure the rate of decay of free oscillations. The larger the damping, the greater will be the rate of decay.

Consider a damped vibration expressed by the general equation (2.3-14)

$$x = Xe^{-\zeta\omega_n t} \sin\left(\sqrt{1 - \zeta^2}\, \omega_n t + \phi\right)$$

which is shown graphically in Fig. 2.4-1. We introduce here a term called *logarithmic decrement* which is defined as the natural logarithm of the ratio of any two successive amplitudes. The expression for the logarithmic decrement then becomes

$$\delta = \ln\frac{x_1}{x_2} = \ln\frac{e^{-\zeta\omega_n t_1} \sin\left(\sqrt{1 - \zeta^2}\, \omega_n t_1 + \phi\right)}{e^{-\zeta\omega_n(t_1 + \tau_d)} \sin\left(\sqrt{1 - \zeta^2}\, \omega_n(t_1 + \tau_d) + \phi\right)} \quad (2.4\text{-}1)$$

and since the values of the sines are equal when the time is increased by

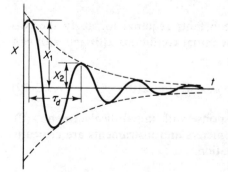

Figure 2.4-1. Rate of decay of oscillation measured by the logarithmic decrement.

Logarithmic Decrement   31

the damped period τ_d, the above relation reduces to

$$\delta = \ln \frac{e^{-\zeta \omega_n t_1}}{e^{-\zeta \omega_n (t_1 + \tau_d)}} = \ln e^{\zeta \omega_n \tau_d} = \zeta \omega_n \tau_d \qquad (2.4\text{-}2)$$

Substituting for the damped period, $\tau_d = 2\pi / \omega_n \sqrt{1 - \zeta^2}$, the expression for the logarithmic decrement becomes

$$\delta = \frac{2\pi \zeta}{\sqrt{1 - \zeta^2}} \qquad (2.4\text{-}3)$$

which is an exact equation.

When ζ is small, $\sqrt{1 - \zeta^2} \cong 1$, and an approximate equation

$$\delta \cong 2\pi \zeta \qquad (2.4\text{-}4)$$

is obtained. Figure 2.4-2 shows a plot of the exact and approximate values of δ as a function of ζ.

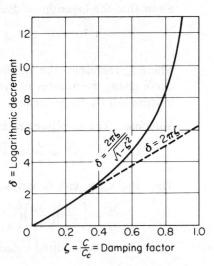

Figure 2.4-2. Logarithmic decrement as function of ζ.

EXAMPLE 2.4-1

The following data are given for a vibrating system with viscous damping: $w = 10$ lb, $k = 30$ lb/in., and $c = 0.12$ lb/in. per sec. Determine the logarithmic decrement and the ratio of any two successive amplitudes.

Solution: The undamped natural frequency of the system in radians per second is

$$\omega_n = \sqrt{\frac{k}{m}} = \sqrt{\frac{30 \times 386}{10}} = 34.0 \text{ rad/sec}$$

The critical damping coefficient c_c and damping factor ζ are

$$c_c = 2m\omega_n = 2 \times \tfrac{10}{386} \times 34.0 = 1.76 \text{ lb/in. per sec}$$

$$\zeta = \frac{c}{c_c} = \frac{0.12}{1.76} = 0.0681$$

The logarithmic decrement, from Eq. (2.4-3), is

$$\delta = \frac{2\pi\zeta}{\sqrt{1 - \zeta^2}} = \frac{2\pi \times 0.0681}{\sqrt{1 - 0.0681^2}} = 0.429$$

The amplitude ratio for any two consecutive cycles is

$$\frac{x_1}{x_2} = e^\delta = e^{0.429} = 1.54$$

EXAMPLE 2.4-2

Show that the logarithmic decrement is also given by the equation

$$\delta = \frac{1}{n}\ln\frac{x_0}{x_n}$$

where x_n represents the amplitude after n cycles have elapsed. Plot a curve giving the number of cycles elapsed against ζ for the amplitude to diminish by 50 per cent.

Solution: The amplitude ratio for any two consecutive amplitudes is

$$\frac{x_0}{x_1} = \frac{x_1}{x_2} = \frac{x_2}{x_3} = \cdots = \frac{x_{n-1}}{x_n} = e^\delta$$

The ratio x_0/x_n can be written as

$$\frac{x_0}{x_n} = \left(\frac{x_0}{x_1}\right)\left(\frac{x_1}{x_2}\right)\left(\frac{x_2}{x_3}\right)\cdots\left(\frac{x_{n-1}}{x_n}\right) = (e^\delta)^n = e^{n\delta}$$

from which the required equation is obtained as

$$\delta = \frac{1}{n}\ln\frac{x_0}{x_n}$$

To determine the number of cycles elapsed for 50 per cent reduction in amplitude, we obtain the following relation from the above equation

$$\delta \cong 2\pi\zeta = \frac{1}{n}\ln 2 = \frac{0.693}{n}$$

$$n\zeta = \frac{0.693}{2\pi} = 0.110$$

The last equation is that of a rectangular hyperbola, and is plotted in Fig. 2.4-3.

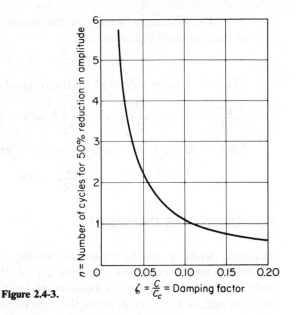

Figure 2.4-3.

EXAMPLE 2.4-3

For small damping, show that the logarithmic decrement is expressible in terms of the vibrational energy U and the energy dissipated per cycle ΔU.

Solution: Figure 2.4-4 shows a damped vibration with consecutive amplitudes $x_1, x_2, x_3, \ldots$. From the definition of the logarithmic decrement $\delta = \ln x_1/x_2$, we can write the ratio of amplitudes in exponential form:

$$\frac{x_2}{x_1} = e^{-\delta} = 1 - \delta + \frac{\delta^2}{2!} - \cdots$$

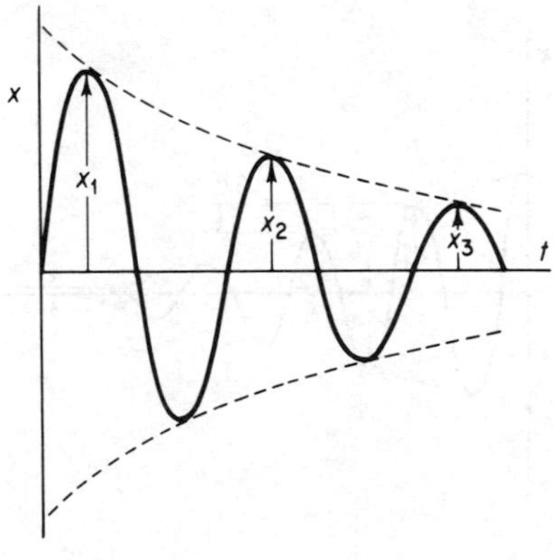

Figure 2.4-4.

The vibrational energy of the system is that stored in the spring at maximum displacement, or

$$U_1 = \tfrac{1}{2}kx_1^2, \qquad U_2 = \tfrac{1}{2}kx_2^2$$

The loss of energy divided by the original energy is

$$\frac{U_1 - U_2}{U_1} = 1 - \frac{U_2}{U_1} = 1 - \left(\frac{x_2}{x_1}\right)^2 = 1 - e^{-2\delta} = 2\delta - \frac{(2\delta)^2}{2!} + \cdots$$

Thus for small δ we obtain the relationship

$$\frac{\Delta U}{U} = 2\delta$$

2.5   COULOMB DAMPING

Coulomb damping results from the sliding of two dry surfaces. The damping force is equal to the product of the normal force and the coefficient of friction μ and is assumed to be independent of the velocity, once the motion is initiated. Since the sign of the damping force is always opposite to that of the velocity, the differential equation of motion for each sign is valid only for half cycle intervals.

To determine the decay of amplitude, we resort to the work-energy principle of equating the work done to the change in kinetic energy. Choosing a half cycle starting at the extreme position with velocity equal to zero and the amplitude equal to X_1, the change in the kinetic energy is zero and the work done on m is also zero.

$$\tfrac{1}{2}k\left(X_1^2 - X_{-1}^2\right) - F_d(X_1 + X_{-1}) = 0$$

or

$$\tfrac{1}{2}k(X_1 - X_{-1}) = F_d$$

where X_{-1} is the amplitude after the half cycle as shown in Fig. 2.5-1.

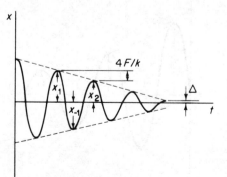

Figure 2.5-1.   Free vibration with coulomb damping.

Repeating this procedure for the next half cycle, a further decrease in amplitude of $2F_d/k$ will be found, so that the decay in amplitude per cycle is a constant and equal to

$$X_1 - X_2 = \frac{4F_d}{k} \tag{2.5-1}$$

The motion will cease, however, when the amplitude becomes less than Δ, at which position the spring force is insufficient to overcome the static friction force, which is generally greater than the kinetic friction force. It can also be shown that the frequency of oscillation is $\omega_\mu = \sqrt{k/m}$, which is the same as that of the undamped system.

Figure 2.5-1 shows the free vibration of a system with Coulomb damping. It should be noted that the amplitudes decay linearly with time.

Table of Spring Stiffness.

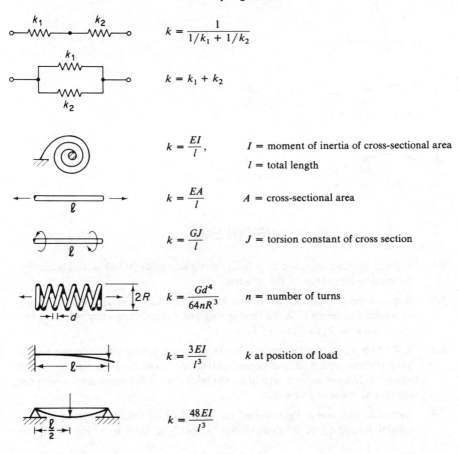

$k = \dfrac{1}{1/k_1 + 1/k_2}$

$k = k_1 + k_2$

$k = \dfrac{EI}{l}$, I = moment of inertia of cross-sectional area

l = total length

$k = \dfrac{EA}{l}$ A = cross-sectional area

$k = \dfrac{GJ}{l}$ J = torsion constant of cross section

$k = \dfrac{Gd^4}{64nR^3}$ n = number of turns

$k = \dfrac{3EI}{l^3}$ k at position of load

$k = \dfrac{48EI}{l^3}$

Table of Spring Stiffness. (*Continued*)

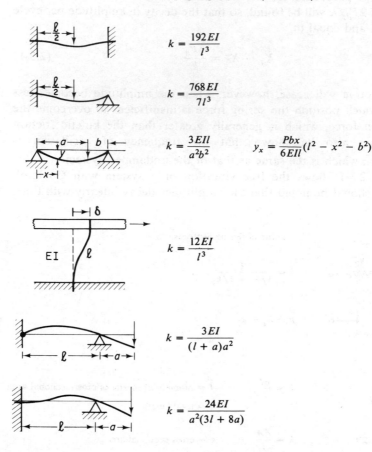

$$k = \frac{192EI}{l^3}$$

$$k = \frac{768EI}{7l^3}$$

$$k = \frac{3EIl}{a^2b^2} \qquad y_x = \frac{Pbx}{6EIl}(l^2 - x^2 - b^2)$$

$$k = \frac{12EI}{l^3}$$

$$k = \frac{3EI}{(l + a)a^2}$$

$$k = \frac{24EI}{a^2(3l + 8a)}$$

PROBLEMS

2-1 A 0.453-kg mass attached to a light spring elongates it 7.87 mm. Determine the natural frequency of the system.

2-2 A spring-mass system k_1, m, has a natural frequency of f_1. If a second spring k_2 is added in series with the first spring, the natural frequency is lowered to $\frac{1}{2}f_1$. Determine k_2 in terms of k_1.

2-3 A 4.53-kg mass attached to the lower end of a spring whose upper end is fixed vibrates with a natural period of 0.45 sec. Determine the natural period when a 2.26-kg mass is attached to the midpoint of the same spring with the upper and lower ends fixed.

2-4 An unknown mass m kg attached to the end of an unknown spring k has a natural frequency of 94 cpm. When a 0.453-kg mass is added to m, the natural frequency is lowered to 76.7 cpm. Determine the unknown mass m and the spring constant k N/m.

2-5   A mass m_1 hangs from a spring k (N/m) and is in static equilibrium. A second mass m_2 drops through a height h and sticks to m_1 without rebound, as shown in Fig. P2-5. Determine the subsequent motion.

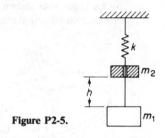

Figure P2-5.

2-6   The ratio k/m of a spring-mass system is given as 4.0. If the mass is deflected 2 cm down, measured from its equilibrium position, and given an upward velocity of 8 cm/sec, determine its amplitude and maximum acceleration.

2-7   A flywheel weighing 70 lb was allowed to swing as a pendulum about a knife-edge at the inner side of the rim as shown in Fig. P2-7. If the measured period of oscillation was 1.22 sec, determine the moment of inertia of the flywheel about its geometric axis.

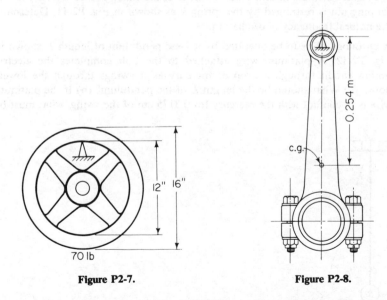

Figure P2-7.   **Figure P2-8.**

2-8   A connecting rod weighing 21.35 N oscillates 53 times in 1 min when suspended as shown in Fig. P2-8. Determine its moment of inertia about its center of gravity, which is located 0.254 m from the point of support.

2-9 A flywheel of mass M is suspended in the horizontal plane by three wires of 1.829 m length equally spaced around a circle of 0.254 m radius. If the period of oscillation about a vertical axis through the center of the wheel is 2.17 sec, determine its radius of gyration.

2-10 A wheel and axle assembly of moment inertia J is inclined from the vertical by an angle α as shown in Fig. P2-10. Determine the frequency of oscillation due to a small unbalance weight w lb at a distance a in. from the axle.

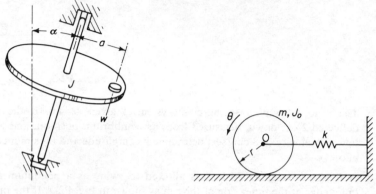

Figure P2-10.   Figure P2-11.

2-11 A cylinder of mass m and mass moment of inertia J_0 is free to roll without slipping but is restrained by the spring k as shown in Fig. P2-11. Determine the natural frequency of oscillation.

2-12 A chronograph is to be operated by a 2-sec pendulum of length L shown in Fig. P2-12. A platinum wire attached to the bob completes the electric timing circuit through a drop of mercury as it swings through the lowest point. (a) What should be the length L of the pendulum? (b) If the platinum wire is in contact with the mercury for 0.3175 cm of the swing, what must be

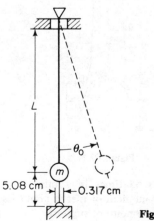

Figure P2-12.

the amplitude θ to limit the duration of contact to 0.01 sec? (Assume that the velocity during contact is constant and that the amplitude of oscillation is small.)

2-13 A hydrometer float, shown in Fig. P2-13, is used to measure the specific gravity of liquids. The mass of the float is 0.0372 kg, and the diameter of the cylindrical section protruding above the surface is 0.0064 m. Determine the period of vibration when the float is allowed to bob up and down in a fluid of specific gravity 1.20.

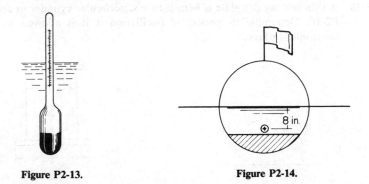

Figure P2-13.   **Figure P2-14.**

2-14 A spherical buoy 3 ft in diameter is weighted to float half out of water as shown in Fig. P2-14. The center of gravity of the buoy is 8 in. below its geometric center, and the period of oscillation in rolling motion is 1.3 sec. Determine the moment of inertia of the buoy about its rotational axis.

2-15 The oscillatory characteristics of ships in rolling motion depends on the position of the metacenter M with respect to the center of gravity G. The metacenter M represents the point of intersection of the line of action of the buoyant force and the center line of the ship, and its distance h measured from G is the metacentric height as shown in Fig. P2-15. The position of M depends on the shape of the hull and is independent of the angular inclination θ of the ship for small values of θ. Show that the period of the

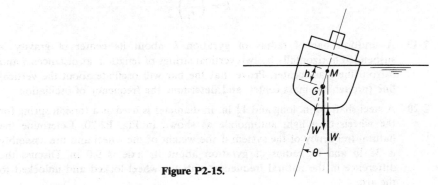

Figure P2-15.

rolling motion is given by

$$\tau = 2\pi\sqrt{\frac{J}{Wh}}$$

where J is the mass moment of inertia of the ship about its roll axis and W is the weight of the ship. In general, the position of the roll axis is unknown and J is obtained from the period of oscillation determined from a model test.

2-16 A thin rectangular plate is bent into a semicircular cylinder as shown in Fig. P2-16. Determine its period of oscillation if it is allowed to rock on a horizontal surface.

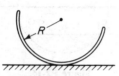

Figure P2-16.

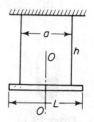

Figure P2-17.

2-17 A uniform bar of length L and weight W is suspended symmetrically by two strings as shown in Fig. P2-17. Set up the differential equation of motion for small angular oscillations of the bar about the vertical axis $O-O$, and determine its period.

2-18 A uniform bar of length L is suspended in the horizontal position by two vertical strings of equal length attached to the ends. If the period of oscillation in the plane of the bar and strings is t_1, and the period of oscillation about a vertical line through the center of gravity of the bar is t_2, show that the radius of gyration of the bar about the center of gravity is given by the expression

$$k = \left(\frac{t_2}{t_1}\right)\frac{L}{2}$$

2-19 A uniform bar of radius of gyration k about its center of gravity is suspended horizontally by two vertical strings of length h, at distances a and b from the mass center. Prove that the bar will oscillate about the vertical line through the mass center, and determine the frequency of oscillation.

2-20 A steel shaft 50 in. long and $1\frac{1}{2}$ in. in diameter is used as a torsion spring for the wheels of a light automobile as shown in Fig. P2-20. Determine the natural frequency of the system if the weight of the wheel and tire assembly is 38 lb and its radius of gyration about its axle is 9.0 in. Discuss the difference in the natural frequency with the wheel locked and unlocked to the arm.

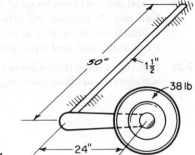

Figure P2-20.

2-21 Using the energy method, show that the natural period of oscillation of the fluid in a *U*-tube manometer shown in Fig. P2-21 is

$$\tau = 2\pi\sqrt{\frac{l}{2g}}$$

where *l* = length of the fluid column.

Figure P2-21.   **Figure P2-22.**

2-22 Figure P2-22 shows a simplified model of a single-story building. The columns are assumed to be rigidly imbedded at the ends. Determine its natural period τ. Refer to the table of stiffness at the end of the chapter.

2-23 Determine the effective mass of the columns of Prob. 2-22 assuming the deflection to be

$$y = \frac{1}{2}y_{max}\left(1 - \cos\frac{\pi x}{l}\right)$$

2-24 Determine the effective mass at point *n* for the system shown in Fig. P2-24 and its natural frequency.

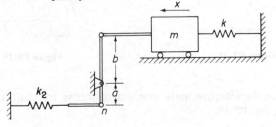

Figure P2-24.

2-25   A uniform cantilever beam of total mass m has a concentrated mass M at its free end. Determine the effective mass of the beam to be added to M assuming the deflection to be that of a massless beam with a concentrated force at the end, and write the equation for its fundamental frequency.

2-26   Determine the effective mass of the rocket engine shown in Fig. P2-26 to be added to the actuator mass m_1.

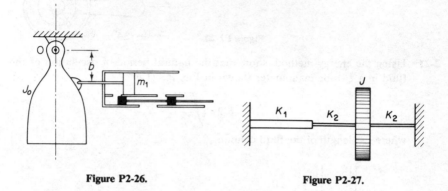

Figure P2-26.   Figure P2-27.

2-27   Determine the effective rotational stiffness of the shaft in Fig. P2-27 and calculate its natural period.

2-28   For purposes of analysis, it is desired to reduce the system of Fig. P2-28 to a simple linear spring-mass system of effective mass m_{eff} and effective stiffness k_{eff}. Determine m_{eff} and k_{eff} in terms of the given quantities.

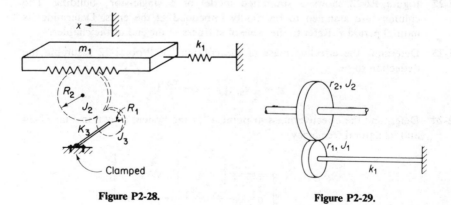

Figure P2-28.   Figure P2-29.

2-29   Determine the effective mass moment of inertia for shaft 1 in the system shown in Fig. P2-29.

2-30 Determine the kinetic energy of the system shown in Fig. P2-30 in terms of $\dot{x}$. Determine the stiffness at m_0 and write the expression for the natural frequency.

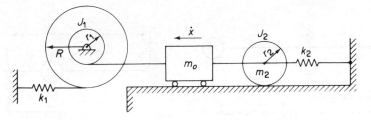

Figure P2-30.

2-31 Tachometers are a reed-type of frequency-measuring instrument consisting of small cantilever beams with weights attached at the ends. When the frequency of vibration corresponds to the natural frequency of one of the reeds, it will vibrate, thereby indicating the frequency. How large a weight must be placed on the end of a reed made of spring steel 0.1016 cm thick, 0.635 cm wide, and 8.890 cm long for a natural frequency of 20 cps?

2-32 A mass of 0.907 kg is attached to the end of a spring with a stiffness of 7.0 N/cm. Determine the critical damping coefficient.

2-33 To calibrate a dashpot, the velocity of the plunger was measured when a given force was applied to it. If a $\frac{1}{2}$ lb weight produced a constant velocity of 1.20 in./sec, determine the damping factor ζ when used with the system of Problem 2-32.

2-34 A vibrating system is started under the following initial conditions: $x = 0$, $\dot{x} = v_0$. Determine the equation of motion when (a) $\zeta = 2.0$, (b) $\zeta = 0.50$, (c) $\zeta = 1.0$. Plot non-dimensional curves for the three cases with $\omega_n t$ as abscissa and $x\omega_n/v_0$ as ordinate.

2-35 A vibrating system consisting of a mass of 2.267 kg and a spring of stiffness 17.5 N/cm is viscously damped such that the ratio of any two consecutive amplitudes is 1.00 and 0.98. Determine (a) the natural frequency of the damped system, (b) the logarithmic decrement, (c) the damping factor, and (d) the damping coefficient.

2-36 A vibrating system consists of a mass of 4.534 kg, a spring of stiffness 35.0 N/cm, and a dashpot with a damping coefficient of 0.1243 N/cm per sec. Find (a) the damping factor, (b) the logarithmic decrement, and (c) the ratio of any two consecutive amplitudes.

2-37 A vibrating system has the following constants: $m = 17.5$ kg, $k = 70.0$ N/cm, and $c = 0.70$ N/cm per sec. Determine (a) the damping factor, (b) the natural frequency of damped oscillation, (c) the logarithmic decrement, and (d) the ratio of any two consecutive amplitudes.

2-38 Set up the differential equation of motion for the system shown in Fig. P2-38. Determine the expression for (a) the critical damping coefficient, and (b) the natural frequency of damped oscillation.

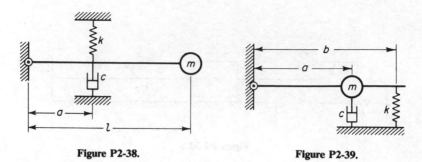

Figure P2-38.   **Figure P2-39.**

2-39 Write the differential equation of motion for the system shown in Fig. P2-39 and determine the natural frequency of damped oscillation and the critical damping coefficient.

2-40 A spring-mass system with viscous damping is displaced from the equilibrium position and released. If the amplitude diminished by 5% each cycle, what fraction of the critical damping does the system have?

2-41 A rigid uniform bar of mass m and length l is pinned at O and supported by a spring and viscous damper as shown in Fig. P2-41. Measuring θ from the static equilibrium position, determine (a) the equation for small θ (the moment of inertia of the bar about 0 is $ml^2/3$), (b) the equation for the undamped natural frequency, and (c) the expression for critical damping.

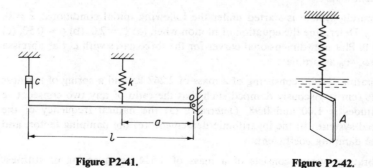

Figure P2-41.   **Figure P2-42.**

2-42 A thin plate of area A and weight W is attached to the end of a spring and is allowed to oscillate in a viscous fluid as shown in Fig. P2-42. If τ_1 is the natural period of undamped oscillation (that is, with the system oscillating in air), and τ_2 the damped period with the plate immersed in the fluid, show that

$$\mu = \frac{2\pi W}{gA\tau_1\tau_2}\sqrt{\tau_2^2 - \tau_1^2}$$

where the damping force on the plate is $F_d = \mu 2Av$, $2A$ is the total surface area of the plate, and v is its velocity.

2-43 A gun barrel weighing 1200 lb has a recoil spring of stiffness 20,000 lb/ft. If the barrel recoils 4 ft on firing, determine (a) the initial recoil velocity of the barrel, (b) the critical damping coefficient of a dashpot which is engaged at the end of the recoil stroke, and (c) the time required for the barrel to return to a position 2 in. from its initial position.

2-44 A piston of mass 4.53 kg is traveling in a tube with a velocity of 15.24 m/s and engages a spring and damper as shown in Fig. P2-44. Determine the maximum displacement of the piston after engaging the spring-damper. How many seconds does it take?

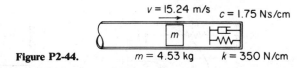

Figure P2-44. $m = 4.53$ kg $k = 350$ N/cm

2-45 A shock absorber is to be designed so that its overshoot is 10% of the initial displacement when released. Determine ζ_1. If ζ is made equal to $\frac{1}{2}\zeta_1$, what will be the overshoot?

2-46 Discuss the limitations of the equation $\Delta U/U = 2\delta$ by considering the case where $x_2/x_1 = \frac{1}{2}$.

2-47 Determine the effective stiffness of the springs shown in Fig. P2-47.

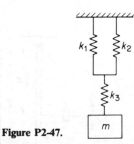

Figure P2-47.

2-48 Determine the flexibility of a simply supported uniform beam of length L at a point $\frac{1}{3}L$ from the end.

2-49 Determine the effective stiffness of the system shown in Fig. P2-49, in terms of the displacement x.

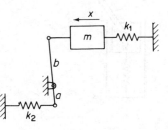

Figure P2-49.

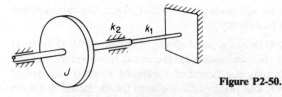

Figure P2-50.

2-50   Determine the effective stiffness of the torsional system shown in Fig. P2-50. The two shafts in series have torsional stiffnesses of k_1 and k_2.

2-51   A spring-mass system m, k, is started with an initial displacement of unity and an initial velocity of zero. Plot $\ln X$ vs. n where X is amplitude at cycle n

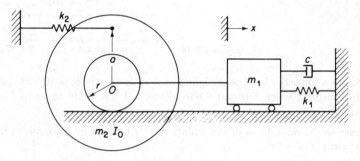

Figure P2-52.

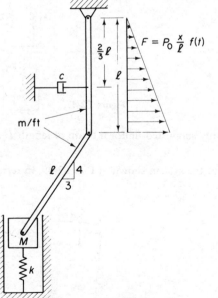

Figure P2-53.

for (a) viscous damping with $\zeta = 0.05$, and (b) Coulomb damping with damping force $F_d = 0.05\ k$. When will the two amplitudes be equal?

2-52 Determine the differential equation of motion and establish the critical damping for the system shown in Fig. P2-52.

2-53 Determine the differential equation of motion for free vibration of the system shown in Fig. P2-53.

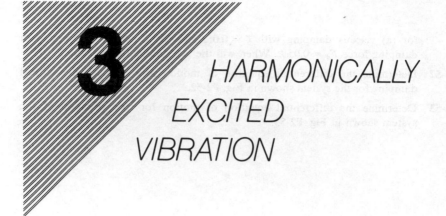

3 HARMONICALLY EXCITED VIBRATION

When a system is subjected to forced harmonic excitation, its vibration response takes place at the same frequency as that of the excitation. Common sources of harmonic excitation are unbalance in rotating machines, forces produced by reciprocating machines, or the motion of the machine itself. These excitations may be undesirable to equipment whose operation may be disturbed or to the safety of the structure if large vibration amplitudes develop. Resonance is to be avoided in most cases, and to prevent large amplitudes from developing, dampers and absorbers are often used. Discussion of their behavior is of importance for their intelligent use. Finally, the theory of vibration measuring instruments is presented as a tool for vibration analysis.

3.1 FORCED HARMONIC VIBRATION

Harmonic excitation is often encountered in engineering systems. It is commonly produced by the unbalance in rotating machinery. Although pure harmonic excitation is less likely to occur than periodic or other types of excitation, understanding the behavior of a system undergoing harmonic excitation is essential in order to comprehend how the system will respond to more general types of excitation. Harmonic excitation may be in the form of a force or displacement of some point in the system.

   We will first consider a single degree of freedom system with viscous damping, excited by a harmonic force $F_0 \sin \omega t$ as shown in Fig. 3.1-1. Its

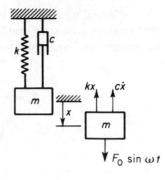

Figure 3.1-1. Viscously damped system with harmonic excitation.

differential equation of motion is found from the free-body diagram to be

$$m\ddot{x} + c\dot{x} + kx = F_0 \sin \omega t \tag{3.1-1}$$

The solution to this equation consists of two parts, the *complimentary function*, which is the solution of the homogeneous equation, and the *particular integral*. The complimentary function, in this case, is a damped free vibration that was discussed in Chapter 2.

The particular solution to the above equation is a steady-state oscillation of the same frequency ω as that of the excitation. We can assume the particular solution to be of the form

$$x = X \sin(\omega t - \phi) \tag{3.1-2}$$

where X is the amplitude of oscillation and ϕ is the phase of the displacement with respect to the exciting force.

The amplitude and phase in the above equation are found by substituting Eq. (3.1-2) into the differential equation Eq. (3.1-1). Remembering that in harmonic motion the phases of the velocity and acceleration are ahead of the displacement by 90° and 180° respectively, the terms of the differential equation can also be displayed graphically as in Fig. 3.1-2. It is easily seen from this diagram that

$$X = \frac{F_0}{\sqrt{(k - m\omega^2)^2 + (c\omega)^2}} \tag{3.1-3}$$

and

$$\phi = \tan^{-1} \frac{c\omega}{k - m\omega^2} \tag{3.1-4}$$

Figure 3.1-2. Vector relationship for forced vibration with damping.

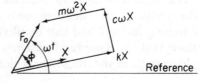

We will now express Eqs. (3.1-3) and (3.1-4) in nondimensional form that enables a concise graphical presentation of these results. Dividing the numerator and denominator of Eqs. (3.1-3) and (3.1-4) by k, we obtain

$$X = \frac{\dfrac{F_0}{k}}{\sqrt{\left(1 - \dfrac{m\omega^2}{k}\right)^2 + \left(\dfrac{c\omega}{k}\right)^2}} \qquad (3.1\text{-}5)$$

and

$$\tan\phi = \frac{\dfrac{c\omega}{k}}{1 - \dfrac{m\omega^2}{k}} \qquad (3.1\text{-}6)$$

The above equations may be further expressed in terms of the following quantities:

$$\omega_n = \sqrt{\frac{k}{m}} \quad = \text{natural frequency of undamped oscillation}$$

$$c_c = 2m\omega_n = \text{critical damping}$$

$$\zeta = \frac{c}{c_c} = \text{damping factor}$$

$$\frac{c\omega}{k} = \frac{c}{c_c}\frac{c_c\omega}{k} = 2\zeta\frac{\omega}{\omega_n}$$

The nondimensional expressions for the amplitude and phase then become

$$\frac{Xk}{F_0} = \frac{1}{\sqrt{\left[1 - \left(\dfrac{\omega}{\omega_n}\right)^2\right]^2 + \left[2\zeta\left(\dfrac{\omega}{\omega_n}\right)\right]^2}} \qquad (3.1\text{-}7)$$

and

$$\tan\phi = \frac{2\zeta\left(\dfrac{\omega}{\omega_n}\right)}{1 - \left(\dfrac{\omega}{\omega_n}\right)^2} \qquad (3.1\text{-}8)$$

These equations indicate that the nondimensional amplitude Xk/F_0 and the phase ϕ are functions only of the frequency ratio ω/ω_n and the damping factor ζ and can be plotted as shown in Fig. 3.1-3. These curves show that the damping factor has a large influence on the amplitude and phase angle in the frequency region near resonance. Further understanding

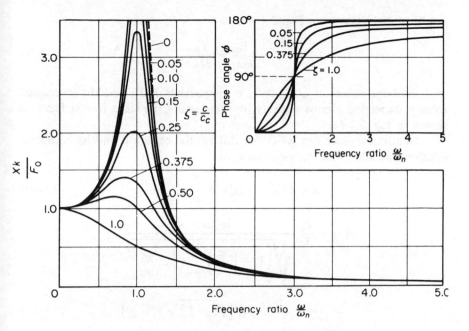

Figure 3.1-3. Plot of Equations (3.1-7) and (3.1-8).

of the behavior of the system may be obtained by studying the force diagram corresponding to Fig. 3.1-2 in the regions ω/ω_n small, $\omega/\omega_n = 1$, and ω/ω_n large.

For small values of $\omega/\omega_n \ll 1$, both the inertia and damping forces are small, which results in a small phase angle ϕ. The magnitude of the impressed force is then nearly equal to the spring force as shown in Fig. 3.1-4a.

For $\omega/\omega_n = 1.0$, the phase angle is 90° and the force diagram appears as in Fig. 3.1-4b. The inertia force, which is now larger, is balanced by the spring force; whereas the impressed force overcomes the damping force. The amplitude at resonance can be found, either from Eqs.

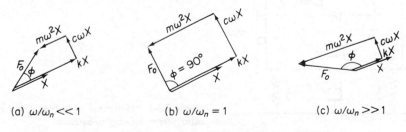

(a) $\omega/\omega_n \ll 1$ (b) $\omega/\omega_n = 1$ (c) $\omega/\omega_n \gg 1$

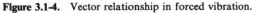

Figure 3.1-4. Vector relationship in forced vibration.

(3.1-5) or (3.1-7) or from Fig. 3.1-4b, to be

$$X = \frac{F_0}{c\omega_n} = \frac{F_0}{2\zeta k} \tag{3.1-9}$$

At large values of $\omega/\omega_n \gg 1$, ϕ approaches $180°$, and the impressed force is expended almost entirely in overcoming the large inertia force as shown in Fig. 3.1-4c.

In summary, we can write the differential equation and its complete solution, including the transient term as

$$\ddot{x} + 2\zeta\omega_n\dot{x} + \omega_n^2 x = \frac{F_0}{m}\sin \omega t \tag{3.1-10}$$

$$x(t) = \frac{F_0}{k} \frac{\sin(\omega t - \phi)}{\sqrt{\left[1 - \left(\frac{\omega}{\omega_n}\right)^2\right]^2 + \left[2\zeta\frac{\omega}{\omega_n}\right]^2}}$$

$$+ X_1 e^{-\zeta\omega_n t} \sin\left(\sqrt{1 - \zeta^2}\,\omega_n t + \phi_1\right) \tag{3.1-11}$$

3.2 ROTATING UNBALANCE

Unbalance in rotating machines is a common source of vibration excitation. We consider here a spring mass system constrained to move in the vertical direction and excited by a rotating machine that is unbalanced, as shown in Fig. 3.2-1. The unbalance is represented by an eccentric mass m with eccentricity e which is rotating with angular velocity ω. Letting x be the displacement of the nonrotating mass $(M - m)$ from the static

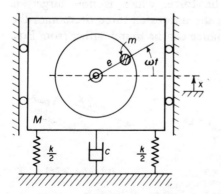

Figure 3.2-1. Harmonic disturbing force resulting from rotating unbalance.

equilibrium position, the displacement of m is

$$x + e \sin \omega t$$

The equation of motion is then

$$(M - m)\ddot{x} + m\frac{d^2}{dt^2}(x + e \sin \omega t) = -kx - c\dot{x}$$

which can be rearranged to

$$M\ddot{x} + c\dot{x} + kx = (me\omega^2) \sin \omega t \qquad (3.2\text{-}1)$$

It is evident, then, that the above equation is identical to Eq. (3.1-1), where F_0 is replaced by $me\omega^2$, and hence the steady-state solution of the previous section can be replaced by

$$X = \frac{me\omega^2}{\sqrt{(k - M\omega^2)^2 + (c\omega)^2}} \qquad (3.2\text{-}2)$$

and

$$\tan \phi = \frac{c\omega}{k - M\omega^2} \qquad (3.2\text{-}3)$$

These can be further reduced to nondimensional form

$$\frac{M}{m}\frac{X}{e} = \frac{\left(\dfrac{\omega}{\omega_n}\right)^2}{\sqrt{\left[1 - \left(\dfrac{\omega}{\omega_n}\right)^2\right]^2 + \left[2\zeta\dfrac{\omega}{\omega_n}\right]^2}} \qquad (3.2\text{-}4)$$

and

$$\tan \phi = \frac{2\zeta\left(\dfrac{\omega}{\omega_n}\right)}{1 - \left(\dfrac{\omega}{\omega_n}\right)^2} \qquad (3.2\text{-}5)$$

and presented graphically as in Fig. 3.2-2. The complete solution is given by

$$x(t) = X_1 e^{-\zeta\omega_n t} \sin\left(\sqrt{1 - \zeta^2}\, \omega_n t + \phi_1\right)$$

$$+ \frac{me\omega^2}{\sqrt{(k - M\omega^2)^2 + (c\omega)^2}} \sin(\omega t - \phi) \qquad (3.2\text{-}6)$$

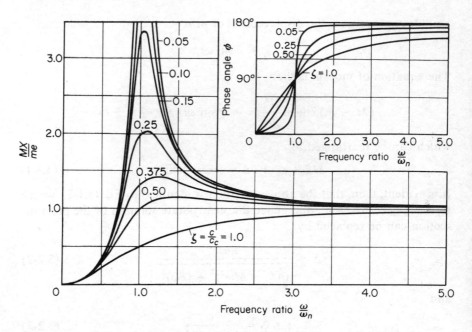

Figure 3.2-2. Plot of Equations (3.2-4) and (3.2-5) for forced vibration with rotating unbalance.

EXAMPLE 3.2-1

A counterrotating eccentric weight exciter is used to produce forced oscillation of a spring-supported mass, as shown in Fig. 3.2-3. By varying the speed of rotation, a resonant amplitude of 0.60 cm was recorded. When the speed of rotation was increased considerably beyond the resonant frequency, the amplitude appeared to approach a fixed value of 0.08 cm. Determine the damping factor of the system.

Solution: From Eq. (3.2-4), the resonant amplitude is

$$X = \frac{\dfrac{me}{M}}{2\zeta} = 0.60 \text{ cm}$$

When ω is very much greater than ω_n, the same equation becomes

$$X = \frac{me}{M} = 0.08 \text{ cm}$$

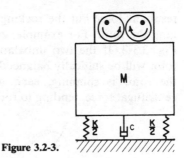

Figure 3.2-3.

Solving the two equations simultaneously, the damping factor of the system is

$$\zeta = \frac{0.08}{2 \times 0.60} = 0.0666$$

3.3 BALANCING OF ROTORS

In Sec. 3.2 the system was idealized to a spring-mass-damper unit with a rotating unbalance acting in a single plane. It is more likely that the unbalance in a rotating wheel or rotor is distributed in several planes. We wish now to distinguish between two types of rotating unbalance and show how they may be corrected.

Static Unbalance.   When the unbalanced masses all lie in a single plane, as in the case of a thin rotor disk, the resultant unbalance is a single radial force. As shown in Fig. 3.3-1, such unbalance can be detected by a static test in which the wheel-axle assembly is placed on a pair of horizontal rails. The wheel will roll to a position where the heavy point is directly below the axle. Since such unbalance can be detected without spinning the wheel, it is called static unbalance.

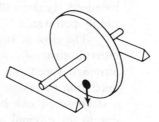

Figure 3.3-1.   System with static unbalance.

Dynamic Unbalance.   When the unbalance appears in more than one plane, the resultant is a force and a rocking moment which is referred to as dynamic unbalance. As previously described, a static test may detect the resultant force but the rocking moment cannot be detected without spinning the rotor. For example, consider a shaft with two disks as shown in Fig. 3.3-2. If the two unbalanced masses are equal and 180° apart, the rotor will be statically balanced about the axis of the shaft. However, when the rotor is spinning, each unbalanced disk would set up a rotating centrifugal force, tending to rock the shaft on its bearings.

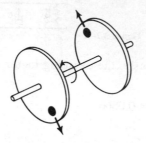

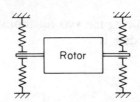

Figure 3.3-2. System with dynamic unbalance.

Figure 3.3-3. A rotor balancing machine.

In general, a long rotor, such as a motor armature or an automobile engine crankshaft, can be considered to be a series of thin disks, each with some unbalance. Such rotors must be spun in order to detect the unbalance. Machines to detect and correct the rotor unbalance are called balancing machines. Essentially the balancing machine consists of supporting bearings which are spring mounted so as to detect the unbalanced forces by their motion, as shown in Fig. 3.3-3. Knowing the amplitude of each bearing and their relative phase, it is possible to determine the unbalance of the rotor and correct for them.

EXAMPLE 3.3-1

Although a thin disk can be balanced statically, it can also be balanced dynamically. We describe one such test which can be simply performed.

The disk is supported on spring restrained bearings that can move horizontally as shown in Fig. 3.3-4. Running at any predetermined speed, the amplitude X_0 and the wheel position "a" at maximum excursion are noted. An accelerometer on the bearing and a stroboscope can be used for this observation. The amplitude X_0, due to the original unbalance m_0, is drawn to scale on the wheel in the direction from o to a.

Next, a trial mass m_1 is added at any point on the wheel and the procedure is repeated at the same speed. The new amplitude X_1 and

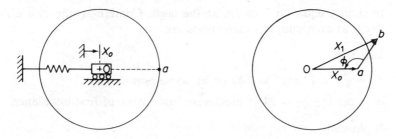

Figure 3.3-4. Experimental balancing of thin disk.

wheel position "*b*," which are due to the original unbalance m_0 and the trial mass m_1, are represented by the vector *ob*. The difference vector *ab* is then the effect of the trial mass m_1 alone. If the position of m_1 is now advanced by the angle ϕ shown in the vector diagram, and the magnitude of m_1 is increased to $m_1 (oa/ab)$, the vector *ab* will become equal and opposite to the vector *oa*. The wheel is now balanced since X_1 is zero.

EXAMPLE 3.3-2

A long rotor can be balanced by the addition or removal of correction weights in any two parallel planes. Generally, the correction is made by drilling holes in the two end planes; i.e., each radial inertia force $me\omega^2$ is replaced by two parallel forces, one in each end plane. With several unbalanced masses treated similarly, the correction to be made is found from their resultant in the two end planes.

Consider the balancing of a 4 in. long rotor shown in Fig. 3.3-5. It has a 3 oz in. unbalance in a plane 1 in. from the left end, and a 2 oz in. unbalance in the middle plane angularly displaced 90° from the first unbalance.

The 3 oz in. unbalance is equivalent to $2\frac{1}{4}$ oz in. at the left end and $\frac{3}{4}$ oz in. at the right end, as shown. The 2 oz in. at the middle is

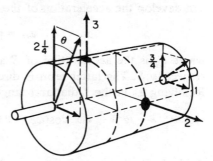

Figure 3.3-5. Correcting a long rotor unbalance in two end planes.

obviously equal to 1 oz in. at the ends. Combining the two unbalances at each end, the corrections are

Left end:

$$C_1 = \sqrt{1^2 + (2.25)^2} = 2.47 \text{ oz in. to be removed}$$

$$\theta_1 = \tan^{-1} \frac{1}{2.25} = 24°0' \text{ clockwise from plane of first unbalance}$$

Right end:

$$C_2 = \sqrt{\left(\tfrac{3}{4}\right)^2 + 1^2} = 1.25 \text{ oz in. to be removed}$$

$$\theta_2 = \tan^{-1} \frac{1}{\left(\tfrac{3}{4}\right)} = 53° \text{ clockwise from plane of first unbalance}$$

3.4   WHIRLING OF ROTATING SHAFTS

Rotating shafts tend to bow out at certain speeds and whirl in a complicated manner. *Whirling* is defined as the rotation of the plane made by the bent shaft and the line of centers of the bearings. The phenomenon results from such various causes as mass unbalance, hysteresis damping in the shaft, gyroscopic forces, fluid friction in bearings, etc. The whirling of the shaft may take place in the same or opposite direction as that of the rotation of the shaft, and the whirling speed may or may not be equal to the rotation speed.

We will consider here a single disk of mass m symmetrically located on a shaft supported by two bearings as shown in Fig. 3.4-1. The center of mass G of the disk is at a distance e (eccentricity) from the geometric center S of the disk. The center line of the bearings intersects the plane of the disk at O, and the shaft center is deflected by $r = OS$.

We will always assume the shaft (i.e., the line $e = SG$) to be rotating at a constant speed ω and in the general case the line $r = OS$ to be whirling at speed $\dot{\theta}$ which is not equal to ω. For the equation of motion, we can develop the acceleration of the mass center as follows;

$$\mathbf{a}_G = \mathbf{a}_S + \mathbf{a}_{G/S} \tag{3.4-1}$$

where $\mathbf{a}_S$ is the acceleration of S and $\mathbf{a}_{G/S}$ is the acceleration of G with respect to S. The latter term is directed from G to S since ω is constant. Resolving $\mathbf{a}_G$ in the radial and tangential directions, we have

$$\mathbf{a}_G = \left[(\ddot{r} - r\dot{\theta}^2) - e\omega^2 \cos(\omega t - \theta) \right]\mathbf{i} + \left[(r\ddot{\theta} + 2\dot{r}\dot{\theta}) - e\omega^2 \sin(\omega t - \theta) \right]\mathbf{j}$$

$$\tag{3.4-2}$$

Aside from the restoring force of the shaft, we will assume a viscous

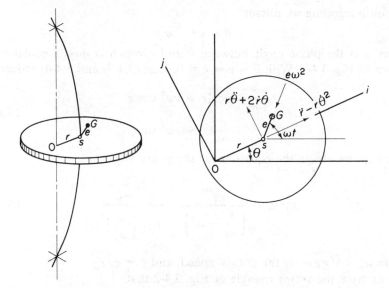

Figure 3.4-1.

damping force to be acting at S. The equations of motion resolved in the radial and tangential directions then become

$$- kr - c\dot{r} = m\left[\ddot{r} - r\dot{\theta}^2 - e\omega^2 \cos(\omega t - \theta)\right]$$

$$- cr\dot{\theta} = m\left[r\ddot{\theta} + 2\dot{r}\dot{\theta} - e\omega^2 \sin(\omega t - \theta)\right]$$

which can be rearranged to

$$\ddot{r} + \frac{c}{m}\dot{r} + \left(\frac{k}{m} - \dot{\theta}^2\right)r = e\omega^2 \cos(\omega t - \theta) \qquad (3.4\text{-}3)$$

$$r\ddot{\theta} + \left(\frac{c}{m}r + 2\dot{r}\right)\dot{\theta} = e\omega^2 \sin(\omega t - \theta) \qquad (3.4\text{-}4)$$

The general case of whirl as described by the above equations comes under the classification of self-excited motion where the exciting forces inducing the motion are controlled by the motion itself. Some aspects of this motion will be treated in Sec. 5.8; in this section we will consider only the simplest case of steady-state synchronous whirl where $\dot{\theta} = \omega$ and $\ddot{\theta} = \ddot{r} = \dot{r} = 0$.

Synchronous Whirl. For the synchronous whirl, the whirling speed $\dot{\theta}$ is equal to the rotation speed ω, which we have assumed to be constant. Thus we have

$$\dot{\theta} = \omega$$

and on integrating we obtain

$$\theta = \omega t - \varphi$$

where φ is the phase angle between e and r which is now a constant as shown in Fig. 3.4-1. With $\ddot{\theta} = \ddot{r} = \dot{r} = 0$, Eqs. (3.4-3) and 3.4-4) reduce to

$$\left(\frac{k}{m} - \omega^2\right)r = e\omega^2 \cos \varphi$$

$$\frac{c}{m}\omega r = e\omega^2 \sin \varphi$$
(3.4-5)

Dividing, we obtain the equation for the phase angle

$$\tan \varphi = \frac{\dfrac{c}{m}\omega}{\left(\dfrac{k}{m} - \omega^2\right)} = \frac{2\zeta\dfrac{\omega}{\omega_n}}{1 - \left(\dfrac{\omega}{\omega_n}\right)^2}$$
(3.4-6)

where $\omega_n = \sqrt{k/m}$ is the critical speed, and $\zeta = c/c_{cr}$. Noting from the vector triangle of Fig. 3.4-2 that

$$\cos \varphi = \frac{\dfrac{k}{m} - \omega^2}{\sqrt{\left(\dfrac{k}{m} - \omega^2\right)^2 + \left(\dfrac{c}{m}\omega\right)^2}}$$

and substituting into the first of Eq. (3.4-5), the amplitude equation becomes

$$r = \frac{me\omega^2}{\sqrt{(k - m\omega^2)^2 + (c\omega)^2}} = \frac{e\left(\dfrac{\omega}{\omega_n}\right)^2}{\sqrt{\left[1 - \left(\dfrac{\omega}{\omega_n}\right)^2\right]^2 + \left[2\zeta\dfrac{\omega}{\omega_n}\right]^2}}$$
(3.4-7)

These equations indicate that the eccentricity line $e = SG$ leads the displacement line $r = OS$ by the phase angle φ which depends on the amount of damping and the rotation speed ratio ω/ω_n. When the rotation

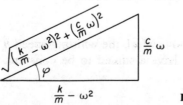

Figure 3.4-2.

speed coincides with the critical speed $\omega_n = \sqrt{k/m}$, or the natural frequency of the shaft in lateral vibration, a condition of resonance is encountered in which the amplitude is restrained only by the damping. Figure 3.4-3 shows the disk-shaft system under three different speed conditions. At very high speeds $\omega \gg \omega_n$, the center of mass G tends to approach the fixed point O and the shaft center S rotates about it in a circle of radius e.

It should be noted that the equations for the steady-state synchronous whirl are identical to those of Sec. 3.2, which is not surprising, since in both cases the exciting force is rotating and equal to $me\omega^2$. The response curves of Fig. 3.2-2 are thus applicable to this section.

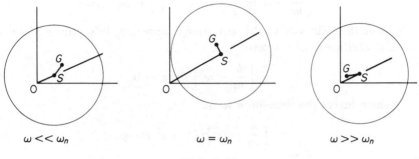

$$\omega \ll \omega_n \qquad\qquad \omega = \omega_n \qquad\qquad \omega \gg \omega_n$$

Figure 3.4-3.

EXAMPLE 3.4-1

Turbines operating above the critical speed must run through dangerous speed at resonance each time they are started or stopped. Assuming the critical speed ω_n to be reached with amplitude r_0, determine the equation for the amplitude build-up with time. Assume zero damping.

Solution: We will assume synchronous whirl as before which makes $\dot{\theta} = \omega = $ constant and $\ddot{\theta} = 0$. However $\ddot{r}$ and $\dot{r}$ terms must be retained unless shown to be zero. With $c = 0$ for the undamped case, the general equations of motion reduce to

$$\ddot{r} + \left(\frac{k}{m} - \omega^2\right)r = e\omega^2 \cos \varphi$$

$$2\dot{r}\omega = e\omega^2 \sin \varphi$$
(a)

The solution of the second equation with initial deflection equal to r_0 is

$$r = \frac{e\omega}{2} t \sin \varphi + r_0$$
(b)

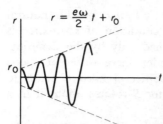

$$r = \frac{e\omega}{2}t + r_0$$

Figure 3.4-4. Amplitude and phase relationship of synchronous whirl with viscous damping.

Differentiating this equation twice we find that $\ddot{r} = 0$; so the first equation with the above solution for r becomes

$$\left(\frac{k}{m} - \omega^2\right)\left(\frac{e\omega}{2}t \sin\varphi + r_0\right) = e\omega^2 \cos\varphi \qquad (c)$$

Since the right side of this equation is constant, it is satisfied only if the coefficient of t is zero

$$\left(\frac{k}{m} - \omega^2\right)\sin\varphi = 0 \qquad (d)$$

which leaves the remaining terms

$$\left(\frac{k}{m} - \omega^2\right)r_0 = e\omega^2 \cos\varphi \qquad (e)$$

With $\omega = \sqrt{k/m}$ the first equation is satisfied, but the second equation is satisfied only if $\cos\varphi = 0$ or $\varphi = \pi/2$. Thus we have shown that at $\omega = \sqrt{k/m}$, or at resonance, the phase angle is $\pi/2$ as before for the damped case, and the amplitude builds up linearly according to the equation which is shown in Fig. 3.4-4.

3.5   SUPPORT MOTION

In many cases the dynamical system is excited by the motion of the support point, as shown in Fig. 3.5-1. We let y be the harmonic displacement of the support point and measure the displacement x of the mass m from an inertial reference.

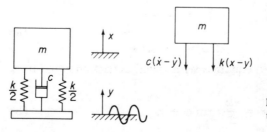

Figure 3.5-1. System excited by motion of support point.

In the displaced position the unbalanced forces are due to the damper and the springs, and the differential equation of motion becomes

$$m\ddot{x} = -k(x-y) - c(\dot{x} - \dot{y}) \qquad (3.5\text{-}1)$$

Making the substitution

$$z = x - y \qquad (3.5\text{-}2)$$

Eq. (3.5-1) becomes

$$m\ddot{z} + c\dot{z} + kz = -m\ddot{y}$$
$$= m\omega^2 Y \sin \omega t \qquad (3.5\text{-}3)$$

where $y = Y \sin \omega t$ has been assumed for the motion of the base. The form of this equation is identical to that of Eq. (3.2-1) where z replaces x and $m\omega^2 Y$ replaces $me\omega^2$. Thus the solution can be immediately written as

$$z = Z \sin(\omega t - \phi)$$

$$Z = \frac{m\omega^2 Y}{\sqrt{(k - m\omega^2)^2 + (c\omega)^2}} \qquad (3.5\text{-}4)$$

$$\tan \phi = \frac{c\omega}{k - m\omega^2} \qquad (3.5\text{-}5)$$

and the curves of Fig. 3.2-2 are applicable with the appropriate change in the ordinate.

If the absolute motion x of the mass is desired, we can solve for $x = z + y$. Using the exponential form of harmonic motion gives

$$y = Ye^{i\omega t}$$
$$z = Ze^{i(\omega t - \phi)} = (Ze^{-i\phi})e^{i\omega t} \qquad (3.5\text{-}6)$$
$$x = Xe^{i(\omega t - \psi)} = (Xe^{-i\psi 4})e^{i\omega t}$$

Substituting into Eq. (3.5-3), we obtain

$$Ze^{-i\phi} = \frac{m\omega^2 Y}{k - m\omega^2 + i\omega c}$$

and

$$x = (Ze^{-i\phi} + Y)e^{i\omega t}$$
$$= \left(\frac{k + i\omega c}{k - m\omega^2 + i\omega c}\right) Ye^{i\omega t} \qquad (3.5\text{-}7)$$

The steady-state amplitude and phase from this equation are

$$\left|\frac{X}{Y}\right| = \sqrt{\frac{k^2 + (\omega c)^2}{(k - m\omega^2)^2 + (c\omega)^2}} \qquad (3.5\text{-}8)$$

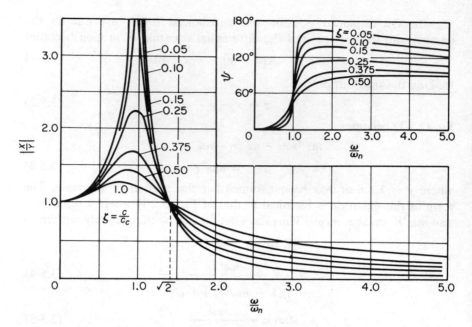

Figure 3.5-2.   Plot of Equations (3.5-8) and (3.5-9).

and

$$\tan \psi = \frac{mc\omega^3}{k(k - m\omega^2) + (\omega c)^2} \tag{3.5-9}$$

which are plotted in Fig. 3.5-2. It should be observed that the amplitude curves for different damping all have the same value of $|X/Y| = 1.0$ at the frequency $\omega/\omega_n = \sqrt{2}$.

3.6   VIBRATION ISOLATION

Vibratory forces generated by machines and engines are often unavoidable; however, their effect on a dynamical system can be reduced substantially by properly designed springs, which are referred to as isolators.

In Fig. 3.6-1 let $F_0 \sin \omega t$ be the exciting force acting on the single degree of freedom system. The transmitted force through the springs and damper is

$$F_T = \sqrt{(kX)^2 + (c\omega X)^2} = kX\sqrt{1 + \left(\frac{c\omega}{k}\right)^2} \tag{3.6-1}$$

Since the amplitude X developed under the force $F_0 \sin \omega t$ is given by Eq.

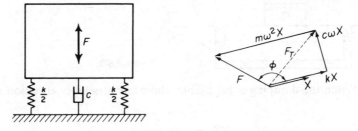

Figure 3.6-1. Disturbing force transmitted through springs and damper.

(3.1-5), the above equation reduces to

$$\frac{F_T}{F_0} = \frac{\sqrt{1 + \left(\dfrac{c\omega}{k}\right)^2}}{\sqrt{\left[1 - \dfrac{m\omega^2}{k}\right]^2 + \left[\dfrac{c\omega}{k}\right]^2}} = \frac{\sqrt{1 + \left(2\zeta\dfrac{\omega}{\omega_n}\right)^2}}{\sqrt{\left[1 - \left(\dfrac{\omega}{\omega_n}\right)^2\right]^2 + \left[2\zeta\dfrac{\omega}{\omega_n}\right]^2}}$$

(3.6-2)

Comparison of Eqs. (3.6-2) and (3.5-8) indicates that $|F_T/F_0|$ is identical to $|X/Y| = |\omega^2 X/\omega^2 Y|$. Thus the problem of isolating a mass from the motion of the support point is identical to that of isolating disturbing forces. Each of these ratios is referred to as transmissibility, and the ordinate of Fig. 3.5-2 can equally represent transmissibility of force or of displacement. These curves show that the transmissibility is less than unity only for $\omega/\omega_n > \sqrt{2}$, thereby establishing the fact that vibration isolation is possible only for $\omega/\omega_n > \sqrt{2}$. As seen from Fig. 3.5-2, in the region $\omega/\omega_n > \sqrt{2}$, an undamped spring is superior to a damped spring in reducing the transmissibility. Some damping is desirable when it is necessary for ω to vary through the resonant region, although the large amplitude at resonance can be limited by stops.

It is possible to reduce the amplitude of vibration by supporting the machine on a large mass M as shown in Fig. 3.6-2. To keep the transmissibility F_T/F_0 the same, k must be increased in the same ratio so that $m + M/k$ remains the same. However, since

$$X = \frac{\dfrac{F_0}{k}}{\sqrt{\left[1 - \dfrac{m\omega^2}{k}\right]^2 + \left[\dfrac{c\omega}{k}\right]^2}}$$

the amplitude X is reduced by the increased value of k.

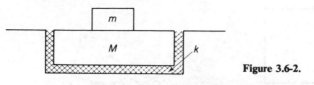

Figure 3.6-2.

When the damping is negligible, the transmissibility equation reduces to

$$TR = \frac{1}{\left(\dfrac{\omega}{\omega_n}\right)^2 - 1}$$ (3.6-3)

where it is understood that the value of ω/ω_n to be used is always greater than $\sqrt{2}$. On further replacing ω_n^2 with g/Δ, where $g = 9.81$ m/s^2 and $\Delta =$ statical deflection in meters, Eq. (3.6-3) may be expressed as

$$TR = \frac{1}{\dfrac{(2\pi f)^2 \Delta}{g} - 1}$$ (3.6-4)

Solving for f, we obtain with Δ in millimeters,

$$f = 15.76\sqrt{\frac{1}{\Delta}\left(\frac{1}{TR} + 1\right)} \text{ Hz}$$ (3.6-5)

Defining the reduction in transmissibility as $R = (1 - TR)$, the above equation may also be written as

$$f = 15.76\sqrt{\frac{1}{\Delta}\left(\frac{2 - R}{1 - R}\right)} \text{ Hz}$$ (3.6-6)

Figure 3.6-3 displays Eq. (3.6-6) for f vs. Δ with R as parameter.

This discussion has been limited to bodies with translation along a single coordinate. In general, a rigid body has six degrees of freedom; namely, translation along and rotation about the three perpendicular coordinate axes. For these more advanced cases the reader is referred to the excellent text on vibration isolation by C. Crede.*

EXAMPLE 3.6-1

A machine of 100 kg mass is supported on springs of total stiffness 700 kN/m and has an unbalanced rotating element which results in a disturbing force of 350 N at a speed of 3000 rpm. Assuming a damping factor of $\zeta = 0.20$, determine (a) its amplitude of motion due to the unbalance, (b) the transmissibility, and (c) the transmitted force.

*C. E. Crede, *Vibration and Shock Isolation* (New York: John Wiley & Sons, 1951).

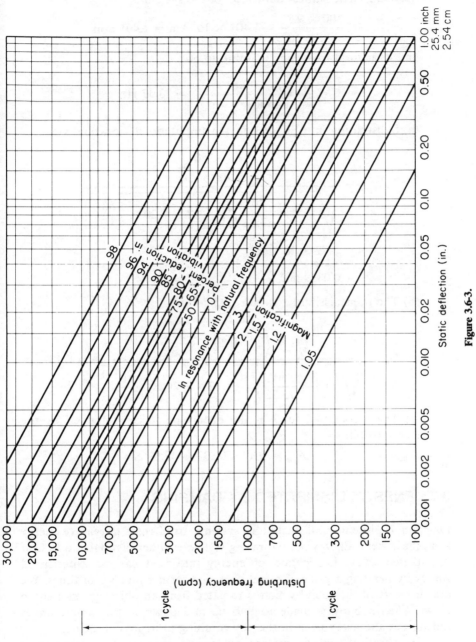

Figure 3.6-3.

Solution: The statical deflection of the system is

$$\frac{100 \times 9.81}{700 \times 10^3} = 1.401 \times 10^{-3} \text{ m} = 1.401 \text{ mm}$$

and its natural frequency is found to be

$$f_n = \frac{1}{2\pi} \sqrt{\frac{9.81}{1.401 \times 10^{-3}}} = 13.32 \text{ Hz}$$

(a) Substituting into Eq. (3.1-5), the amplitude of vibration is

$$X = \frac{\left(\dfrac{350}{700 \times 10^3}\right)}{\sqrt{\left[1 - \left(\dfrac{50}{13.32}\right)^2\right]^2 + \left[2 \times 0.20 \times \dfrac{50}{13.32}\right]^2}}$$

$$= 3.79 \times 10^{-5} \text{ m}$$

$$= 0.0379 \text{ mm}$$

(b) The transmissibility from Eq. (3.6-2) is

$$TR = \frac{\sqrt{1 + \left(2 \times 0.20 \times \dfrac{50}{13.32}\right)^2}}{\sqrt{\left[1 - \left(\dfrac{50}{13.32}\right)^2\right]^2 + \left[2 \times 0.20 \times \dfrac{50}{13.32}\right]^2}} = 0.137$$

(c) The transmitted force is the disturbing force multiplied by the transmissibility.

$$F_{TR} = 350 \times 0.137 = 47.89 \text{ N}$$

3.7 ENERGY DISSIPATED BY DAMPING

Damping is present in all oscillatory systems. Its effect is to remove energy from the system. Energy in a vibrating system is either dissipated into heat or radiated away. Dissipation of energy into heat can be experienced simply by bending a piece of metal back and forth a number of times. We are all aware of the sound which is radiated from an object given a sharp blow. When a buoy is made to bob up and down in the water, waves radiate out and away from it, thereby resulting in its loss of energy.

In vibration analysis, we are generally concerned with damping in terms of system response. The loss of energy from the oscillatory system results in the decay of amplitude of free vibration. In steady-state forced vibration, the loss of energy is balanced by the energy which is supplied by the excitation.

A vibrating system may encounter many different types of damping forces, from internal molecular friction to sliding friction and fluid resistance. Generally their mathematical description is quite complicated and not suitable for vibration analysis. Thus simplified damping models have been developed that in many cases are found to be adequate in evaluating the system response. For example, we have already used the viscous damping model, designated by the dashpot, which leads to manageable mathematical solutions.

Energy dissipation is usually determined under conditions of cyclic oscillations. Depending on the type of damping present, the force-displacement relationship when plotted may differ greatly. In all cases, however, the force-displacement curve will enclose an area, referred to as the *hysteresis loop*, that is proportional to the energy lost per cycle. The energy lost per cycle due to a damping force F_d is computed from the general equation

$$W_d = \oint F_d \, dx \qquad (3.7\text{-}1)$$

In general, W_d will depend on many factors, such as temperature, frequency, or amplitude.

We will consider in this section the simplest case of energy dissipation, that of a spring-mass system with viscous damping. The damping force in this case is $F_d = c\dot{x}$. With the steady-state displacement and velocity

$$x = X \sin(\omega t - \phi)$$
$$\dot{x} = \omega X \cos(\omega t - \phi)$$

the energy dissipated per cycle, from Eq. (3.7-1), becomes

$$W_d = \oint c\dot{x} \, dx = \oint c\dot{x}^2 \, dt$$
$$= c\omega^2 X^2 \int_0^{2\pi/\omega} \cos^2(\omega t - \phi) \, dt = \pi c\omega X^2 \qquad (3.7\text{-}2)$$

Of particular interest is the energy dissipated in forced vibration at resonance. Substituting $\omega_n = \sqrt{k/m}$ and $c = 2\zeta\sqrt{km}$, the above equation at resonance becomes

$$W_d = 2\zeta\pi k X^2 \qquad (3.7\text{-}3)$$

The energy dissipated per cycle by the damping force can be represented graphically as follows. Writing the velocity in the form

$$\dot{x} = \omega X \cos(\omega t - \phi) = \pm\omega X\sqrt{1 - \sin^2(\omega t - \phi)}$$
$$= \pm\omega\sqrt{X^2 - x^2}$$

the damping force becomes

$$F_d = c\dot{x} = \pm c\omega\sqrt{X^2 - x^2} \tag{3.7-4}$$

Rearranging the above equation to

$$\left(\frac{F_d}{c\omega X}\right)^2 + \left(\frac{x}{X}\right)^2 = 1 \tag{3.7-5}$$

we recognize it as that of an ellipse with F_d and x plotted along the vertical and horizontal axes, as shown in Fig. 3.7-1a. The energy dissipated per cycle is then given by the area enclosed by the ellipse. If we add to F_d the force kx of the lossless spring, the hysteresis loop is rotated as shown in Fig. 3.7-1b. This representation then conforms to the *Voigt model*, which consists of a dashpot in parallel with a spring.

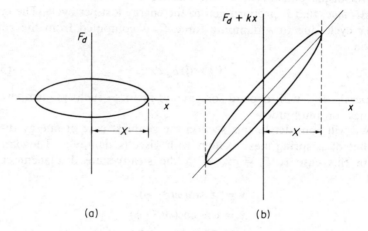

Figure 3.7-1. Energy dissipated by viscous damping.

Damping properties of materials are listed in many different ways depending on the technical areas to which they are applied. Of these we list two relative energy units which have wide usage. First of these is *specific damping capacity* defined as the energy loss per cycle W_d divided by the peak potential energy U.

$$\frac{W_d}{U} \tag{3.7-6}$$

The second quantity is the *loss coefficient* defined as the ratio of damping energy loss per radian $W_d/2\pi$ divided by the peak potential or strain energy U.

$$\eta = \frac{W_d}{2\pi U} \tag{3.7-7}$$

For the case of linear damping where the energy loss is proportional to the square of the strain or amplitude, the hysteresis curve is an ellipse. The loss coefficient for most materials varies between 0.001 to unity depending on the material and conditions under which the tests are performed. When the damping loss is not a quadratic function of the strain or amplitude, the hysteresis curve is no longer an ellipse. Again, the loss coefficient may vary from 0.001 to approximately 0.2.

EXAMPLE 3.7-1

Determine the expression for the power developed by a force $F = F_0 \sin(\omega t + \phi)$ acting on a displacement $x = X_0 \sin \omega t$.

Solution: Power is the rate of doing work which is the product of the force and velocity.

$$P = F \frac{dx}{dt} = (\omega X_0 F_0) \sin(\omega t + \phi) \cos \omega t$$

$$= (\omega X_0 F_0) \left[\cos \phi \cdot \sin \omega t \cos \omega t + \sin \phi \cdot \cos^2 \omega t \right]$$

$$= \tfrac{1}{2} \omega X_0 F_0 \left[\sin \phi + \sin(2\omega t + \phi) \right]$$

The first term is a constant, representing the steady flow of work per unit time. The second term is a sine wave of twice the frequency that represents the fluctuating component of power, the average value of which is zero over any interval of time that is a multiple of the period.

EXAMPLE 3.7-2

A force $F = 10 \sin \pi t$ Newtons acts on a displacement of $x = 2 \sin(\pi t - \pi/6)$ meters. Determine (a) the work done during the first 6 sec; (b) the work done during the first $\tfrac{1}{2}$ sec.

Solution: Rewriting Eq. (3.7-1) as $W = \int F \dot{x} \, dt$, and substituting $F = F_0 \sin \omega t$ and $x = X \sin(\omega t - \phi)$, the work done per cycle becomes

$$W = \pi F_0 X \sin \phi$$

For the force and displacement given in this problem, $F_0 = 10$ N, $X = 2$ m, $\phi = \pi/6$, and the period $\tau = 2$ seconds. Thus in the 6 seconds specified in (a), three complete cycles take place, and the work done is

$$W = 3(\pi F_0 X \sin \phi) = 3\pi \times 10 \times 2 \times \sin 30° = 94.2 \text{ Nm}$$

The work done in part (b) is determined by integrating the expression for work between the limits 0 and $\frac{1}{2}$ sec.

$$W = \omega F_0 X_0 \left[\cos 30° \int_0^{1/2} \sin \pi t \cos \pi t \, dt + \sin 30° \int_0^{1/2} \sin^2 \pi t \, dt \right]$$

$$= \pi \times 10 \times 2 \left[-\frac{0.866}{4\pi} \cos 2\pi t + 0.50 \left(\frac{t}{2} - \frac{\sin 2\pi t}{4\pi} \right) \right]_0^{1/2}$$

$$= 16.51 \text{ Nm}$$

3.8   EQUIVALENT VISCOUS DAMPING

The primary influence of damping on oscillatory systems is that of limiting the amplitude of response at resonance. As seen from the response curves of Fig. 3.1-3, damping has little influence on the response in the frequency regions away from resonance.

In the case of viscous damping, the amplitude at resonance, Eq. (3.1-9), was found to be

$$X = \frac{F_0}{c\omega_n} \tag{3.8-1}$$

For other types of damping, no such simple expression exists. It is possible, however, to approximate the resonant amplitude by substituting an equivalent damping c_{eq} in the above equation.

The equivalent damping c_{eq} is found by equating the energy dissipated by the viscous damping to that of the nonviscous damping force with assumed harmonic motion. From Eq. (3.7-2)

$$\pi c_{eq} \omega X^2 = W_d \tag{3.8-2}$$

where W_d must be evaluated from the particular type of damping force.

EXAMPLE 3.8-1

Bodies moving with moderate speed (3 to 20 m/s) in fluids such as water or air, are resisted by a damping force which is proportional to the square of the speed. Determine the equivalent damping for such forces acting on an oscillatory system, and find its resonant amplitude.

Solution: Let the damping force be expressed by the equation

$$F_d = \pm a \dot{x}^2$$

where the negative sign must be used when $\dot{x}$ is positive, and vice versa. Assuming harmonic motion with the time measured from the position of extreme negative displacement

$$x = -X \cos \omega t$$

the energy dissipated per cycle is

$$W_d = 2 \int_{-x}^{x} a\dot{x}^2 \, dx = 2a\omega^2 X^3 \int_0^\pi \sin^3 \omega t \, d(\omega t)$$

$$= \tfrac{8}{3} a\omega^2 X^3$$

The equivalent viscous damping from Eq. (3.8-2) is then

$$c_{eq} = \frac{8}{3\pi} a\omega X$$

The amplitude at resonance is found by substituting $c = c_{eq}$ in Eq. (3.8-1) with $\omega = \omega_n$

$$X = \sqrt{\frac{3\pi F_0}{8a\omega_n^2}}$$

EXAMPLE 3.8-2

An oscillatory system forced to vibrate by an exciting force $F_0 \sin \omega t$ is known to be acted upon by several different forms of damping. Develop the equation for the equivalent damping and indicate the procedure for determining the amplitude at resonance.

Solution: Let U_1, U_2, U_3, etc., be the energy dissipated per cycle by the various damping forces. Equating the total energy dissipated to that of equivalent viscous damping

$$\pi c_{eq} \omega X^2 = U_1 + U_2 + U_3 + \cdots$$

The equivalent viscous damping coefficient is found to be

$$c_{eq} = \frac{\sum U}{\pi \omega X^2}$$

To determine the amplitude, it is necessary to obtain expressions for U_1, U_2, U_3, etc., which will contain X raised to various powers. Substituting c_{eq} into the expression

$$X = \frac{F_0}{c_{eq} \omega}$$

the equation with $\omega = \omega_n$ is solved for X.

EXAMPLE 3.8-3

Find the equivalent viscous damping for Coulomb damping.

Solution: We assume that under forced sinusoidal excitation the displacement of the system with Coulomb damping is sinusoidal and equal to $x = X \sin \omega t$. The equivalent viscous damping can then be found from Eq. (3.8-2) by noting that the work done per cycle by the Coulomb force F_d is equal to $W_d = F_d \times 4X$. Its substitution into Eq. (3.8-2) gives

$$\pi c_{eq} \omega X^2 = 4 F_d X$$

$$c_{eq} = \frac{4 F_d}{\pi \omega X}$$

The amplitude of forced vibration can be found by substituting c_{eq} into Eq. (3.1-3).

$$X = \frac{F_0}{\sqrt{(k - m\omega^2)^2 + \left(\dfrac{4 F_d \omega}{\pi \omega X}\right)^2}}$$

Solving for X, we obtain

$$|X| = \frac{\sqrt{F_0^2 - \left(\dfrac{4 F_d}{\pi}\right)^2}}{k - m\omega^2} = \frac{F_0}{k} \frac{\sqrt{1 - \left(\dfrac{4 F_d}{\pi F_0}\right)^2}}{1 - \left(\dfrac{\omega}{\omega_n}\right)^2}$$

We note here that unlike the system with viscous damping X/δ_{st} goes to ∞ when $\omega = \omega_n$. For the numerator to remain real, the term $4 F_d / \pi F_0$ must be less than 1.0.

3.9   STRUCTURAL DAMPING

When materials are cyclicly stressed, energy is dissipated internally within the material itself. Experiments by several investigators[*,†] indicate that for most structural metals, such as steel or aluminum, the energy dissipated per cycle is independent of the frequency over a wide frequency range, and proportional to the square of the amplitude of vibration. Internal damping fitting this classification is called *solid damping* or *structural damping*. With the energy dissipation per cycle proportional to the square of the vibration amplitude, the loss coefficient is a constant and the shape of the hysteresis curve remains unchanged with amplitude and independent of the strain rate.

Energy dissipated by structural damping may be written as

$$W_d = \alpha X^2 \tag{3.9-1}$$

---

[*] A. L. Kimball, "Vibration Damping, Including the Case of Solid Damping," Trans. ASME, APM 51–52 (1929).

[†] B. J. Lazan, *Damping of Materials and Members in Structural Mechanics* (Elmsford, N.Y.: Pergamon Press, 1968).

where α is a constant with units of force/displacement. Using the concept of equivalent viscous damping, Eq. (3.8-2) gives

$$\pi c_{eq} \omega X^2 = \alpha X^2$$

or

$$c_{eq} = \frac{\alpha}{\pi \omega} \qquad (3.9\text{-}2)$$

Substitution of c_{eq} for c, the differential equation of motion for a system with structural damping may be written as

$$m\ddot{x} + \left(\frac{\alpha}{\pi \omega}\right)\dot{x} + kx = F(t) \qquad (3.9\text{-}3)$$

Complex Stiffness. In the calculation of the flutter speeds of airplane wings and tail surfaces, the concept of *complex stiffness* is used. It is arrived at by assuming the oscillations to be harmonic, which enables Eq. (3.9-3) to be written as

$$m\ddot{x} + \left(k + i\frac{\alpha}{\pi}\right)x = F_0 e^{i\omega t}$$

By factoring out the stiffness k and letting $\gamma = \alpha/\pi k$, the above equation becomes

$$m\ddot{x} + k(1 + i\gamma)x = F_0 e^{i\omega t} \qquad (3.9\text{-}4)$$

The quantity $k(1 + i\gamma)$ is called the *complex stiffness* and γ is the *structural damping factor*.

Using the concept of complex stiffness for problems in structural vibrations is advantageous in that one needs only to multiply the stiffness terms in the system by $(1 + i\gamma)$. The method is justified, however, only for harmonic oscillations. With the solution $x \equiv \overline{X} e^{i\omega t}$, the steady state amplitude from Eq. (3.9-4) becomes

$$\overline{X} = \frac{F_0}{(k - m\omega^2) + i\gamma k} \qquad (3.9\text{-}5)$$

The amplitude at resonance is then

$$|X| = \frac{F_0}{\gamma k} \qquad (3.9\text{-}6)$$

Comparing this with the resonant response of a system with viscous damping

$$|X| = \frac{F_0}{2\zeta k}$$

we conclude that with equal amplitudes at resonance, the structural damping factor is equal to twice the viscous damping factor.

3.10 SHARPNESS OF RESONANCE

In forced vibration there is a quantity Q related to damping that is a measure of the sharpness of resonance. To determine this quantity, we will assume viscous damping and start with Eq. (3.1-7).

When $\omega/\omega_n = 1$, the resonant amplitude is $X_{res} = (F_0/k)/2\zeta$. We now seek the two frequencies on either side of resonance (often referred to as *sidebands*), where X is $0.707X_{res}$. These points are also referred to as the *half-power points* and are shown in Fig. 3.10-1.

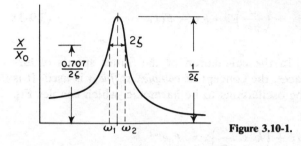

Figure 3.10-1.

Letting $X = 0.707X_{res}$ and squaring Eq. (3.1-7), we obtain

$$\frac{1}{2}\left(\frac{1}{2\zeta}\right)^2 = \frac{1}{\left[1 - \left(\frac{\omega}{\omega_n}\right)^2\right]^2 + \left[2\zeta\left(\frac{\omega}{\omega_n}\right)\right]^2}$$

or

$$\left(\frac{\omega}{\omega_n}\right)^4 - 2(1 - 2\zeta^2)\left(\frac{\omega}{\omega_n}\right)^2 + (1 - 8\zeta^2) = 0 \qquad (3.10\text{-}1)$$

Solving for $(\omega/\omega_n)^2$ we have

$$\left(\frac{\omega}{\omega_n}\right)^2 = (1 - 2\zeta^2) \pm 2\zeta\sqrt{1 - \zeta^2} \qquad (3.10\text{-}2)$$

Assuming $\zeta \ll 1$ and neglecting higher order terms of ζ, we arrive at the result

$$\left(\frac{\omega}{\omega_n}\right)^2 = 1 \pm 2\zeta \qquad (3.10\text{-}3)$$

Letting the two frequencies corresponding to the roots of Eq. (3.10-3) be ω_1 and ω_2, we obtain

$$4\zeta = \frac{\omega_2^2 - \omega_1^2}{\omega_n^2} \cong 2\left(\frac{\omega_2 - \omega_1}{\omega_n}\right)$$

76

The quantity Q is then defined as

$$Q = \frac{\omega_n}{\omega_2 - \omega_1} = \frac{f_n}{f_2 - f_1} = \frac{1}{2\zeta} \qquad (3.10\text{-}4)$$

Here, again, equivalent damping can be used to define Q for systems with other forms of damping. Thus, for structural damping, Q is equal to

$$Q = \frac{1}{\gamma} \qquad (3.10\text{-}5)$$

3.11   RESPONSE TO PERIODIC FORCES

Any periodic force can be resolved into a series of harmonic components by Fourier analysis. In place of the Fourier series of Sec. 1.2, the sine and cosine series can be combined into an alternative series, and the periodic force may be written in the form

$$F(t) = \sum_n F_n \sin(n\omega_1 t - \phi_n) \qquad (3.11\text{-}1)$$

If this force is applied to a SDF system, the steady-state response becomes the superposition of the harmonic components, one of which is

$$x_n(t) = \frac{F_n \sin(\omega_n t - \psi_n)}{\sqrt{(k - m\omega_n^2)^2 + (c\omega_n)^2}} \qquad (3.11\text{-}2)$$

The steady-state response is then given by the series

$$x(t) = \sum_n x_n(t) \qquad (3.11\text{-}3)$$

To this must be added the free vibration which in general dies down due to damping.

EXAMPLE 3.11-1

Determine the steady-state response of a single degree of freedom system to the square wave of Prob. 1-9.

Solution: The Fourier series for the rectangular wave of amplitude P is

$$F(t) = \frac{4P}{\pi} \left(\sin \omega_1 t + \frac{1}{3} \sin 3\omega_1 t + \frac{1}{5} \sin 5\omega_1 t + \cdots \right)$$

which contains only odd harmonics. The steady-state response is then the sum of the following terms:

$$x(t) = \frac{4P}{\pi}\left\{\frac{\sin(\omega_1 t - \phi_1)}{\sqrt{(k - m\omega_1^2)^2 + (c\omega_1)^2}}\right.$$

$$+ \frac{\frac{1}{3}\sin(3\omega_1 t - \phi_3)}{\sqrt{[k - m(3\omega_1)^2]^2 + [c3\omega_1]^2}}$$

$$\left. + \frac{\frac{1}{5}\sin(5\omega_1 t - \phi_5)}{\sqrt{[k - m(5\omega_1)^2]^2 + [c5\omega_1]^2}} + \cdots\right\}$$

$$\tan\phi_1 = \frac{c\omega_1}{k - m\omega_1^2}, \qquad \tan\phi_3 = \frac{c3\omega_1}{k - m(3\omega_1)^2}, \qquad \text{etc.}$$

3.12 VIBRATION MEASURING INSTRUMENTS

The basic element of many vibration measuring instruments is the seismic unit of Fig. 3.12-1. Depending on the frequency range utilized, displacement, velocity, or acceleration is indicated by the relative motion of the suspended mass with respect to the case.

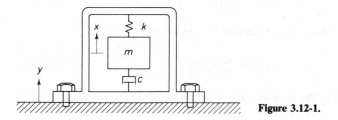

Figure 3.12-1.

To determine the behavior of such instruments we consider the equation of motion of m which is

$$m\ddot{x} = -c(\dot{x} - \dot{y}) - k(x - y) \qquad (3.12\text{-}1)$$

where x and y are the displacement of the seismic mass and the vibrating body, respectively, both measured with respect to an inertial reference. Letting the relative displacement of the mass m and the case attached to the vibrating body be

$$z = (x - y) \qquad (3.12\text{-}2)$$

and assuming sinusoidal motion $y = Y \sin \omega t$ of the vibrating body, we obtain the equation

$$m\ddot{z} + c\dot{z} + kz = m\omega^2 Y \sin \omega t \qquad (3.12\text{-}3)$$

This equation is identical in form to Eq. (3.2-1) with z and $m\omega^2 Y$ replacing x and $me\omega^2$, respectively. The steady-state solution $z = Z \sin(\omega t - \phi)$ is then available from inspection to be

$$Z = \frac{m\omega^2 Y}{\sqrt{(k - m\omega^2)^2 + (c\omega)^2}} = \frac{Y\left(\dfrac{\omega}{\omega_n}\right)^2}{\sqrt{\left[1 - \left(\dfrac{\omega}{\omega_n}\right)^2\right]^2 + \left[2\zeta\dfrac{\omega}{\omega_n}\right]^2}}$$

$$(3.12\text{-}4)$$

and

$$\tan \phi = \frac{\omega c}{k - m\omega^2} = \frac{2\zeta\dfrac{\omega}{\omega_n}}{1 - \left(\dfrac{\omega}{\omega_n}\right)^2} \qquad (3.12\text{-}5)$$

It is evident then that the parameters involved are the frequency ratio ω/ω_n and the damping factor ζ. Figure 3.12-2 shows a plot of these equations

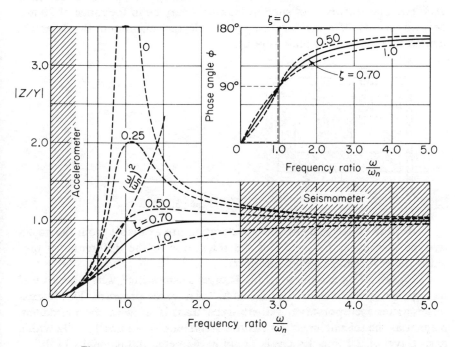

Figure 3.12-2. Response of a vibration measuring instrument.

and is identical to Fig. 3.2-2 except that Z/Y replaces MX/me. The type of instrument is determined by the useful range of frequencies with respect to the natural frequency ω_n of the instrument.

Seismometer—Instrument with Low Natural Frequency. When the natural frequency ω_n of the instrument is low in comparison to the vibration frequency ω to be measured, the ratio ω/ω_n approaches a large number, and the relative displacement Z approaches Y regardless of the value of damping ζ, as indicated in Fig. 3.12-2. The mass m then remains stationary while the supporting case moves with the vibrating body. Such instruments are called *seismometers*.

One of the disadvantages of the seismometer is its large size. Since $Z = Y$, the relative motion of the seismic mass must be of the same order of magnitude as that of the vibration to be measured.

The relative motion z is usually converted to an electric voltage by making the seismic mass a magnet moving relative to coils fixed in the case as shown in Fig. 3.12-3. Since the voltage generated is proportional to the rate of cutting of the magnetic field, the output of the instrument will be proportional to the velocity of the vibrating body. Such instruments are called velometers. A typical instrument of this kind may have a natural frequency between 1 Hz to 5 Hz and a useful frequency range of 10 Hz to 2000 Hz. The sensitivity of such instruments may be in the range of 20 mV per cm/sec to 350 mV per cm/sec with the maximum displacement limited to about 0.5 cm peak to peak.

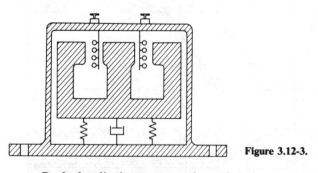

Figure 3.12-3.

Both the displacement and acceleration are available from the velocity-type transducer by means of the integrator or the differentiator provided in most signal conditioner units.

Figure 3.12-4 is a photo of the Ranger seismometer, which because of its high sensitivity was used in the U.S. lunar space program. The Ranger seismometer incorporates a velocity-type transducer with the permanent magnet as the seismic mass. Its natural frequency is nominally 1 Hz with a mass travel of ± 1 mm. Its size is 15 cm in diameter and it weighs 11 lb.

Figure 3.12-4.   Ranger seismometer. (*Courtesy of Kinemetrics, Inc., Pasadena, California.*)

Accelerometer—Instrument with High Natural Frequency.   When the natural frequency of the instrument is high compared to that of the vibration to be measured, the instrument indicates acceleration. Examination of Eq. (3.12-4) shows that the factor

$$\sqrt{\left[1 - \left(\frac{\omega}{\omega_n}\right)^2\right]^2 + \left[2\zeta\frac{\omega}{\omega_n}\right]^2}$$

approaches unity for $\omega/\omega_n \to 0$, so that

$$Z = \frac{\omega^2 Y}{\omega_n^2} = \frac{\text{(acceleration)}}{\omega_n^2} \tag{3.12-6}$$

Thus Z becomes proportional to the acceleration of the motion to be measured with a factor $1/\omega_n^2$. The useful range of the accelerometer can be

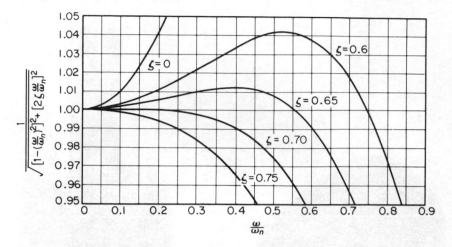

Figure 3.12-5.   Acceleration error vs. frequency with ζ as parameter.

seen from Fig. 3.12-5 which is a magnified plot of

$$
\frac{1}{\sqrt{\left[1 - \left(\dfrac{\omega}{\omega_n}\right)^2\right]^2 + \left[2\zeta\dfrac{\omega}{\omega_n}\right]^2}}
$$

for various values of damping ζ. The diagram shows that the useful frequency range of the undamped accelerometer is somewhat limited. However, with $\zeta = 0.7$ the useful frequency range is $0 \le \omega/\omega_n \le 0.20$ with a maximum error less than 0.01 percent. Thus an instrument with a natural frequency of 100 Hz has a useful frequency range between 0 Hz to 20 Hz with negligible error. Electromagnetic-type accelerometers generally utilize damping around $\zeta = 0.7$, which not only extends the useful frequency range but also prevents phase distortion for complex waves, as will be shown later. On the other hand, very high natural frequency instruments, such as the piezoelectric crystal accelerometers, have almost zero damping and operate without distortion up to frequencies of $0.06 f_n$.

Several different accelerometers are in use today. The seismic mass accelerometer is often used for low frequency vibration, and the supporting springs may be four electric strain gage wires connected in a bridge circuit. A more accurate variation of this accelerometer is one in which the seismic mass is servo-controlled to have zero relative displacement; the force necessary to accomplish this becomes a measure of the acceleration. Both of these instruments require an external source of electric power.

The piezoelectric properties of crystals like quartz or barium titanate are utilized in accelerometers for higher frequency measurements. The

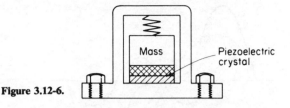

Figure 3.12-6.

crystals are mounted so that under acceleration they are either compressed or bent to generate an electric charge. Figure 3.12-6 shows one such arrangement. The natural frequency of such accelerometers can be made very high, in the 50,000 Hz range, which enables acceleration measurements to be made up to 3000 Hz. The size of the crystal accelerometer is very small, approximately 1 cm in diameter and height, and it is remarkably rugged and can stand shocks as high as 10,000 g's.

The sensitivity of the crystal accelerometer is given either in terms of charge (picocoulombs = pC = 10^{-12} coulombs) per g, or in terms of voltage (millivolts = mV = 10^{-3} V) per g. Since the voltage is related to the charge by the equation $E = Q/C$, the capacitance of the crystal, including the shunt capacitance of the connecting cable, must be specified. Typical sensitivity for a crystal accelerometer is 25 pC/g with crystal capacitance of 500 pF (picofarads). The equation $E = Q/C$ then gives $25/500 = 0.050$ V/g = 50 mV/g for the sensitivity in terms of voltage. If the accelerometer is connected to a vacuum tube voltmeter through a 3-meter length of cable of capacitance 300 pF, the open circuit output voltage of the accelerometer will be reduced to

$$50 \times \frac{500}{500 + 300} = 31.3 \text{ mV/g}$$

This severe loss of signal can be avoided by using a charge amplifier, in which case the capacitance of the cable has no effect.

Phase Distortion. To reproduce a complex wave such as the one shown in Fig. 3.12-7, without changing its shape, the phase of all harmonic components must remain unchanged with respect to the fundamental. This requires that the phase angle be zero or that all the harmonic components must be shifted equally. The first case of zero phase shift corresponds to $\zeta = 0$ for $\omega/\omega_n < 1$. The second case of equal timewise shift of all harmonics is nearly satisfied for $\zeta = 0.70$ for $\omega/\omega_n < 1$. As shown in Fig. 3.12-2, when $\zeta = 0.70$, the phase for $\omega/\omega_n < 1$ can be expressed by the equation

$$\phi \cong \frac{\pi}{2} \frac{\omega}{\omega_n}$$

Thus for $\zeta = 0$, or 0.70, the phase distortion is practically eliminated.

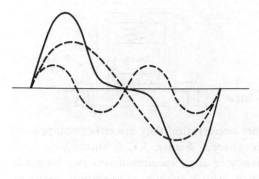

Figure 3.12-7.

EXAMPLE 3.12

Investigate the output of an accelerometer with damping $\zeta = 0.70$ when used to measure a periodic motion with the displacement given by the equation

$$y = Y_1 \sin \omega_1 t + Y_2 \sin \omega_2 t$$

Solution: For $\zeta = 0.70$, $\phi \cong \pi/2 \times \omega/\omega_n$, so that $\phi_1 = \pi/2 \times \omega_1/\omega_n$ and $\phi_2 = \pi/2 \times \omega_2/\omega_n$. The output of the accelerometer is then

$$z = Z_1 \sin(\omega_1 t - \phi_1) + Z_2 \sin(\omega_2 t - \phi_2)$$

Substituting for Z_1 and Z_2 from Eq. (3.12-6), the output of the instrument is

$$z = \frac{1}{\omega_n^2} \left\{ \omega_1^2 Y_1 \sin \omega_1 \left(t - \frac{\pi}{2\omega_n} \right) + \omega_2^2 Y_2 \sin \omega_2 \left(t - \frac{\pi}{2\omega_n} \right) \right\}$$

Since the time function $(t - \pi/2\omega_n)$ in both terms is equal, the shift of both components along the time axis is equal. Thus the instrument faithfully reproduces the acceleration y without distortion. It is obvious that if ϕ_1 and ϕ_2 are both zero, we again obtain zero phase distortion.

PROBLEMS

3-1 A machine part of 1.95 kg mass vibrates in a viscous medium. Determine the damping coefficient when a harmonic exciting force of 24.46 N results in a resonant amplitude of 1.27 cm with a period of 0.20 sec.

3-2 If the system of Prob. 3-1 is excited by a harmonic force of frequency 4 cps, what will be the percentage increase in the amplitude of forced vibration when the dashpot is removed?

3-3 A weight attached to a spring of stiffness 525 N/m has a viscous damping device. When the weight is displaced and released, the period of vibration is found to be 1.80 sec, and the ratio of consecutive amplitudes is 4.2 to 1.0. Determine the amplitude and phase when a force $F = 2 \cos 3t$ acts on the system.

3-4 Show that for the damped spring-mass system, the peak amplitude occurs at a frequency ratio given by the expression

$$\left(\frac{\omega}{\omega_n}\right)_p = \sqrt{1 - 2\zeta^2}$$

3-5 A spring-mass system is excited by a force $F_0 \sin \omega t$. At resonance the amplitude is measured to be 0.58 cm. At 0.80 resonant frequency, the amplitude is measured to be 0.46 cm. Determine the damping factor ζ of the system.

3-6 Arrive at Eqs. (3.1-3) and (3.1-4) by substituting the general steady-state solution $x = C_1 \sin \omega t + C_2 \cos \omega t$ into the differential equation and solving for C_1 and C_2.

3-7 For a system shown in Fig. P3-7, set up the equation of motion and solve for the steady-state amplitude and phase angle by using complex algebra.

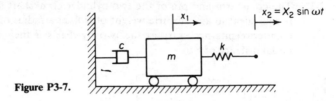

Figure P3-7.

3-8 Shown in Fig. P3-8 is a cylinder of mass m connected to a spring of stiffness k excited through viscous friction c to a piston with motion $y = A \sin \omega t$. Determine the amplitude of the cylinder motion and its phase with respect to the piston.

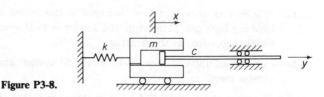

Figure P3-8.

3-9 A counterrotating eccentric mass exciter shown in Fig. P3-9 is used to determine the vibrational characteristics of a structure of mass 181.4 kg. At a speed of 900 rpm, a stroboscope shows the eccentric masses to be at the top at the instant the structure is moving upward through its static equilibrium position, and the corresponding amplitude is 21.6 mm. If the unbalance of each wheel of the exciter is 0.0921 kg m, determine (a) the natural frequency of the structure, (b) the damping factor of the structure, (c) the

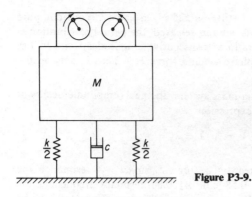

Figure P3-9.

amplitude at 1200 rpm, and (d) the angular position of the eccentrics at the instant the structure is moving upward through its equilibrium position.

3-10 A circular disk rotating about its geometric axis has two holes A and B drilled through it. The diameter and position of hole A are $d_A = 10$ mm, $r_A = 30$ cm, $\theta_A = 0°$. The diameter and position at hole B are $d_B = 5$ mm, $r_B = 20$ cm, $\theta_B = 90°$. Determine the diameter and position of a third hole at 10 cm radius that will balance the disk.

3-11 The crank arm and pin of the two-cylinder crankshaft shown in Fig. P3-11 is equivalent to an eccentric weight of w lb at a radius of r in. Determine the counterweights necessary at the two flywheels if they are also placed at a radial distance of r in.

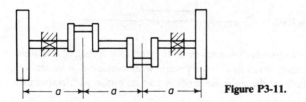

Figure P3-11.

3-12 A solid disk of weight 10 lb is keyed to the center of a $\frac{1}{2}$ in. steel shaft 2 ft between bearings. Determine the lowest critical speed. (Assume shaft to be simply supported at the bearings.)

3-13 Convert all units in Prob. 3-12 to the SI system and recalculate the lowest critical speed.

3-14 The rotor of a turbine 13.6 kg in mass is supported at the midspan of a shaft with bearings 0.4064 m apart, as shown in Fig. P3-14. The rotor is known to have an unbalance of 0.2879 kg cm. Determine the forces exerted on the bearings at a speed of 6000 rpm if the diameter of the steel shaft is 2.54 cm. Compare this result with that of the same rotor mounted on a steel shaft of diameter 1.905 cm. (Assume the shaft to be simply supported at the bearings.)

3-15 For turbines operating above the critical speed, stops are provided to limit the amplitude as it runs through the critical speed. In the turbine of Prob.

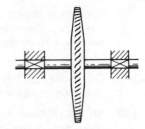

Figure P3-14.

3-14, if the clearance between the 2.54-cm shaft and the stops is 0.0508 cm, and if the eccentricity is 0.0212 cm, determine the time required for the shaft to hit the stops. Assume that the critical speed is reached with zero amplitude.

3-16 Figure P3-16 represents a simplified diagram of a spring-supported vehicle traveling over a rough road. Determine the equation for the amplitude of W as a function of the speed and determine the most unfavorable speed.

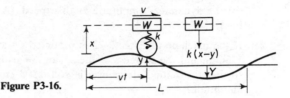

Figure P3-16.

3-17 The springs of an automobile trailer are compressed 10.16 cm under its weight. Find the critical speed when the trailer is traveling over a road with a profile approximated by a sine wave of amplitude 7.62 cm and wave length of 14.63 m. What will be the amplitude of vibration at 64.4 km/h? (Neglect damping.)

3-18 The point of suspension of a simple pendulum is given a harmonic motion $x_0 = X_0 \sin \omega t$ along a horizontal line, as shown in Fig. P3-18. Write the differential equation of motion for small amplitude of oscillation, using the

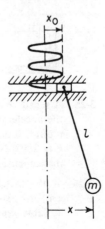

Figure P3-18.

coordinates shown. Determine the solution for x/x_0 and show that when $\omega = \sqrt{2}\,\omega_n$, the node is found at the mid-point of l. Show that in general the distance h from the mass to the node is given by the relation $h = l(\omega_n/\omega)^2$ where $\omega_n = \sqrt{g/l}$.

3-19   Derive Eqs. (3.5-8) and (3.5-9) for the amplitude and phase by letting $y = Y \sin \omega t$ and $x = X \sin(\omega t - \phi)$ in the differential equation (3.5-1).

3-20   An aircraft radio weighing 106.75 N is to be isolated from engine vibrations ranging in frequencies from 1600 cpm to 2200 cpm. What statical deflection must the isolators have for 85% isolation?

3-21   A refrigerator unit weighing 65 lb is to be supported by three springs of stiffness k lb/in. each. If the unit operates at 580 rpm, what should be the value of the spring constant k if only 10 per cent of the shaking force of the unit is to be transmitted to the supporting structure?

3-22   An industrial machine of mass 453.4 kg is supported on springs with a statical deflection of 0.508 cm. If the machine has a rotating unbalance of 0.2303 kg m, determine (a) the force transmitted to the floor at 1200 rpm and (b) the dynamical amplitude at this speed. (Assume damping to be negligible.)

3-23   If the machine of Prob. 3-22 is mounted on a large concrete block of mass 1136 kg and the stiffness of the springs or pads under the block is increased so that the statical deflection is still 0.508 cm, what will be the dynamical amplitude?

3-24   An electric motor of mass 68 kg is mounted on an isolator block of mass 1200 kg and the natural frequency of the total assembly is 160 cpm with a damping factor of $\zeta = 0.10$ (see Fig. P3-24). If there is an unbalance in the motor that results in a harmonic force of $F = 100 \sin 31.4t$, determine the amplitude of vibration of the block and the force transmitted to the floor.

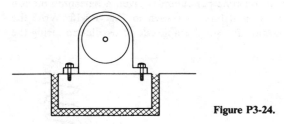

Figure P3-24.

3-25   A sensitive instrument with mass 113 kg is to be installed at a location where the acceleration is 15.24 cm/sec^2 at a frequency of 20 Hz. It is proposed to mount the instrument on a rubber pad with the following properties: $k = 2802$ N/cm and $\zeta = 0.10$. What acceleration is transmitted to the instrument?

3-26   If the instrument of Prob. 3-25 can only tolerate an acceleration of 2.03 cm/sec^2, suggest a solution assuming that the same rubber pad is the only isolator available. Give numerical values to substantiate your solution.

3-27 For the system shown in Fig. P3-27, verify that the transmissibility $TR = |x/y|$ is the same as that for force. Plot the transmissibility in decibels, $20 \log |TR|$ vs. ω/ω_n between $\omega/\omega_n = 1.50$ to 10 with $\zeta = 0.02, 0.04 \ldots 0.10$.

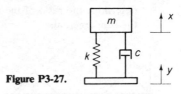

Figure P3-27.

3-28 Show that the energy dissipated per cycle for viscous friction can be expressed as

$$W_d = \frac{\pi F_0^2}{k} \frac{2\zeta (\omega/\omega_n)}{\left[1 - (\omega/\omega_n)^2\right]^2 + [2\zeta(\omega/\omega_n)]^2}$$

3-29 Show that for viscous damping, the loss factor η is independent of the amplitude and proportional to the frequency.

3-30 Express the equation for the free vibration of a single degree of freedom system in terms of the loss factor η at resonance.

3-31 Show that τ_n/τ_d plotted against ζ is a quarter circle where $\tau_d =$ damped natural period and $\tau_n =$ undamped natural period.

3-32 For small damping, the energy dissipated per cycle divided by the peak potential energy is equal to 2δ and also to $1/Q$. [See Eq. (3.7-6).] For viscous damping show that

$$\delta = \frac{\pi c \omega_n}{k}$$

3-33 In general, the energy loss per cycle is a function of both amplitude and frequency. State under what condition the logarithmic decrement δ is independent of the amplitude.

3-34 Coulomb damping between dry surfaces is a constant D always opposed to the motion. Determine the equivalent viscous damping.

3-35 Using the result of Prob. 3-34, determine the amplitude of motion of a spring mass system with Coulomb damping when excited by a harmonic force $F_0 \sin \omega t$. Under what condition can this motion be maintained?

3-36 Plot the results of Prob. 3-35 in the permissible range.

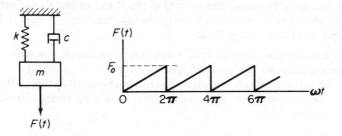

Figure P3-37.

3-37 In Fig. P 3-37 determine the steady-state response of the spring-mass damper system to the excitation of Prob. 1-12.

3-38 If the periodic force shown in Fig. P3-38 is applied to a spring-mass system, determine the ratio of the response to the various harmonics compared with the fundamental.

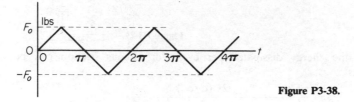

Figure P3-38.

3-39 If the excitation given in Prob. 3-38 is the displacement of the support point of the spring-mass system, determine the equation for (a) the relative motion and (b) the absolute motion of the mass. Assume structural damping $\gamma = 0.05$.

3-40 The shaft of a torsiograph, shown in Fig. P 3-40, undergoes harmonic torsional oscillation $\theta_0 \sin \omega t$. Determine the expression for the relative amplitude of the outer wheel with respect to (a) the shaft, (b) a fixed reference.

Figure P3-40.

3-41 A commercial-type vibration pickup has a natural frequency of 4.75 cps and a damping factor $\zeta = 0.65$. What is the lowest frequency that can be measured with (a) 1 percent error, (b) 2 percent error?

3-42 An undamped vibration pickup having a natural frequency of 1 cps is used to measure a harmonic vibration of 4 cps. If the amplitude indicated by the pickup (relative amplitude between pickup mass and frame) is 0.052 cm, what is the correct amplitude?

3-43 A manufacturer of vibration measuring instruments gives the following specifications for one of its vibration pickups:

    Frequency range: Velocity response flat from 10 cps to 1000 cps.

    Sensitivity: 0.096 V/cm/sec, both volts and velocity in root-mean-square values

Amplitude range: Almost no lower limit to maximum stroke between stops of 0.60 in.

(a) This instrument was used to measure the vibration of a machine with a known frequency of 30 cps. If a reading of 0.024 V is indicated, determine the root-mean-square amplitude.

(b) Could this instrument be used to measure the vibration of a machine with known frequency of 12 cps and double amplitude of 0.80 cm? Give reasons.

3-44 A vibration pickup has a sensitivity of 40 mV/(cm/sec) between $f = 10$ Hz to 2000 Hz. If 1 g acceleration is maintained over this frequency range, what will be the output voltage at (a) 10 Hz and (b) at 2000 Hz.

3-45 Using the equations of harmonic motion, obtain the relationship for the velocity vs. frequency applicable to the velocity pickup.

3-46 A vibration pickup has a sensitivity of 30 mV/cm/sec. Assuming that 3 mV (rms) is the accuracy limit of the instrument, determine the upper frequency limit of the instrument for 1g excitation. What voltage would be generated at 200 Hz?

3-47 The sensitivity of a certain crystal accelerometer is given as 18 pC/g, with its capacitance equal to 450 pF. It is used with a vacuum tube voltmeter and its connecting cable is 5 m long with a capacitance of 50 pF/m. Determine its voltage output per g.

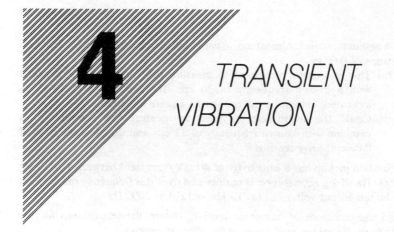

4 TRANSIENT VIBRATION

When a dynamical system is excited by a suddenly applied nonperiodic excitation $F(t)$, the response to such excitation is called *transient response* since steady state oscillations are generally not produced. Such oscillations take place at the natural frequencies of the system with the amplitude varying in a manner dependent on the type of excitation.

We first study the response of a spring-mass system to an impulse excitation because this case is important in the understanding of the more general problem of transients.

4.1 IMPULSE EXCITATION

Impulse is the time integral of the force, and we designate it by the notation $\hat{F}$

$$\hat{F} = \int F(t) \, dt \qquad (4.1\text{-}1)$$

We frequently encounter a force of very large magnitude which acts for a very short time, but with a time integral which is finite. Such forces are called *impulsive*.

Figure 4.1-1 shows an impulsive force of magnitude $\hat{F}/\epsilon$ with a time duration of ϵ. As ϵ approaches zero, such forces tend to become infinite; however, the impulse defined by its time integral is $\hat{F}$ which is considered to be finite. When $\hat{F}$ is equal to unity, such force in the limiting case $\epsilon \to 0$

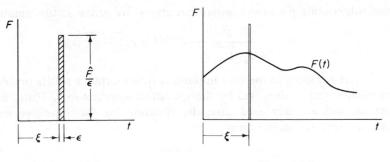

Figure 4.1-1.   Figure 4.1-2.

is called the *unit impulse* or the *delta function*. A delta function at $t = \xi$ is identified by the symbol $\delta(t - \xi)$ and has the following properties

$$\delta(t - \xi) = 0 \qquad \text{for all } t \neq \xi$$
$$= \text{greater than any assumed value for } t = \xi \quad (4.1\text{-}2)$$
$$\int_0^\infty \delta(t - \xi)\, dt = 1.0 \qquad 0 < \xi < \infty$$

If $\delta(t - \xi)$ is multiplied by any time function $f(t)$ as shown in Fig. 4.1-2, the product will be zero everywhere except at $t = \xi$, and its time integral will be

$$\int_0^\infty f(t)\delta(t - \xi)\, dt = f(\xi) \qquad 0 < \xi < \infty \qquad (4.1\text{-}3)$$

Since $F dt = m\, dv$, the impulse $\hat{F}$ acting on the mass will result in a sudden change in its velocity equal to $\hat{F}/m$ without an appreciable change in its displacement. Under free vibration we found that the undamped spring-mass system with initial conditions $x(0)$ and $\dot{x}(0)$ behaved according to the equation

$$x = \frac{\dot{x}(0)}{\omega_n} \sin \omega_n t + x(0) \cos \omega_n t$$

Hence the response of a spring-mass system initially at rest and excited by an impulse $\hat{F}$ is

$$x = \frac{\hat{F}}{m\omega_n} \sin \omega_n t \qquad (4.1\text{-}4)$$

where

$$\omega_n = \sqrt{\frac{k}{m}}$$

When damping is present we can start with the free vibration equation

$$x = X e^{-\zeta \omega_n t} \sin\left(\sqrt{1 - \zeta^2}\, \omega_n t - \phi\right)$$

and substituting the above initial conditions, we arrive at the equation

$$x = \frac{\hat{F}}{m\omega_n\sqrt{1-\zeta^2}}\, e^{-\zeta\omega_n t}\, \sin\sqrt{1-\zeta^2}\,\omega_n t \qquad (4.1\text{-}5)$$

The response to the unit impulse is of importance to the problems of transients, and is identified by the special designation $h(t)$. Thus, in either the damped or undamped case, the equation for the impulsive response can be expressed in the form

$$x = \hat{F}h(t) \qquad (4.1\text{-}6)$$

where the right side of the equation is given by either Eq. (4.1-4) or (4.1-5).

4.2   ARBITRARY EXCITATION

Having the response $h(t)$ to a unit impulse excitation, it is possible to establish the equation for the response of the system excited by an arbitrary force $f(t)$. For this development, we consider the arbitrary force to be a series of impulses as shown in Fig. 4.2-1. If we examine one of the impulses (shown crosshatched) at time $t = \xi$, its strength is

$$\hat{F} = f(\xi)\,\Delta\xi$$

and its contribution to the response at time t is dependent upon the

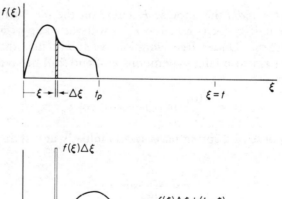

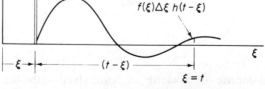

Figure 4.2-1.

elapsed time $(t - \xi)$, or

$$f(\xi) \, \Delta\xi h(t - \xi)$$

Since the system we are considering is linear, the principle of superposition holds. Thus, by combining all such contributions, the response to the arbitrary excitation $f(t)$ is represented by the integral

$$x(t) = \int_0^t f(\xi)h(t - \xi) \, d\xi \qquad (4.2\text{-}1)$$

The above integral is called the *Convolution integral*, or is sometimes referred to as the *superposition integral*.

Another form of this equation is found by letting $\tau = (t - \xi)$. Then $\xi = t - \tau$, $d\xi = - d\tau$, and we obtain

$$x(t) = \int_0^t f(t - \tau)h(\tau) \, d\tau \qquad (4.2\text{-}2)$$

When t is greater than the pulse time, say t_p, the upper limit of the general equation, Eq. (4.2-1), remains at t_p because the integral can then be written as

$$x(t) = \int_0^{t_p} f(\xi)h(t - \xi) \, d\xi + \int_{t_p}^t f(\xi)h(t - \xi) \, d\xi$$

$$= \int_0^{t_p} f(\xi)h(t - \xi) \, d\xi, \qquad t > t_p \qquad (4.2\text{-}3)$$

Here the second integral is zero since $f(\xi) = 0$ for $\xi > t_p$.

Base Excitation. Often the support of the dynamical system is subjected to a sudden movement specified by its displacement, velocity, or acceleration. The equation of motion can then be expressed in terms of the relative displacement $z = x - y$ as follows

$$\ddot{z} + 2\zeta\omega_n\dot{z} + \omega_n^2 z = -\ddot{y} \qquad (4.2\text{-}4)$$

and hence all of the results for the force-excited system apply to the base-excited system for z when the term F_0/m is replaced by $-\ddot{y}$ or the negative of the base acceleration.

For an undamped system initially at rest, the solution for the relative displacement becomes

$$z = -\frac{1}{\omega_n} \int_0^t \ddot{y}(\xi) \sin \omega_n(t - \xi) \, d\xi \qquad (4.2\text{-}5)$$

EXAMPLE 4.2-1

Determine the response of a single degree of freedom system to the step excitation shown in Fig. 4.2-2.

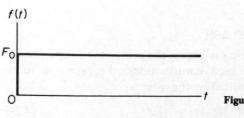

Figure 4.2-2. Step function excitation.

Solution: Considering the undamped system, we have

$$h(t) = \frac{1}{m\omega_n} \sin \omega_n t$$

Substituting into Eq. (4.2-1) the response of the undamped system is

$$x(t) = \frac{F_0}{m\omega_n} \int_0^t \sin \omega_n(t - \xi)\, d\xi$$

$$= \frac{F_0}{k}(1 - \cos \omega_n t)$$

This result indicates that the peak response to the step excitation of magnitude F_0 is equal to twice the statical deflection.

For a damped system the procedure can be repeated with

$$h(t) = \frac{e^{-\zeta\omega_n t}}{m\omega_n\sqrt{1 - \zeta^2}} \sin \sqrt{1 - \zeta^2}\, \omega_n t$$

or, alternatively, we can simply consider the differential equation

$$\ddot{x} + 2\zeta\omega_n\dot{x} + \omega_n^2 x = \frac{F_0}{m}$$

whose solution is the sum of the solutions to the homogeneous equation and that of the particular solution, which for this case is $F_0/m\omega_n^2$. Thus the equation

$$x(t) = Xe^{-\zeta\omega_n t} \sin\left(\sqrt{1 - \zeta^2}\, \omega_n t - \phi\right) + \frac{F_0}{m\omega_n^2}$$

fitted to the initial conditions of $x(0) = \dot{x}(0) = 0$ will result in the solution which is given as

$$x = \frac{F_0}{k}\left[1 - \frac{e^{-\zeta\omega_n t}}{\sqrt{1 - \zeta^2}} \cos\left(\sqrt{1 - \zeta^2}\, \omega_n t - \psi\right)\right]$$

where

$$\tan \psi = \frac{\zeta}{\sqrt{1 - \zeta^2}}$$

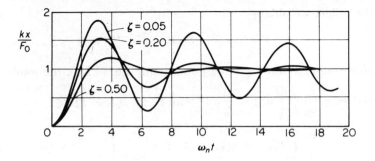

Figure 4.2-3. Response to a unit step function.

Figure 4.2-3 shows a plot of xk/F_0 versus $\omega_n t$ with ζ as parameter, and it is evident that the peak response is less than $2F_0/k$ when damping is present.

EXAMPLE 4.2-2

Consider an undamped spring-mass system where the motion of the base is specified by a velocity pulse of the form

$$\dot{y}(t) = v_0 e^{-t/t_0} u(t)$$

where $u(t)$ is a unit step function. The velocity together with its time rate of change is shown in Fig. 4.2-4.

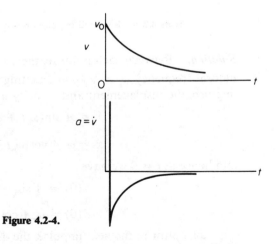

Figure 4.2-4.

Solution: The velocity pulse at $t = 0$ has a sudden jump from zero to v_0, and its rate of change (or acceleration) is infinite. Differentiating $\dot{y}(t)$ and recognizing that $(d/dt)u(t) = \delta(t)$, a delta function at the origin, we obtain

$$\ddot{y} = v_0 e^{-t/t_0} \delta(t) - \frac{v_0}{t_0} e^{-t/t_0} u(t)$$

Substituting $\ddot{y}$ into Eq. (4.2-5) and noting Eq. (4.2-3), the result is

$$z(t) = -\frac{v_0}{\omega_n} \int_0^t \left[\delta(\xi) - \frac{1}{t_0} e^{-\xi/t_0} u(\xi) \right] \sin \omega_n(t - \xi) \, d\xi$$

$$= -\frac{v_0}{\omega_n} \int_0^t \delta(\xi) \sin \omega_n(t - \xi) \, d\xi + \frac{v_0}{\omega_n t_0} \int_0^t e^{-\xi/t_0} \sin \omega_n(t - \xi) \, d\xi$$

$$= \frac{v_0 t_0}{1 + (\omega_n t_0)^2} \left[e^{-t/t_0} - \omega_n t_0 \sin \omega_n t - \cos \omega_n t \right]$$

EXAMPLE 4.2-3

A mass m attached to a spring of stiffness k is subjected to repeated impulse $\hat{F}$ of negligible duration at intervals of τ_i, as shown in Fig. 4.2-5. Determine the steady-state response.

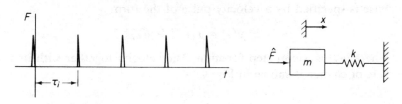

Figure 4.2-5. Repeated impulse on a spring-mass system.

Solution: Between each impulse, the system is in free vibration at its natural frequency $\omega_n = \sqrt{k/m}$. Letting $t = 0$ immediately after the impulse, the displacement and velocity may be expressed as

$$x = A \sin(\omega_n t + \phi) \tag{a}$$

$$\dot{x} = \omega_n A \cos(\omega_n t + \phi) \tag{b}$$

and hence at $t = 0$ we have

$$x(0) = A \sin \phi \tag{c}$$

$$\dot{x}(0) = \omega_n A \cos \phi \tag{d}$$

Just prior to the next impulse, the displacement and velocity are

$$x(\tau_i) = A \sin(\omega_n \tau_i + \phi) \tag{e}$$

$$\dot{x}(\tau_i) = \omega_n A \cos(\omega_n \tau_i + \phi) \tag{f}$$

where τ_i is the time interval of the impulses. The impulse acting at this time increases the velocity suddenly by $\hat{F}/m$ although the displacement remains essentially unchanged.

If steady state is attained, the displacement and velocity after each cycle must repeat themselves. Thus we can write

$$A \sin \phi = A \sin(\omega_n \tau_i + \phi) \tag{g}$$

$$\omega_n A \cos \phi = \omega_n A \cos(\omega_n \tau_i + \phi) + \frac{\hat{F}}{m} \tag{h}$$

Rearranging these equations to

$$\sin(\omega_n \tau_i + \phi) - \sin \phi = 0 \tag{i}$$

$$\cos(\omega_n \tau_i + \phi) - \cos \phi = -\frac{\hat{F}}{\omega_n m A} \tag{j}$$

we note that these equations may be rewritten as

$$\sin \frac{\omega_n \tau_i}{2} \cos\left(\frac{\omega_n \tau_i}{2} + \phi\right) = 0 \tag{i'}$$

$$\sin \frac{\omega_n \tau_i}{2} \sin\left(\frac{\omega_n \tau_i}{2} + \phi\right) = \frac{\hat{F}}{2\omega_n m A} \tag{j'}$$

Since $\sin \omega_n \tau_i / 2$ cannot be zero for arbitrary τ_i, Eq. (i') can be satisfied only if

$$\cos\left(\frac{\omega_n \tau_i}{2} + \phi\right) = 0$$

or

$$\sin\left(\frac{\omega_n \tau_i}{2} + \phi\right) = 1$$

Equation (j') then becomes

$$\sin \frac{\omega_n \tau_i}{2} = \frac{\hat{F}}{2\omega_n m A} \tag{j''}$$

from which the amplitude is available as

$$A = \frac{\hat{F}}{2\omega_n m \sin \dfrac{\omega_n \tau_i}{2}} \tag{k}$$

The maximum spring force $F_s = kA$ may be of interest, in which case Eq. (k) takes the form

$$\frac{\tau_i F_s}{\hat{F}} = \frac{\dfrac{\omega_n \tau_i}{2}}{\sin \dfrac{\omega_n \tau_i}{2}} \tag{l}$$

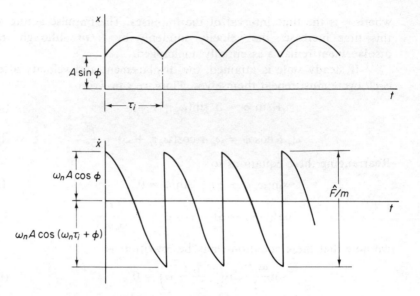

Figure 4.2-6.   Displacement and velocity.

Thus the amplitude or spring force becomes infinite when

$$\frac{\omega_n \tau_i}{2} = \frac{\pi f_n}{f_i} = 0, \pi, 2\pi, 3\pi, \ldots$$

Equation (l) also shows that the maximum spring force F_s is a minimum when

$$\frac{\pi f_n}{f_i} = \frac{\pi}{2}, \frac{3\pi}{2}, \frac{5\pi}{2}, \ldots$$

The time variation of the displacement and velocity may appear as in Fig. 4.2-6.

When damping is included, a similar procedure can be applied, although the numerical work is increased considerably.

4.3   LAPLACE TRANSFORM FORMULATION

The Laplace transform method of solving the differential equation provides a complete solution, yielding both transient and forced vibration. For those unfamiliar with the method, a brief presentation of the Laplace transform theory is given in the Appendix. In this section we will illustrate its use by some simple examples.

EXAMPLE 4.3-1

Formulate the Laplace transform solution of a viscously damped spring-mass system with initial conditions $x(0)$ and $\dot{x}(0)$.

Solution: The equation of motion for the system excited by an arbitrary force $F(t)$ is

$$m\ddot{x} + c\dot{x} + kx = F(t)$$

Taking its Laplace transform, we find

$$m\left[s^2\bar{x}(s) - x(0)s - \dot{x}(0) \right] + c\left[s\bar{x}(s) - x(0) \right] + k\bar{x}(s) = \bar{F}(s)$$

Solving for $\bar{x}(s)$ we obtain the *subsidiary equation*

$$\bar{x}(s) = \frac{\bar{F}(s)}{ms^2 + cs + k} + \frac{(ms + c)x(0) + m\dot{x}(0)}{ms^2 + cs + k} \qquad \text{(a)}$$

The response $x(t)$ is found from the inverse of Eq. (a); the first term represents the forced vibration and the second term represents the transient solution due to the initial conditions.

For the more general case, the subsidiary equation can be written in the form

$$\bar{x}(s) = \frac{A(s)}{B(s)} \qquad \text{(b)}$$

where $A(s)$ and $B(s)$ are polynomials and $B(s)$, in general, is of higher order than $A(s)$.

If only the forced solution is considered, we can define the *impedance transform* as

$$\frac{\bar{F}(s)}{\bar{x}(s)} = z(s) = ms^2 + cs + k \qquad \text{(c)}$$

Its reciprocal is the admittance transform

$$H(s) = \frac{1}{z(s)} \qquad \text{(d)}$$

Frequently a block diagram is used to denote input and output as shown in Fig. 4.3-1. The admittance transform $H(s)$ then can also be considered as the *system transfer function*, defined as the ratio in the subsidiary plane of the output over the input with all initial conditions equal to zero.

Figure 4.3-1. Block diagram.   Input $\bar{F}(s) \longrightarrow \boxed{H(s)} \longrightarrow$ Output $\bar{x}(s)$

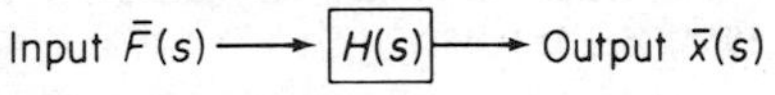

EXAMPLE 4.3-2

The question of how far a body can be dropped without incurring damage is of frequent interest. Such considerations are of paramount importance in the landing of airplanes or the cushioning of packaged articles.* In this example we shall discuss some of the elementary aspects of this problem by idealizing the mechanical system in terms of linear spring-mass components.

Consider the spring-mass system of Fig. 4.3-2 dropped through a height h. If x is measured from the position of m at the instant $t = 0$ when the spring first contacts the floor, the differential equation of motion for m applicable as long as the spring remains in contact with the floor is

$$m\ddot{x} + kx = mg \tag{a}$$

Taking the Laplace transform of this equation with the initial conditions $x(0) = 0$ and $\dot{x}(0) = \sqrt{2gh}$, we can write the subsidiary equation as

$$\bar{x}(s) = \frac{\sqrt{2gh}}{s^2 + \omega_n^2} + \frac{g}{s(s^2 + \omega_n^2)} \tag{b}$$

where $\omega_n = \sqrt{k/m}$ is the natural frequency of the system. From the inverse transformation of $\bar{x}(s)$ the displacement equation becomes

$$x(t) = \frac{\sqrt{2gh}}{\omega_n} \sin \omega_n t + \frac{g}{\omega_n^2}(1 - \cos \omega_n t)$$

$$\tag{c}$$

$$= \sqrt{\frac{2gh}{\omega_n^2} + \left(\frac{g}{\omega_n^2}\right)^2} \ \sin(\omega_n t - \phi) \qquad x(t) > 0$$

where the relationship is shown in Fig. 4.3-3. By differentiation, the velocity and acceleration are

$$\dot{x}(t) = \omega_n \sqrt{\frac{2gh}{\omega_n^2} + \left(\frac{g}{\omega_n^2}\right)^2} \ \cos(\omega_n t - \phi)$$

$$\ddot{x}(t) = -\omega_n^2 \sqrt{\frac{2gh}{\omega_n^2} + \left(\frac{g}{\omega_n^2}\right)^2} \ \sin(\omega_n t - \phi)$$

*R. D. Mindlin, "Dynamics of Package Cushioning," *Bell Syst. Tech. Jour.* (July, 1954), pp. 353–461.

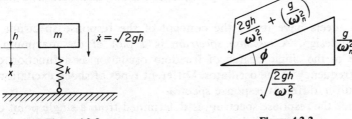

Figure 4.3-2.   Figure 4.3-3.

We recognize here that $g/\omega^2 = \delta_{st}$ and that the maximum displacement and acceleration occur at $\sin(\omega_n t - \phi) = 1.0$. Thus the maximum acceleration in terms of gravity is found to depend only on the ratio of the distance dropped to the statical deflection as given by the equation

$$\frac{\ddot{x}}{g} = -\sqrt{\frac{2h}{\delta_{st}} + 1}$$ (d)

A plot of this equation is shown in Fig. 4.3-4.

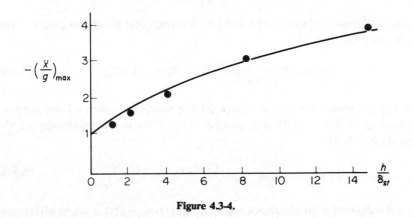

Figure 4.3-4.

4.4 RESPONSE SPECTRUM

A shock represents a sudden application of a force or other form of disruption which results in a transient response of a system. The maximum value of the response is a good measure of the severity of the shock and is, of course, dependent upon the dynamic characteristics of the system. In order to categorize all types of shock excitations, a single degree of freedom undamped oscillator (spring-mass system) is chosen as a standard system.

Engineers have found the concept of the response spectrum to be useful in design. A *response spectrum* is a plot of the maximum peak response of the single degree of freedom oscillator as a function of the natural frequency of the oscillator. Different types of shock excitations will then result in different response spectra.

Since the response spectrum is determined from a single point on the time response curve, which is itself an incomplete bit of information, it does not uniquely define the shock input. In fact, it is possible for two different shock excitations to have very similar response spectra. In spite of this limitation, the response spectrum is a useful concept that is extensively used.

The response of a system to arbitrary excitation $f(t)$ was expressed in terms of the impulse response $h(t)$ by Eq. (4.2-1).

$$x(t) = \int_0^t f(\xi)h(t - \xi)\, d\xi \qquad (4.4\text{-}1)$$

For the undamped single degree of freedom oscillator, we have

$$h(t) = \frac{1}{m\omega_n}\sin \omega_n t \qquad (4.4\text{-}2)$$

so that the peak response to be used in the response spectrum plot is given by the equation

$$x(t)_{max} = \left| \frac{1}{m\omega_n} \int_0^t f(\xi) \sin \omega_n(t - \xi)\, d\xi \right|_{max} \qquad (4.4\text{-}3)$$

In the case where the shock is due to the sudden motion of the support point, $f(t)$ in Eq. (4.4-3) is replaced by $-\ddot{y}(t)$, the acceleration of the support point, or

$$z(t)_{max} = \left| \frac{-1}{\omega_n} \int_0^t \ddot{y}(\xi) \sin \omega_n(t - \xi)\, d\xi \right|_{max} \qquad (4.4\text{-}4)$$

Associated with the shock excitation $f(t)$ or $-\ddot{y}(t)$ is some characteristic time t_1, such as the duration of the shock pulse. With τ as the natural period of the oscillator, the maximum value of $x(t)$ or $z(t)$ is plotted as a function of t_1/τ.

Figures 4.4-1, 4.4-2, and 4.4-3 represent response spectra for three different excitations. The horizontal scale is equal to the ratio t_1/τ, while the vertical scale is a nondimensional number which is a measure of the dynamic effect over that of a statically applied load. The dynamic factor of a shock is then generally less than two.

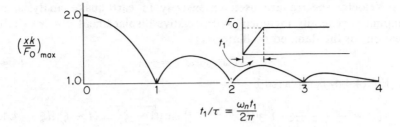

$$t_1/\tau = \frac{\omega_n t_1}{2\pi}$$

Figure 4.4-1.

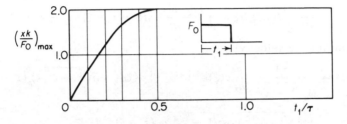

Figure 4.4-2.

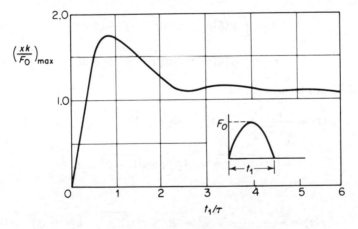

Figure 4.4-3.

Pseudo Response Spectra. In ground shock situations, it is often convenient to express the response spectra in terms of the *velocity spectra*. The displacement and acceleration spectra then can be expressed in terms of the velocity spectra by dividing or multiplying by ω_n. Such results are called *pseudo spectra* since they are exact only if the peak response occurs after the shock pulse has passed, in which case the motion is harmonic.

Velocity spectra are used extensively in earthquake analysis, and damping is generally included. With relative displacement $z = x - y$, the equation for the damped oscillator is

$$\ddot{z} + 2\zeta\omega_n\dot{z} + \omega_n^2 z = -\ddot{y} \tag{4.4-5}$$

and Eq. (4.4-4) is replaced by

$$z(t) = \frac{-1}{\omega_n\sqrt{1 - \zeta^2}} \int_0^t \ddot{y}(\xi)e^{-\zeta\omega_n(t-\xi)} \sin\sqrt{1 - \zeta^2}\,\omega_n(t - \xi)\,d\xi \tag{4.4-6}$$

Differentiating, using the equation

$$\frac{d}{dt}\int_0^t f(t,\xi)\,d\xi = \int_0^t \frac{\partial f(t, \xi)}{dt}\,d\xi + f(t, \xi) \qquad \xi = t \tag{4.4-7}$$

we obtain for the velocity

$$\dot{z}(t) = \frac{-1}{\omega_n\sqrt{1 - \zeta^2}} \int_0^t \ddot{y}(\xi)e^{-\zeta\omega_n(t-\xi)} \Big[-\zeta\omega_n \sin\sqrt{1 - \zeta^2}\,\omega_n(t - \xi)$$

$$+\omega_n\sqrt{1 - \zeta^2}\,\cos\sqrt{1 - \zeta^2}\,\omega_n(t - \xi) \Big]\,d\xi \tag{4.4-8}$$

Letting

$$A = \int_0^t \ddot{y}(\xi)e^{\zeta\omega_n\xi} \cos\sqrt{1 - \zeta^2}\,\omega_n\xi\,d\xi \tag{4.4-9}$$

$$B = \int_0^t \ddot{y}(\xi)e^{\zeta\omega_n\xi} \sin\sqrt{1 - \zeta^2}\,\omega_n\xi\,d\xi \tag{4.4-10}$$

Eq. (4.4-8) can be written as

$$\dot{z}(t) = \frac{e^{-\zeta\omega_n t}}{\sqrt{1 - \zeta^2}} \Big\{ \Big[A\zeta - B\sqrt{1 - \zeta^2} \Big] \sin\sqrt{1 - \zeta^2}\,\omega_n t$$

$$+ \Big[A\sqrt{1 - \zeta^2} + B\zeta \Big] \cos\sqrt{1 - \zeta^2}\,\omega_n t \Big\} \tag{4.4-11}$$

or

$$\dot{z}(t) = \frac{e^{-\zeta\omega_n t}}{\sqrt{1 - \zeta^2}} \sqrt{A^2 + B^2} \sin\left(\sqrt{1 - \zeta^2}\,\omega_n t - \phi\right) \tag{4.4-12}$$

If Eq. (4.4-12) is plotted against time, it would appear as an amplitude modulated wave, as shown in Fig. 4.4-4. Thus the peak velocity response S_v or the velocity spectrum is given with sufficient accuracy by the peak value of the envelope

$$S_v = |\dot{z}(t)|_{\max} = \left| \frac{e^{-\zeta\omega_n t}}{\sqrt{1 - \zeta^2}} \sqrt{A^2 + B^2} \right|_{\max} \tag{4.4-13}$$

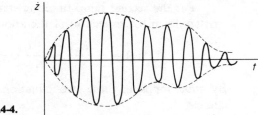

Figure 4.4-4.

Approximate relations for the peak displacement and acceleration, known as *pseudo spectra*, are then

$$|x - y|_{\max} = \frac{S_v}{\omega_n} \qquad (4.4\text{-}14)$$

$$|\ddot{z}|_{\max} = \omega_n S_v \qquad (4.4\text{-}15)$$

EXAMPLE 4.4-1

Determine the undamped response spectrum for a step function with a rise time t_1, shown in Fig. 4.4-5.

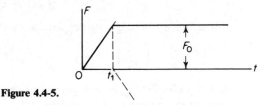

Figure 4.4-5.

Solution: The input can be considered to be the sum of two ramp functions $F_0(t/t_1)$, the second of which is negative and delayed by the time t_1. For the first ramp function the terms of the convolution integral are

$$f(t) = F_0(t/t_1)$$

$$h(t) = \frac{1}{m\omega_n} \sin \omega_n t = \frac{\omega_n}{k} \sin \omega_n t$$

and the response becomes

$$x(t) = \frac{\omega_n}{k} \int_0^t \frac{F_0 \xi}{t_1} \sin \omega_n (t - \xi) \, d\xi$$

$$= \frac{F_0}{k} \left(\frac{t}{t_1} - \frac{\sin \omega_n t}{\omega_n t_1} \right), \qquad t < t_1$$

For the second ramp function starting at t_1, the solution can be written down by inspection of the above equation as

$$x(t) = -\frac{F_0}{k}\left[\frac{(t - t_1)}{t_1} - \frac{\sin \omega_n(t - t_1)}{\omega_n t_1}\right]$$

By superimposing these two equations the response for $t > t_1$ becomes

$$x(t) = \frac{F_0}{k}\left[1 - \frac{\sin \omega_n t}{\omega_n t_1} + \frac{1}{\omega_n t_1}\sin \omega_n(t - t_1)\right] \qquad t > t_1$$

Differentiating and equating to zero, the peak time is obtained as

$$\tan \omega_n t_p = \frac{1 - \cos \omega_n t_1}{\sin \omega_n t_1}$$

Since $\omega_n t_p$ must be greater than π, we also obtain

$$\sin \omega_n t_p = -\sqrt{\tfrac{1}{2}(1 - \cos \omega_n t_1)}$$

$$\cos \omega_n t_p = \frac{-\sin \omega_n t_1}{\sqrt{2(1 - \cos \omega_n t_1)}}$$

Substituting these quantities into $x(t)$, the peak amplitude is found as

$$\left(\frac{xk}{F_0}\right)_{\max} = 1 + \frac{1}{\omega_n t_1}\sqrt{2(1 - \cos \omega_n t_1)}$$

Letting $\tau = 2\pi/\omega_n$ be the period of the oscillator, the above equation is plotted against t_1/τ as in Fig. 4.4-1.

EXAMPLE 4.4-2

Determine the response spectrum for the base velocity input, $\dot{y}(t) = v_0 e^{-t/t_0}$ of Example 4.2-2.

Solution: The relative displacement $z(t)$ was found in Example 4.2-2 to be

$$z(t) = \frac{v_0 t_0}{1 + (\omega_n t_0)^2} \times \left(e^{-t/t_0} - \omega_n t_0 \sin \omega_n t - \cos \omega_n t\right)$$

To determine the peak value z_p, the usual procedure is to differentiate the equation with respect to time t, set it equal to zero, and substitute this time back into the equation for $z(t)$. It is evident that for this problem this results in a transcendental equation which must be solved by plotting. To avoid this numerical task, we will consider a different approach as follows.

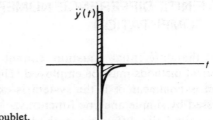

Figure 4.4-6. Impulsive doublet.

For a very stiff system, which corresponds to large ω_n, the peak response will certainly occur at small t, and we would obtain for the time varying part of the equation the peak value

$$(1 - \omega_n t_0 - 1) = -\omega_n t_0$$

Thus for large ω_n the peak value will be nearly equal to

$$|z_p| \cong \frac{v_0 t_0}{1 + (\omega_n t_0)^2}(\omega_n t_0) \cong \frac{v_0 t_0}{\omega_n t_0}$$

so that $z_p/v_0 t_0$ plots against $\omega_n t_0$ as a rectangular hyperbola.

For small ω_n, or a very soft spring, the duration of the input would be small compared to the period of the system. Hence the input would appear as an impulsive doublet shown in Fig. 4.4-6 with the equation $v_0 t_0 \delta'(t)$. The solution for $z(t)$ is then

$$z(t) = v_0 t_0 \cos \omega_n t$$

and its peak value is

$$|z_p| \cong v_0 t_0$$

With these extreme conditions evaluated, we can now fill in the response spectrum which is shown in Fig. 4.4-7.

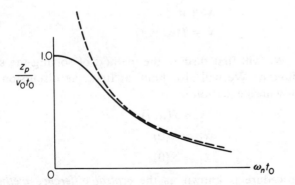

Figure 4.4-7. Response spectrum for the base velocity input $y(t) = v_0 e^{-t/t_0}$.

4.5 FINITE DIFFERENCE NUMERICAL COMPUTATION

When the differential equation cannot be integrated in closed form, numerical methods must be employed. This may well be the case when the system is nonlinear or if the system is excited by a force that cannot be expressed by simple analytic functions.

In the finite difference method the continuous variable t is replaced by the discrete variable t_i and the differential equation is solved progressively in time increments $h = \Delta t$ starting from known initial conditions. The solution is approximate, but with a sufficiently small time increment the solution of acceptable accuracy is obtainable.

Although there are a number of different finite difference procedures available, in this chapter we consider only two methods chosen for their simplicity. Merits of the various methods are associated with the accuracy, stability, and length of computation, which are discussed in a number of texts on numerical analysis listed at the end of the chapter.

The differential equation of motion for a dynamical system, which may be linear or nonlinear, can be expressed in the following general form:

$$\ddot{x} = f(x, \dot{x}, t)$$
$$x_1 = x(0) \qquad\qquad (4.5\text{-}1)$$
$$\dot{x}_1 = \dot{x}(0)$$

where the initial conditions x_1 and $\dot{x}_1$ are presumed to be known. (The subscript 1 is chosen to correspond to $t = 0$ since most computer languages do not allow sub-zero.)

In the first method the second-order equation is integrated without change in form; in the second method the second-order equation is reduced to two first-order equations before integration. The equation then takes the form

$$\dot{x} = y$$
$$\dot{y} = f(x, y, t) \qquad\qquad (4.5\text{-}2)$$

Method 1: We will first discuss the method of solving the second-order equation directly. We will also limit, at first, the discussion to the undamped system whose equation is

$$\ddot{x} = f(x, t)$$
$$x_1 = x(0) \qquad\qquad (4.5\text{-}3)$$
$$\dot{x}_1 = \dot{x}(0)$$

The following procedure is known as the *central difference method*, the basis of which can be developed from the Taylor expansion of x_{i+1} and

x_{i-1} about the pivotal point i

$$x_{i+1} = x_i + h\dot{x}_i + \frac{h^2}{2}\ddot{x}_i + \frac{h^3}{6}\dddot{x}_i + \cdots$$

$$x_{i-1} = x_i - h\dot{x}_i + \frac{h^2}{2}\ddot{x}_i - \frac{h^3}{6}\dddot{x}_i + \cdots$$

(4.5-4)

where the time interval is $h = \Delta t$. Subtracting and ignoring higher-order terms, we obtain

$$\dot{x}_i = \frac{1}{2h}(x_{i+1} - x_{i-1}) \qquad (4.5\text{-}5)$$

Adding, we find

$$\ddot{x}_i = \frac{1}{h^2}(x_{i-1} - 2x_i + x_{i+1}) \qquad (4.5\text{-}6)$$

In both Eqs. (4.4-5) and (4.5-6) the ignored terms are of order h^2. Substituting from the differential equation, Eq. (4.5-3), Eq. (4.5-6) can be rearranged to

$$x_{i+1} = 2x_i - x_{i-1} + h^2 f(x_i, t_i) \qquad i \geq 2 \qquad (4.5\text{-}7)$$

which is known as the *recurrence formula*.

(Starting the Computation). If we let $i = 2$ in the recurrence equation, we note that it is not self-starting, i.e., x_1 is known, but we need x_2 to find x_3. Thus to start the computation we need another equation for x_2. This is supplied by the first of Taylor's series, Eq. (4.5-4), ignoring higher-order terms, which gives

$$x_2 = x_1 + h\dot{x}_1 + \frac{h^2}{2}\ddot{x}_1 = x_1 + h\dot{x}_1 + \frac{h^2}{2}f(x_1, t) \qquad (4.5\text{-}8)$$

Thus Eq. (4.5-8) enables one to find x_2 in terms of the initial conditions, after which $x_3, x_4 \ldots$ are available from Eq. (4.5-7).

In this development we have ignored higher-order terms that introduce what is known as *truncation errors*. Other errors, such as round-off errors, are introduced due to loss of significant figures. These are all related to the time increment $h = \Delta t$ in a rather complicated way which is beyond the scope of this text. In general, better accuracy is obtained by choosing a smaller Δt, but the number of computations will then increase together with errors generated in the computation. A safe rule to use in Method 1 is to choose $h \leq \tau/10$ where τ is the natural period of the system.

A flow diagram for the digital calculation is shown in Fig. 4.5-1. From the given data in block Ⓐ we proceed to block Ⓑ which is the differential equation. Going to Ⓒ for the first time, I is not greater than 1, and hence we proceed to the left where x_2 is calculated. Increasing I by 1

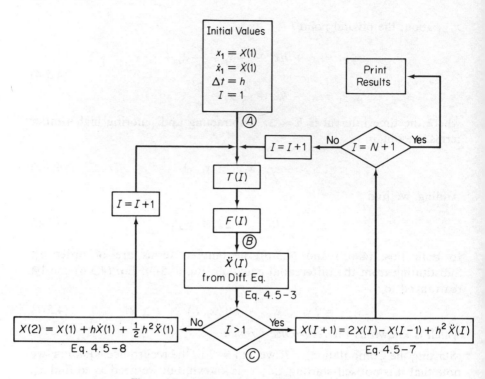

Figure 4.5-1. Flow diagram (undamped system).

we complete the left loop Ⓑ and Ⓒ where I is now equal to 2, so we proceed to the right to calculate x_3. Assuming N intervals of Δt, the path is to the NO direction and the right loop is repeated N times until $I = N + 1$, at which time the results are printed out.

EXAMPLE 4.5-1

Solve numerically the differential equation

$$4\ddot{x} + 2000x = F(t)$$

with initial conditions

$$x_1 = \dot{x}_1 = 0$$

and forcing function as shown in Fig. 4.5-2.

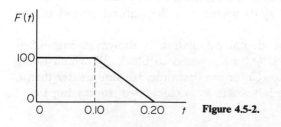

Figure 4.5-2.

Solution: The natural period of the system is first found as

$$\omega = \frac{2\pi}{\tau} = \sqrt{\frac{2000}{4}} = 22.36 \text{ rad/sec}$$

$$\tau = \frac{2\pi}{22.36} = 0.281 \text{ sec}$$

According to the rule $h \le \tau/10$ and for convenience for representing $F(t)$, we will choose $h = 0.020$ sec.
From the differential equation we have

$$\ddot{x} = f(x, t) = \frac{1}{4}F(t) - 500x$$

Eq. (4.5-8) gives $x_2 = \frac{1}{2}(25)(0.02)^2 = 0.005$. x_3 is then found from Eq. (4.5-7).

$$x_3 = 0.005 - 0 + (0.02)^2(25 - 500 \times 0.005) = 0.0190$$

The following values of x_4, x_5, etc. are now available from Eq. (4.5-7).
Figure 4.5-3 shows the computed values compared with the exact solution. The latter was obtained by the superposition of the solutions for the step function and the ramp function in the following

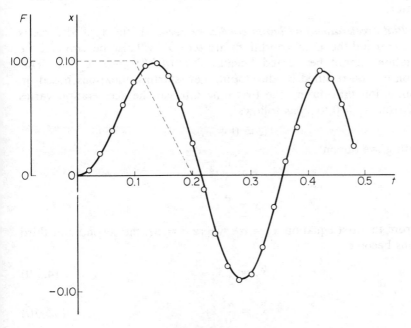

Figure 4.5-3.

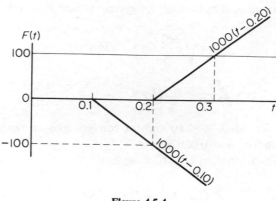

Figure 4.5-4.

manner. Figure 4.5-4 shows the superposition of forces. The equations to be superimposed for the exact solution are

$$x_1 = 0.05(1 - \cos 22.36t) \qquad 0 \le t \le 0.1$$

$$x_2 = -\left[\frac{1}{2}(t - 0.1) - 0.02236 \sin 22.36(t - 0.10)\right] \qquad \text{add at } t = 0.1$$

$$x_3 = +\left[\frac{1}{2}(t - 0.2) - 0.02236 \sin 22.36(t - 0.2)\right] \qquad \text{add at } t = 0.2$$

Both computations were carried out on a programmable hand calculator.

Initial acceleration and initial conditions zero. If the applied force is zero at $t = 0$ and the initial conditions are zero, $\ddot{x}_1$ will also be zero and the computation cannot be started because Eq. (4.5-8) gives $x_2 = 0$. This condition can be rectified by developing new starting equations based on the assumption that during the first-time interval the acceleration varies linearly from $\ddot{x}_1 = 0$ to $\ddot{x}_2$ as follows:

$$\ddot{x} = 0 + \alpha t$$

Integrating, we obtain

$$\dot{x} = \frac{\alpha}{2} t^2$$

$$x = \frac{\alpha}{6} t^3$$

Since from the first equation $\ddot{x}_2 = \alpha h$ where $h = \Delta t$, the second and third equations become

$$\dot{x}_2 = \frac{h}{2} \ddot{x}_2 \qquad (4.5\text{-}9)$$

$$x_2 = \frac{h^2}{6} \ddot{x}_2 \qquad (4.5\text{-}10)$$

Substituting these equations into the differential equation at time $t_2 = h$ enables one to solve for $\ddot{x}_2$ and x_2. Example 4.5-2 illustrates the situation encountered here.

EXAMPLE 4.5-2

Solve by the digital computer the problem of a spring-mass system excited by a triangular pulse. The differential equation of motion and the initial conditions are given as

$$0.5\ddot{x} + 8\pi^2 x = F(t)$$

$$x_1 = \dot{x}_1 = 0$$

The triangular force is defined in Fig. 4.5-5.

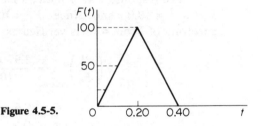

Figure 4.5-5.

Solution: The natural period of the system is

$$\tau = \frac{2\pi}{\omega} = \frac{2\pi}{4\pi} = 0.50$$

The time increment will be chosen as $h = 0.05$, and the differential equation is reorganized as

$$\ddot{x} = f(x, t) = 2F(t) - 16\pi^2 x$$

This equation is to be solved together with the recurrence equation, Eq. (4.5-7),

$$x_{i+1} = 2x_i - x_{i-1} + h^2 f(x, t)$$

Since the force and the acceleration are zero at $t = 0$, it is necessary to start the computational process with Eqs. (4.5-9) and (4.5-10) and the differential equation.

$$x_2 = \tfrac{1}{6}\ddot{x}_2(.05)^2 = .000417\ddot{x}_2$$

$$\ddot{x}_2 = 2F(.05) - 16\pi^2 x_2 = 50 - 158 x_2$$

Their simultaneous solution leads to

$$x_2 = \frac{(.05)^2 F(.05)}{3 + 8\pi^2(.05)^2} = .0195$$

$$\ddot{x}_2 = 46.91$$

The flow diagram for the computation is shown in Fig. 4.5-6. With $h = 0.05$, the time duration for the force must be divided into regions $I = 1$ to 5, $I = 6$ to 9 and $I > 9$. The index I controls the computation path on the diagram.

The Fortran program can be written in many ways, one of which is shown in Fig. 4.5-7, and the results, Fig. 4.5-8, can also be plotted by the computer, as presented in Fig. 4.5-9. A smaller Δt would have resulted in a smoother plot.

The response x vs. t indicates a maximum at $x \simeq 1.97$ in. Since $k = 8\pi^2 = 79.0$, and since $F_0 = 100$, a point on the response spectrum of Prob. 4-21 is verified as

$$\left(\frac{xk}{F_0}\right)_{\text{max}} = \frac{1.97 \times 79}{100} = 1.54$$

$$\frac{t_1}{\tau} = \frac{0.4}{0.5} = 0.80$$

Damped System.   When damping is present, the differential equation contains an additional term $\dot{x}_i$ and Eq. (4.5-7) is replaced by

$$x_{i+1} = 2x_i - x_{i-1} + h^2 f(x_i, \dot{x}_i, t_i) \qquad i \geq 2 \qquad (4.5\text{-}7')$$

We now need to calculate the velocity at each step as well as the displacement.

Considering again the first three terms of the Taylor series, Eq. (4.5-4), x_2 is available from the expansion of x_{i+1} with $i = 1$

$$x_2 = x_1 + \dot{x}_1 h + \frac{h^2}{2} f(x_1, \dot{x}_1, t_1)$$

The quantity $\dot{x}_2$ is found from the second equation for x_{i-1} with $i = 2$

$$x_1 = x_2 - \dot{x}_2 h + \frac{h^2}{2} f(x_2, \dot{x}_2, t_2)$$

With these results, x_3 can be calculated from Eq. (4.5-7'). The procedure is thus repeated for other values of x_i and $\dot{x}_i$ using the Taylor series.

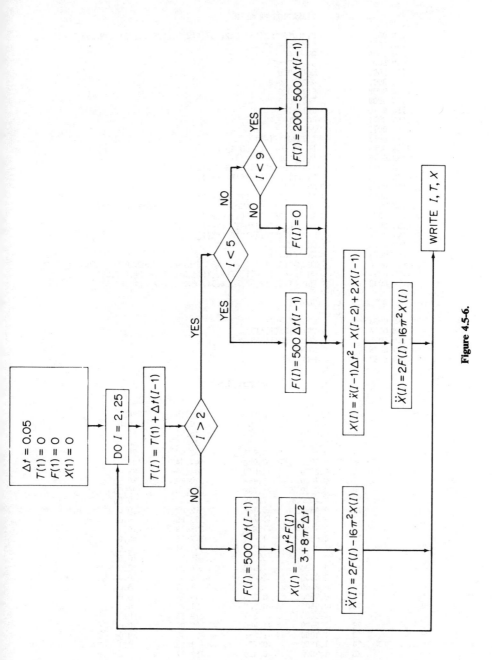

Figure 4.5-6.

117

```
C
C
C      VIBRATION PROBLEM
C      .
ISN 0002      DIMENSION X(25),DX2(25),F(25),T(25),J(25),VAR(25)
ISN 0003      PI2=3.1415**2
ISN 0004      DT=0.05
ISN 0005      DT2=DT**2
ISN 0006      X(1)=0.0
ISN 0007      DX2(1)=0.0
ISN 0008      F(1)=0.0
ISN 0009      T(1)=0.0
ISN 0010      J(1)=1
ISN 0011      DO 1 I=2,25
ISN 0012      J(I)=I
ISN 0013      T(I)=DT*(I-1)
ISN 0014      IF (I .GT. 2) GO TO 2
ISN 0016      F(I)=500*DT*(I-1)
ISN 0017      X(I)=(DT2*F(1) )/(3+8*PI2*DT2)
ISN 0018      DX2(I)=2*F(I)-16*PI2*X(I)
ISN 0019      GO TO 1
ISN 0020    2 IF(I .LE. 5) F(I)=500*DT*(I-1)
ISN 0022      IF (I .GT. 5 .AND. I .LT. 9) F(I)=200-500*DT*(I-1)
ISN 0024      IF (I .GE. 9) F(I)=0.0
ISN 0026      X(I)=DX2(I-1)*DT2-X(I-2)+2*X(I-1)
ISN 0027      DX2(I)=2*F(I)-16*PI2*X(I)
ISN 0028    1 CONTINUE
ISN 0029      WRITE(6,3)
ISN 0030    3 FORMAT(41H1   J   TIME   DISPL   ACCLRTN   FORCE)
ISN 0031      WRITE(6,4) (J(I),T(I),X(I),DX2(I),F(I),I=1,25)
ISN 0032    4 FORMAT(3X,I2,2X,F6.4,3X,F6.3,3X,F7.2,4X,F7.2)
C
C      PLOTTING
C
ISN 0033      DO 5 I=1,25
ISN 0034    5 VAR(I)=X(I)*10
ISN 0035      CALL PLOT1(VAR,25)
ISN 0036      STOP
ISN 0037      END
```

Figure 4.5-7.

J	TIME	DISPL	ACCLRTN	FORCE
1	0.0	0.0	0.0	0.0
2	0.0500	0.020	46.91	25.00
3	0.1000	0.156	75.31	50.00
4	0.1500	0.481	73.97	75.00
5	0.2000	0.992	43.44	100.00
6	0.2500	1.610	-104.25	75.00
7	0.3000	1.968	-210.78	50.00
8	0.3500	1.799	-234.10	25.00
9	0.4000	1.045	-165.01	0.00
10	0.4500	-0.122	19.22	0.0
11	0.5000	-1.240	195.86	0.0
12	0.5500	-1.869	295.19	0.0
13	0.6000	-1.760	277.98	0.0
14	0.6500	-0.957	151.04	0.0
15	0.7000	0.225	-35.52	0.0
16	0.7500	1.318	-208.06	0.0
17	0.8000	1.890	-298.47	0.0
18	0.8500	1.717	-271.05	0.0
19	0.9000	0.865	-136.64	0.0
20	0.9500	-0.328	51.72	0.0
21	1.0000	-1.391	219.66	0.0
22	1.0500	-1.906	300.89	0.0
23	1.1000	-1.668	263.33	0.0
24	1.1500	-0.772	121.83	0.0
25	1.2000	0.429	-67.77	0.0

Figure 4.5-8.

118

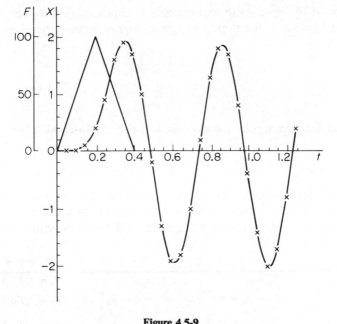

Figure 4.5-9.

4.6 RUNGE-KUTTA METHOD—METHOD 2

The Runge-Kutta computation procedure is popular since it is self-starting and results in good accuracy. A brief discussion of its basis is presented here.

In the Runge-Kutta method the second-order differential equation is first reduced to two first-order equations. As an example, consider the differential equation for the single degree of freedom system, which may be written as

$$\ddot{x} = \frac{1}{m}\left[F(t) - kx - c\dot{x}\right] = f(x, \dot{x}, t) \qquad (4.6\text{-}1)$$

By letting $\dot{x} = y$, the above equation is reduced to the following two first-order equations:

$$\dot{x} = y$$
$$\dot{y} = f(x, y, t) \qquad (4.6\text{-}2)$$

Both x and y in the neighborhood of x_i and y_i can be expressed in terms of the Taylor series. Letting the time increment be $h = \Delta t$

$$x = x_i + \left(\frac{dx}{dt}\right)_i h + \left(\frac{d^2x}{dt^2}\right)_i \frac{h^2}{2} + \cdots$$

$$y = y_i + \left(\frac{dy}{dt}\right)_i h + \left(\frac{d^2y}{dt^2}\right)_i \frac{h^2}{2} + \cdots \qquad (4.6\text{-}3)$$

Instead of using these expressions, it is possible to replace the first derivative by an average slope and ignore higher order derivatives

$$x = x_i + \left(\frac{dx}{dt}\right)_{i\,av} h$$

$$y = y_i + \left(\frac{dy}{dt}\right)_{i\,av} h \tag{4.6-4}$$

If we used Simpson's rule, the average slope in the interval h becomes, i.e.

$$\left(\frac{dy}{dt}\right)_{i\,av} = \frac{1}{6}\left[\left(\frac{dy}{dt}\right)_{t_i} + 4\left(\frac{dy}{dt}\right)_{t_i+h/2} + \left(\frac{dy}{dt}\right)_{t_i+h}\right]$$

The Runge-Kutta method is very similar to the above computations, except that the center term of the above equation is split into two terms and four values of t, x, y, and f are computed for each point i as follows

t	x	$y = \dot{x}$	$f = \dot{y} = \ddot{x}$
$T_1 = t_i$	$X_1 = x_i$	$Y_1 = y_i$	$F_1 = f(T_1, X_1, Y_1)$
$T_2 = t_i + \dfrac{h}{2}$	$X_2 = x_i + Y_1\dfrac{h}{2}$	$Y_2 = y_i + F_1\dfrac{h}{2}$	$F_2 = f(T_2, X_2, Y_2)$
$T_3 = t_i + \dfrac{h}{2}$	$X_3 = x_i + Y_2\dfrac{h}{2}$	$Y_3 = y_i + F_2\dfrac{h}{2}$	$F_3 = f(T_3, X_3, Y_3)$
$T_4 = t_i + h$	$X_4 = x_i + Y_3 h$	$Y_4 = y_i + F_3 h$	$F_4 = f(T_4, X_4, Y_4)$

These quantities are then used in the following recurrence formula

$$x_{i+1} = x_i + \frac{h}{6}\left[Y_1 + 2Y_2 + 2Y_3 + Y_4\right] \tag{4.6-5}$$

$$y_{i+1} = y_i + \frac{h}{6}\left[F_1 + 2F_2 + 2F_3 + F_4\right] \tag{4.6-6}$$

where it is recognized that the four values of Y divided by 6 represent an average slope dx/dt and the four values of F divided by 6 results in an average of dy/dt as defined by Eqs. (4.6-4).

EXAMPLE 4.6-1

Solve Example 4.5-1 by the Runge-Kutta method.

Solution: The differential equation of motion is

$$\ddot{x} = \frac{1}{4}F(t) - 500x$$

Let $y = \dot{x}$; then

$$\dot{y} = f(x, t) = \frac{1}{4}F(t) - 500x$$

With $h = 0.02$, the following table is calculated:

	t	x	$y = \dot{x}$	f
$t_1 =$	0	0	0	25
	0.01	0	0.25	25
	0.01	0.0025	0.25	23.75
$t_2 =$	0.02	0.0050	0.475	22.50

The calculation for x_2 and y_2 follows.

$$x_2 = 0 + \frac{0.02}{6}(0 + 0.50 + 0.50 + 0.475) = 0.00491667$$

$$y_2 = 0 + \frac{0.02}{6}(25 + 50 + 47.50 + 22.50) = 0.4833333$$

To continue to point 3 we repeat the above table

$t_2 =$	0.02	0.00491667	0.4833333	22.541665
	0.03	0.0097500	0.70874997	20.12500
	0.03	0.01200417	0.6845833	18.997915
$t_3 =$	0.04	0.01860834	0.8632913	15.695830

and calculate x_3 and y_3

$$x_3 = 0.00491667$$

$$+ \frac{0.02}{6}(0.483333 + 1.4174999 + 1.3691666 + 0.8632913)$$

$$= 0.00491667 + 0.01377764 = 0.01869431$$

$$y_3 = 0.483333 + 0.38827775 = 0.87161075$$

To complete the calculation, the problem was programmed on a digital computer, and the results showed excellent accuracy. Table 4.6-1 gives the numerical values for the central difference and the Runge-Kutta methods compared with the analytical solution. It is seen that the Runge-Kutta method gives greater accuracy than the central difference method.

Although the Runge-Kutta method does not require the evaluation of derivatives beyond the first, its higher accuracy is achieved by four evaluations of the first derivatives to obtain agreement with the Taylor series solution through terms of order h^4. Moreover, the versatility of the Runge-Kutta method is evident in the fact that by replacing the variable by a vector, the same method is applicable to a system of differential

TABLE 4.6-1

COMPARISON OF METHODS FOR PROBLEM 4.5-1

Time t	Exact Solution	Central Difference	Runge-Kutta
0	0	0	0
0.02	0.00492	0.00500	0.00492
0.04	0.01870	0.01900	0.01869
0.06	0.03864	0.03920	0.03862
0.08	0.06082	0.06159	0.06076
0.10	0.08086	0.08167	0.08083
0.12	0.09451	0.09541	0.09447
0.14	0.09743	0.09807	0.09741
0.16	0.08710	0.08712	0.08709
0.18	0.06356	0.06274	0.06359
0.20	0.02949	0.02782	0.02956
0.22	−0.01005	−0.01267	−0.00955
0.24	−0.04761	−0.05063	−0.04750
0.26	−0.07581	−0.07846	−0.07571
0.28	−0.08910	−0.09059	−0.08903
0.30	−0.08486	−0.08461	−0.08485
0.32	−0.06393	−0.06171	−0.06400
0.34	−0.03043	−0.02646	−0.03056
0.36	0.00906	0.01407	0.00887
0.38	0.04677	0.05180	0.04656
0.40	0.07528	0.07916	0.07509
0.42	0.08898	0.09069	0.08886
0.44	0.08518	0.08409	0.08516
0.46	0.06463	0.06066	0.06473
0.48	0.03136	0.02511	0.03157

equations. For example, the first-order equation of one variable is

$$\dot{x} = f(x, t)$$

For two variables, x and y, as in this problem, we can let $Z = \{^x_y\}$ and write the two first-order equations as

$$\left\{ \begin{matrix} \dot{x} \\ \dot{y} \end{matrix} \right\} = \left\{ \begin{matrix} y \\ f(x, y, t) \end{matrix} \right\} = F(x, y, t)$$

or

$$\dot{z} = F(x, y, t)$$

Thus the above vector equation is identical in form to the equation in one variable and can be treated in the same manner.

REFERENCES

[1] CRANDALL, S. H. "Engineering Analysis," *A Survey of Numerical Procedures*. New York: McGraw-Hill Book Company, 1956.

[2] RALSTON, A., and WILF, H. S. *Mathematical Methods for Digital Computers*, Vols. I and II. New York: John Wiley & Sons, Inc., 1968.

[3] SALVADORI, M. G., and BARON, M. L. *Numerical Methods in Engineering*. Englewood Cliffs, N. J.: Prentice-Hall, Inc., 1952.

[4] BATHE, K-J., and WILSON, E. L. *Numerical Methods in Finite Element Analysis*. Englewood Cliffs, N. J.: Prentice-Hall, Inc., 1976.

PROBLEMS

4-1 Show that the time t_p corresponding to the peak response for the impulsively excited spring-mass system is given by the equation

$$\tan \sqrt{1 - \zeta^2}\, \omega_n t_p = \sqrt{1 - \zeta^2}\,/\zeta$$

4-2 Determine the peak displacement for the impulsively excited spring-mass system, and show that it can be expressed in the form

$$\frac{x_{\text{peak}} \sqrt{km}}{\hat{F}} = \exp\left(-\frac{\zeta}{\sqrt{1 - \zeta^2}} \tan^{-1} \frac{\sqrt{1 - \zeta^2}}{\zeta}\right)$$

Plot this result as a function of ζ.

4-3 Show that the time t_p corresponding to the peak response of the damped spring-mass system excited by a step force F_0 is $\omega_n t_p = \pi/\sqrt{1 - \zeta^2}$.

4-4 For the system of Prob. 4-3, show that the peak response is equal to

$$\left(\frac{xk}{F_0}\right)_{\text{max}} = 1 + \exp\left(-\frac{\zeta\pi}{\sqrt{1 - \zeta^2}}\right)$$

4-5 A rectangular pulse of height F_0 and duration t_0 is applied to an undamped spring-mass system. Considering the pulse to be the sum of two step pulses, as shown in Fig. P4-5, determine its response for $t > t_0$ by the superposition of the undamped solutions.

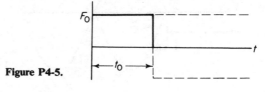

Figure P4-5.

123

4-6   If an arbitrary force $f(t)$ is applied to an undamped oscillator which has initial conditions other than zero, show that the solution must be in the form

$$x(t) = x_0 \cos \omega_n t + \frac{v_0}{\omega_n} \sin \omega_n t + \frac{1}{m\omega_n} \int_0^t f(\xi) \sin \omega_n(t - \xi)\, d\xi$$

4-7   Show that the response to a unit step function, designated by $g(t)$, is related to the impulsive response $h(t)$ by the equation $h(t) = \dot{g}(t)$.

4-8   Show that the convolution integral can also be written in terms of $g(t)$ as

$$x(t) = f(0)g(t) + \int_0^t \dot{f}(\xi)g(t - \xi)\, d\xi$$

where $g(t)$ is the response to a unit step function.

4-9   In Sec. 4.3 the subsidiary equation for the viscously damped spring-mass system was given by Eq. (a). Evaluate the second term due to initial conditions by the inverse transforms.

4-10   An undamped spring-mass system is given a base excitation of $\ddot{y}(t) = 20(1 - 5t)$. If the natural frequency of the system is $\omega_n = 10$ sec^{-1}, determine the maximum relative displacement.

4-11   A sinusoidal pulse is considered to be the superposition of two sine waves as shown in Fig. P4-11. Show that its solution is

$$\left(\frac{xk}{F_0}\right) = \frac{1}{(\tau/2t_1 - 2t_1/\tau)}\left(\sin \frac{2\pi t}{\tau} - \frac{2t_1}{\tau}\sin \frac{\pi t}{t_1}\right) \qquad t < t_1$$

$$\left(\frac{xk}{F_0}\right) = \frac{1}{(\tau/2t_1 - 2t_1/\tau)}\left[\left(\sin \frac{2\pi t}{\tau} - \frac{2t_1}{\tau}\sin \frac{\pi t}{t_1}\right)\right.$$

$$\left. + \left(\sin 2\pi \frac{t - t_1}{\tau} - \frac{2t_1}{\tau}\sin \pi \frac{t - t_1}{t_1}\right)\right], \qquad t > t_1$$

where $\tau = 2\pi/\omega$.

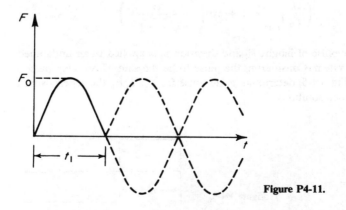

Figure P4-11.

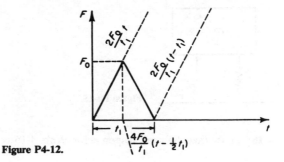

Figure P4-12.

4-12 For the triangular pulse shown in Fig. P4-12 show that the response is

$$x = \frac{2F_0}{k}\left(\frac{t}{t_1} - \frac{\tau}{2\pi t_1}\sin 2\pi\frac{t}{\tau}\right), \qquad 0 < t < \tfrac{1}{2}t_1$$

$$x = \frac{2F_0}{k}\left\{1 - \frac{t}{t_1} + \frac{\tau}{2\pi t_1}\left[2\sin\frac{2\pi}{\tau}\left(t - \frac{1}{2}t_1\right) - \sin 2\pi\frac{t}{\tau}\right]\right\},$$

$$\tfrac{1}{2}t_1 < t < t_1$$

$$x = \frac{2F_0}{k}\left\{\frac{\tau}{2\pi t_1}\left[2\sin\frac{2\pi}{\tau}\left(t - \frac{1}{2}t_1\right) - \sin\frac{2\pi}{\tau}(t - t_1) - \sin 2\pi\frac{t}{\tau}\right]\right\},$$

$$t > t_1$$

4-13 A spring-mass system slides down a smooth 30° inclined plane as shown in Fig. P4-13. Determine the time elapsed from first contact of the spring until it breaks contact again.

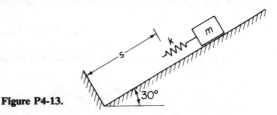

Figure P4-13.

4-14 A 38.6 lb weight is supported on several springs whose combined stiffness is 6.40 lb/in. If the system is lifted so that the bottom of the springs are just free and released, determine the maximum displacement of m, and the time for maximum compression.

4-15 A spring-mass system of Fig. P4-15 has a Coulomb damper which exerts a constant friction force f. For a base excitation, show that the solution is

$$\frac{\omega_n z}{v_0} = \frac{1}{\omega_n t_1}\left(1 - \frac{ft_1}{mv_0}\right)(1 - \cos \omega_n t) - \sin \omega_n t$$

where the base velocity of Prob. 4-24 is assumed.

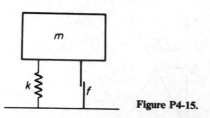

Figure P4-15.

4-16 Show that the peak response for Prob. 4-15 is

$$
\frac{\omega_n z_{\max}}{v_0} = \frac{1}{\omega_n t_1}\left(1 - \frac{ft_1}{mv_0}\right)\left\{1 - \frac{\dfrac{1}{\omega_n t_1}\left(1 - \dfrac{ft_1}{mv_0}\right)}{\sqrt{1 + \left[\dfrac{1}{\omega_n t_1}\left(1 - \dfrac{ft_1}{mv_0}\right)\right]^2}}\right\}
$$

$$
- \frac{1}{\sqrt{1 + \left[\dfrac{1}{\omega_n t_1}\left(1 - \dfrac{ft_1}{mv_0}\right)\right]^2}}
$$

By dividing by $\omega_n t_1$, the quantity $z_{\max}/v_0 t_1$ can be plotted as a function of $\omega_n t_1$ with ft_1/mv_0 as parameter.

4-17 In Prob. 4-16 the maximum force transmitted to m is

$$
F_{\max} = f + |kz_{\max}|
$$

To plot this quantity in nondimensional form, multiply by t_1/mv_0 to obtain

$$
\frac{F_{\max}t_1}{mv_0} = \frac{ft_1}{mv_0} + (\omega_n t_1)^2\left(\frac{z_{\max}}{v_0 t_1}\right)
$$

which again can be plotted as a function of ωt_1 with parameter ft_1/mv_0. Plot $|\omega_n z_{\max}/v_0|$ and $|z_{\max}/v_0 t_1|$ as function of $\omega_n t_1$ for ft_1/mv_0 equal to 0, 0.20, and 1.0.

4-18 Show that the response spectrum for the rectangular pulse of time duration t_0 shown in Fig. P4-18 is given by

$$
\left(\frac{xk}{F_0}\right)_{\max} = 2\sin\frac{\pi t_0}{\tau}, \qquad \frac{t_0}{\tau} < 0.50
$$

$$
= 2 \qquad \frac{t_0}{\tau} > 0.50
$$

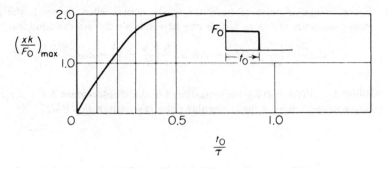

Figure P4-18.

where

$$\tau = \frac{2\pi}{\omega_n}$$

4-19 Shown in Fig. P4-19 is the response spectrum for the sine pulse. Show that for small values of t_1/τ the peak response occurs in the region $t > t_1$. Determine t_p/t_1 when $t_1/\tau = \frac{1}{2}$.

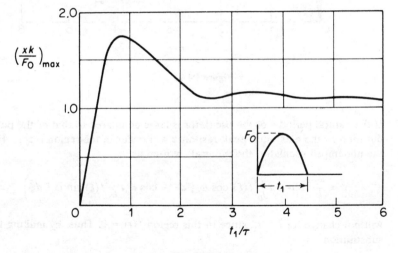

Figure P4-19.

4-20 An undamped spring-mass system with $w = 16.1$ lb has a natural period of 0.5 seconds. It is subjected to an impulse of 2.0 lb-sec which has a triangular shape with time duration of 0.40 sec. Determine the maximum displacement of the mass.

4-21   For a triangular pulse of duration t_1, show that when $t_1/\tau = \frac{1}{2}$, the peak response occurs at $t = t_1$, which can be established from the equation

$$2 \cos \frac{2\pi t_1}{\tau}\left(\frac{t_p}{t_1} - 0.5\right) - \cos 2\pi \frac{t_1}{\tau}\left(\frac{t_p}{t_1} - 1\right) - \cos \frac{2\pi t_1}{\tau}\frac{t_p}{t_1} = 0$$

found by differentiating the equation for the displacement for $t > t_1$. The response spectrum for the triangular pulse is shown in Fig. P4-21.

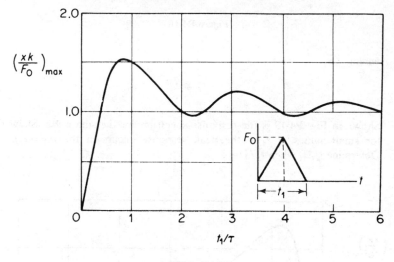

Figure P4-21.

4-22   If the natural period τ of the oscillator is large compared to that of the pulse duration t_1, the maximum peak response will occur in the region $t > t_1$. For the undamped oscillator, the integrals written as

$$x = \frac{\omega_n}{k}\left\{\sin \omega_n t \int_0^t f(\xi) \cos \omega_n \xi \, d\xi - \cos \omega_n t \int_0^t f(\xi) \sin \omega_n \xi \, d\xi\right\}$$

will not change for $t > t_1$, since in this region $f(t) = 0$. Thus, by making the substitution

$$A \cos \phi = \omega_n \int_0^{t_1} f(\xi) \cos \omega_n \xi \, d\xi$$

$$A \sin \phi = \omega_n \int_0^{t_1} f(\xi) \sin \omega_n \xi \, d\xi$$

the response for $t > t_1$, is a simple harmonic motion with amplitude A. Discuss the nature of the response spectrum for this case.

4-23 An undamped spring-mass system, m, k, is given a force excitation $F(t)$ as shown in Fig. P4-23. Show that for $t < t_0$

$$\frac{kx(t)}{F_0} = \frac{1}{\omega_n t_0}(\omega_n t - \sin \omega_n t)$$

and for $t > t_0$

$$\frac{kx(t)}{F_0} = \frac{1}{\omega_n t_0}[\sin \omega_n(t - t_0) - \sin \omega_n t] + \cos \omega_n(t - t_0)$$

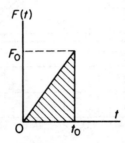

Figure P4-23.

4-24 The base of an undamped spring-mass system, m, k, is given a velocity pulse as shown in Fig. P4-24. Show that if the peak occurs at $t < t_1$, the response spectrum is given by the equation

$$\frac{\omega_n z_{max}}{v_0} = \frac{1}{\omega_n t_1} - \frac{1}{\omega_n t_1 \sqrt{1 + (\omega_n t_1)^2}} - \frac{\omega_n t_1}{\sqrt{1 + (\omega_n t_1)^2}}$$

Plot this result.

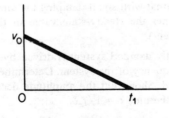

Figure P4-24.

4-25 In Prob. 4-24 if $t > t_1$, show that the solution is

$$\frac{\omega_n z}{v_0} = -\sin \omega_n t + \frac{1}{\omega_n t_1}[\cos \omega_n(t - t_1) - \cos \omega_n t]$$

4-26 Determine the time response for Prob. 4-10 using numerical integration.

4-27 Determine the time response for Prob. 4-20 using numerical integration.

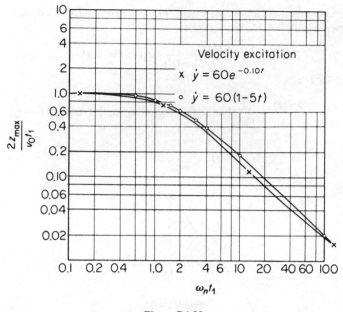

Figure P4-28.

4-28 Figure P4-28 shows the response spectra for the undamped spring-mass system under two different base velocity excitations. Solve the problem for the base velocity excitation of $\dot{y}(t) = 60e^{-0.10t}$ and verify a few of the points on the spectra.

4-29 A spring-mass system with viscous damping is initially at rest with zero displacement. If the system is activated by a harmonic force of frequency $\omega = \omega_n = \sqrt{k/m}$, determine the equation for its motion.

4-30 In Prob. 4-29 show that with small damping the amplitude will build up to a value $(1 - e^{-1})$ times the steady-state value in the time $t = 1/f_n\delta$. ($\delta =$ logarithmic decrement).

4-31 Assume that a lightly damped system is driven by a force $F_0 \sin \omega_n t$ where ω_n is the natural frequency of the system. Determine the equation if the force is suddenly removed. Show that the amplitude decays to a value e^{-1} times the initial value in the time $t = 1/f_n\delta$.

4-32 Set up a computer program for Example 4.5-1.

4-33 Draw a general flow diagram for the damped system with zero initial conditions excited by a force with zero initial value.

4-34 Draw a flow diagram for the damped system excited by base motion $y(t)$ with initial conditions $x(0) = X_1$ and $\dot{x}(0) = V_1$.

4-35 Write a Fortran program for Prob. 4-34, where the base motion is a half sine wave.

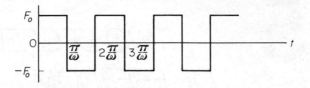

Figure P4-36.

4-36 Determine the response of an undamped spring-mass system to the alternating square wave of force shown in Fig. P4-36 by superimposing the solution
to the step function and matching the displacement and velocity at each
transition time. Plot the result and show that the peaks of the response will
increase as straight lines from the origin.

4-37 For the central difference method, supply the first higher-order term left out
in the recurrence formula for $\ddot{x}_i$ and verify that its error is $O(h^2)$.

4-38 Consider a curve $x = t^3$ and determine x_i at $t = 0.8, 0.9, 1.0, 1.1$, and 1.2.
Calculate $\dot{x}_{1.0}$ by using $\dot{x}_i = 1/2h(x_{i+1} - x_{i-1})$ with $h = 0.20$ and $h = 0.10$,
and show that the error is approximately $O(h^2)$.

4-39 Repeat Prob. 4-38 with $\dot{x}_i = 1/h(x_i - x_{i-1})$ and show that the error is
approximately $O(h)$.

4-40 Verify the correctness of the superimposed exact solution in Example 4.5-1.

4-41 Calculate the problem in Example 4.5-2 by using the Runge-Kutta method.

5 TWO DEGREES OF FREEDOM SYSTEM

When a system requires two coordinates to describe its motion, it is said to have two degrees of freedom. Such a system offers a simple introduction to the behavior of systems with several degrees of freedom.

A two degrees of freedom system will have two natural frequencies. When free vibration takes place at one of these natural frequencies, a definite relationship exists between the amplitudes of the two coordinates, and the configuration is referred to as the *normal mode*. The two degrees of freedom system will then have two normal mode vibrations corresponding to the two natural frequencies. Free vibration initiated under any condition will in general be the superposition of the two normal mode vibrations. However, forced harmonic vibration will take place at the frequency of the excitation, and the amplitude of the two coordinates will tend to a maximum at the two natural frequencies.

5.1 NORMAL MODE VIBRATION

Consider the undamped system of Fig. 5.1-1. Using coordinates x_1 and x_2 measured from inertial reference, the differential equations of motion for the system become

$$m\ddot{x}_1 = -k(x_1 - x_2) - kx_1$$
$$2m\ddot{x}_2 = k(x_1 - x_2) - kx_2 \qquad (5.1\text{-}1)$$

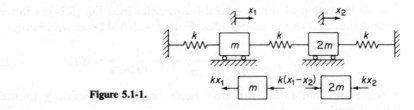

Figure 5.1-1.

We now define a normal mode oscillation as one in which each mass undergoes harmonic motion of the same frequency, passing simultaneously through the equilibrium position. For such motion we can let

$$x_1 = A_1 e^{i\omega t}$$
$$x_2 = A_2 e^{i\omega t}$$

$(5.1\text{-}2)$

Substituting these into the differential equations gives

$$(2k - \omega^2 m)A_1 - kA_2 = 0$$
$$- kA_1 + (2k - 2\omega^2 m)A_2 = 0$$

$(5.1\text{-}3)$

which are satisfied for any A_1 and A_2 if the following determinant is zero

$$\begin{vmatrix} (2k - \omega^2 m) & -k \\ -k & (2k - 2\omega^2 m) \end{vmatrix} = 0$$

$(5.1\text{-}4)$

Letting $\omega^2 = \lambda$, the above determinant leads to the *characteristic equation*

$$\lambda^2 - \left(3\frac{k}{m}\right)\lambda + \frac{3}{2}\left(\frac{k}{m}\right)^2 = 0$$

$(5.1\text{-}5)$

The two roots of this equation are

$$\lambda_1 = \left(\frac{3}{2} - \frac{1}{2}\sqrt{3}\right)\frac{k}{m} = 0.634\frac{k}{m}$$

and

$$\lambda_2 = \left(\frac{3}{2} + \frac{1}{2}\sqrt{3}\right)\frac{k}{m} = 2.366\frac{k}{m}$$

and the *natural frequencies* of the system are found to be

$$\omega_1 = \lambda_1^{1/2} = \sqrt{0.634\frac{k}{m}}$$

$(5.1\text{-}6)$

and

$$\omega_2 = \lambda_2^{1/2} = \sqrt{2.366\frac{k}{m}}$$

Substitution of these natural frequencies into Eq. (5.1-3) enables one to find the ratio of the amplitudes. For $\omega_1^2 = 0.634 k/m$, we obtain

$$\left(\frac{A_1}{A_2}\right)^{(1)} = \frac{k}{2k - \omega_1^2 m} = \frac{1}{2 - 0.634} = 0.731 \qquad (5.1\text{-}7)$$

which is the amplitude ratio or *mode shape* corresponding to the first normal mode.

Similarly, using $\omega_2^2 = 2.366 k/m$, we obtain

$$\left(\frac{A_1}{A_2}\right)^{(2)} = \frac{k}{2k - \omega_2^2 m} = \frac{1}{2 - 2.366} = -2.73 \qquad (5.1\text{-}8)$$

for the mode shape corresponding to the second normal mode. We can display the two normal modes graphically as in Fig. 5.1-2. In the first normal mode, the two masses move in phase; in the second normal mode the masses move in opposition, or out of phase, with each other. For the normal mode shape function, we will find the following notation useful for later chapters:

$$\phi_1(x) = \left\{ \begin{array}{c} 0.731 \\ 1.00 \end{array} \right\} \qquad \phi_2(x) = \left\{ \begin{array}{c} -2.73 \\ 1.00 \end{array} \right\}$$

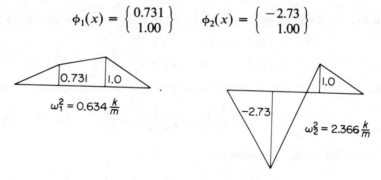

Figure 5.1-2. Normal modes of the system shown in Figure 5.1-1.

EXAMPLE 5.1-1

For the system of Fig. 5.1-1, let the coupling spring at the center equal nk and compute the natural frequencies and mode shapes.

Solution: Let $k/m = \omega_{11}^2$, in which case the characteristic equation becomes

$$\omega^4 - \tfrac{3}{2}\omega_{11}^2(1 + n)\omega^2 + \tfrac{1}{2}\omega_{11}^4(1 + 2n) = 0$$

The two normal mode frequencies in terms of ω_{11}^2 then become

$$\omega_{1,\,2}^2/\omega_{11}^2 = \frac{3}{4}(1 + n) \pm \sqrt{\frac{9}{16}(1 + n)^2 - \frac{1}{2}(1 + 2n)}$$

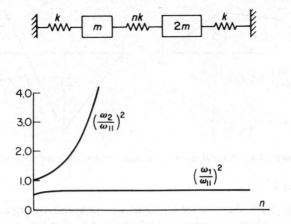

Figure 5.1-3. Natural frequencies as function of coupling n.

On varying the value of n, the following numerical values for $(\omega_1/\omega_{11})^2$ and $(\omega_2/\omega_{11})^2$ are found and plotted in Fig. 5.1-3. Note that $(\omega_1/\omega_{11})^2$ remains nearly constant.

NORMAL MODE FREQUENCIES AS FUNCTION OF n

n	$(\omega_1/\omega_{11})^2$	$(\omega_2/\omega_{11})^2$
0	0.50	1.0
0.5	0.611	1.641
1.0	0.634	2.366
2.0	0.650	3.850
4.0	0.660	6.840
10.0	0.666	15.83
100.0	0.666	150.8
∞	0.666	∞

The mode shapes can now be found for any n as

$$\left(\frac{A_1}{A_2}\right)^{(1)} = \frac{1 + n - 2(\omega_1/\omega_{11})^2}{n}$$

$$\left(\frac{A_1}{A_2}\right)^{(2)} = \frac{1 + n - 2(\omega_2/\omega_{11})^2}{n}$$

For example, if $n = 4$, the two natural modes are as shown in Fig. 5.1-4.

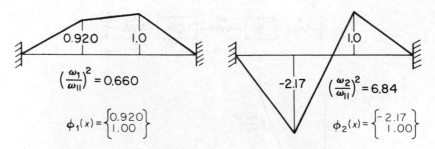

Figure 5.1-4.   Normal modes of system shown in Figure 5.1-3 for $n = 4$.

EXAMPLE 5.1-2

In Fig. 5.1-5 the two pendulums are coupled by means of a weak spring k, which is unstrained when the two pendulum rods are in the vertical position. Determine the normal mode vibrations.

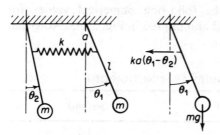

Figure 5.1-5.   Coupled pendulum.

Solution: Assuming the counterclockwise angular displacements to be positive, and taking moments about the points of suspension, we obtain the following equations of motion for small oscillations

$$ml^2\ddot{\theta}_1 = -mgl\theta_1 - ka^2(\theta_1 - \theta_2)$$

$$ml^2\ddot{\theta}_2 = -mgl\theta_2 + ka^2(\theta_1 - \theta_2)$$

Assuming the normal mode solutions as

$$\theta_1 = A_1 \cos \omega t$$

$$\theta_2 = A_2 \cos \omega t$$

the natural frequencies and mode shapes are found to be

$$\omega_1 = \sqrt{\frac{g}{l}} \qquad\qquad \omega_2 = \sqrt{\frac{g}{l} + 2\frac{k}{m}\frac{a^2}{l^2}}$$

$$\left(\frac{A_1}{A_2}\right)^{(1)} = 1.0 \qquad\qquad \left(\frac{A_1}{A_2}\right)^{(2)} = -1.0$$

Thus in the first mode the two pendulums move in phase and the spring remains unstretched. In the second mode the two pendulums move in opposition and the coupling spring is actively involved with a node at its midpoint. Consequently, the natural frequency is higher.

EXAMPLE 5.1-3

If the coupled pendulum of Example 5.1-2 is set into motion with initial conditions differing from those of the normal modes, the oscillations will contain both normal modes simultaneously. For example, if the initial conditions are $\theta_1(0) = A$ and $\theta_2(0) = 0$, the equations of motion will be

$$\theta_1(t) = \tfrac{1}{2}A \cos \omega_1 t + \tfrac{1}{2}A \cos \omega_2 t$$

$$\theta_2(t) = \tfrac{1}{2}A \cos \omega_1 t - \tfrac{1}{2}A \cos \omega_2 t$$

Consider the case where the coupling spring is very weak, and show that a beating phenomena takes place between the two pendulums.

Solution: The equations above can be rewritten as follows

$$\theta_1(t) = A \cos\left(\frac{\omega_1 - \omega_2}{2}\right)t \cos\left(\frac{\omega_1 + \omega_2}{2}\right)t$$

$$\theta_2(t) = -A \sin\left(\frac{\omega_1 - \omega_2}{2}\right)t \sin\left(\frac{\omega_1 + \omega_2}{2}\right)t$$

Since $(\omega_1 - \omega_2)$ is very small, $\theta_1(t)$ and $\theta_2(t)$ will behave like $\cos(\omega_1 + \omega_2)t/2$ and $\sin(\omega_1 + \omega_2)t/2$ with slowly varying amplitudes as shown in Fig. 5.1-6. Since the system is conservative, energy is transferred from one pendulum to the other.

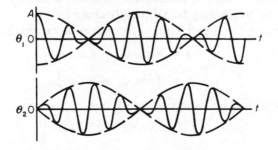

Figure 5.1-6.   Exchange of energy between pendulums.

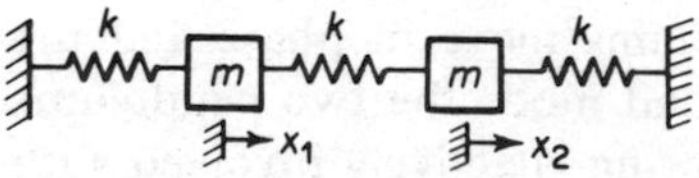

Figure 5.1-7.

EXAMPLE 5.1-4

If the masses and springs of the system shown in Fig. 5.1-7 are made equal to m and k as shown, the normal modes become

$$\omega_1^2 = \frac{k}{m} \qquad \omega_2^2 = \frac{3k}{m}$$

$$\frac{A_1}{A_2} = 1 \qquad \frac{A_1}{A_2} = -1$$

Determine the free vibration of the system when the initial conditions are

$$x_1(0) = 5 \qquad x_2(0) = 0$$

$$\dot{x}_1(0) = 0 \qquad \dot{x}_2(0) = 0$$

Solution: Any free vibration can be considered to be the superposition of its normal modes. Thus the two displacements can be written as

$$x_1 = A \sin(\omega_1 t + \psi_1) - B \sin(\omega_2 t + \psi_2)$$

$$x_2 = A \sin(\omega_1 t + \psi_1) + B \sin(\omega_2 t + \psi_2)$$

(a)

It should be noted here that the first terms on the right correspond to the first normal mode at the natural frequency ω_1. Its amplitude ratio is also $A_1/A_2 = A/A = 1$, which is the first normal mode shape. The second terms oscillate at frequency ω_2 with amplitude ratio $B_1/B_2 = -B/B = -1$, in conformity with the second normal mode vibration. The phase ψ_1 and ψ_2 simply allows the freedom of shifting the time origin and does not alter the character of the normal modes. The constants A, B, ψ_1 and ψ_2 are sufficient to satisfy the four initial conditions, which may be arbitrarily chosen.

Letting $t = 0$ and $x_1(0) = 5$, $x_2(0) = 0$, we obtain

$$5 = A \sin \psi_1 - B \sin \psi_2$$

$$0 = A \sin \psi_1 + B \sin \psi_2$$

Thus by adding and subtracting we find

$$A \sin \psi_1 = 2.5$$

$$B \sin \psi_2 = -2.5$$

Differentiating Eq. (a) for the velocity and letting $t = 0$ we obtain two other equations

$$0 = \omega_1 A \cos \psi_1 - \omega_2 B \cos \psi_2$$

$$0 = \omega_1 A \cos \psi_1 + \omega_2 B \cos \psi_2$$

from which we find

$$\cos \psi_1 = 0 \qquad \text{or} \qquad \psi_1 = 90°$$

$$\cos \psi_2 = 0 \qquad \text{or} \qquad \psi_2 = 90°$$

The solution is then easily seen to be

$$x_1 = 2.5 \cos\sqrt{\frac{k}{m}}\, t + 2.5 \cos\sqrt{\frac{3k}{m}}\, t$$

$$x_2 = 2.5 \cos\sqrt{\frac{k}{m}}\, t - 2.5 \cos\sqrt{\frac{3k}{m}}\, t$$

which may be written in the matrix form

$$\begin{Bmatrix} x_1 \\ x_2 \end{Bmatrix} = 2.5 \begin{Bmatrix} 1 \\ 1 \end{Bmatrix} \cos\sqrt{\frac{k}{m}}\, t - 2.5 \begin{Bmatrix} -1 \\ 1 \end{Bmatrix} \cos\sqrt{\frac{3k}{m}}\, t$$

5.2 COORDINATE COUPLING

The differential equation of motion for the two degrees of freedom system are in general *coupled*, in that both coordinates appear in each equation. In the most general case the two equations for the undamped system have the form

$$m_{11}\ddot{x}_1 + m_{12}\ddot{x}_2 + k_{11}x_1 + k_{12}x_2 = 0$$

$$m_{21}\ddot{x}_1 + m_{22}\ddot{x}_2 + k_{21}x_1 + k_{22}x_2 = 0$$

These equations can be expressed in matrix form (See Appendix C) as

$$\begin{bmatrix} m_{11} & m_{12} \\ m_{21} & m_{22} \end{bmatrix} \begin{Bmatrix} \ddot{x}_1 \\ \ddot{x}_2 \end{Bmatrix} + \begin{bmatrix} k_{11} & k_{12} \\ k_{21} & k_{22} \end{bmatrix} \begin{Bmatrix} x_1 \\ x_2 \end{Bmatrix} = \begin{Bmatrix} 0 \\ 0 \end{Bmatrix}$$

which immediately reveals the type of coupling present. Mass of *dynamical coupling* exists if the mass matrix is nondiagonal, whereas stiffness or *static coupling* exists if the stiffness matrix is nondiagonal.

It is also possible to establish the type of coupling from the expressions for the kinetic and potential energies. Cross products of coordinates in either expression denote coupling, dynamic or static, depending on whether they are found in T or U. The choice of coordinates establishes the type of coupling, and both dynamic and static coupling may be present.

It is possible to find a coordinate system which has neither form of coupling. The two equations are then decoupled and each equation may be solved independently of the other. Such coordinates are called *principal coordinates* (also called *normal coordinates*).

Although it is always possible to decouple the equations of motion for the undamped system, this is not always the case for a damped system. The following matrix equations show a system which has zero dynamic and static coupling, but the coordinates are coupled by the damping matrix.

$$\begin{bmatrix} m_{11} & 0 \\ 0 & m_{22} \end{bmatrix} \begin{Bmatrix} \ddot{x}_1 \\ \ddot{x}_2 \end{Bmatrix} + \begin{bmatrix} c_{11} & c_{12} \\ c_{21} & c_{22} \end{bmatrix} \begin{Bmatrix} \dot{x}_1 \\ \dot{x}_2 \end{Bmatrix} + \begin{bmatrix} k_{11} & 0 \\ 0 & k_{22} \end{bmatrix} \begin{Bmatrix} x_1 \\ x_2 \end{Bmatrix} = \begin{Bmatrix} 0 \end{Bmatrix}$$

If in the above equation $c_{12} = c_{21} = 0$, then the damping is said to be *proportional* (proportional to the stiffness or mass matrix), and the system equations become uncoupled.

EXAMPLE 5.2-1

Figure 5.2-1 shows a rigid bar with its center of mass not coinciding with its geometric center, i.e., $l_1 \neq l_2$, and supported by two springs, k_1, k_2. It represents a two degree of freedom system since two coordinates are necessary to describe its motion. The choice of the coordinates will define the type of coupling which can be immediately determined from the mass and stiffness matrices. Mass or *dynamical coupling* exists if the mass matrix is nondiagonal, whereas stiffness or *static coupling* exists if the stiffness matrix is nondiagonal. It is also possible to have both forms of coupling.

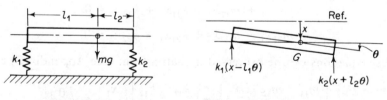

Figure 5.2-1.   **Figure 5.2-2.** Coordinates leading to static coupling.

Static Coupling. Choosing coordinates x and θ, shown in Fig. 5.2-2, where x is the linear displacement of the center of mass, the system will have static coupling as shown by the matrix equation

$$\begin{bmatrix} m & 0 \\ 0 & J \end{bmatrix} \begin{Bmatrix} \ddot{x} \\ \ddot{\theta} \end{Bmatrix} + \begin{bmatrix} (k_1 + k_2) & (k_2 l_2 - k_1 l_1) \\ (k_2 l_2 - k_1 l_1) & (k_1 l_1^2 + k_2 l_2^2) \end{bmatrix} \begin{Bmatrix} x \\ \theta \end{Bmatrix} = \begin{Bmatrix} 0 \end{Bmatrix}$$

If $k_1 l_1 = k_2 l_2$, the coupling disappears, and we obtain uncoupled x and θ vibrations.

Dynamic Coupling. There is some point C along the bar where a force applied normal to the bar produces pure translation; i.e., $k_1 l_3 = k_2 l_4$. (See Fig. 5.2-3.) The equations of motion in terms of x_c and θ can be shown to be

$$\begin{bmatrix} m & me \\ me & J_c \end{bmatrix} \begin{Bmatrix} \ddot{x}_c \\ \ddot{\theta} \end{Bmatrix} + \begin{bmatrix} (k_1 + k_2) & 0 \\ 0 & (k_1 l_3^2 + k_2 l_4^2) \end{bmatrix} \begin{Bmatrix} x_c \\ \theta \end{Bmatrix} = \begin{Bmatrix} 0 \\ 0 \end{Bmatrix}$$

which shows that the coordinates chosen eliminated the static coupling and introduced dynamic coupling.

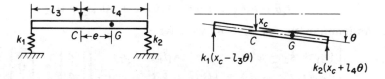

Figure 5.2-3. Coordinates leading to dynamic coupling.

Static and Dynamic Coupling. If we choose $x = x_1$ at the end of the bar, as shown in Fig. 5.2-4, the equations of motion become

$$\begin{bmatrix} m & ml_1 \\ ml_1 & J_1 \end{bmatrix} \begin{Bmatrix} \ddot{x}_1 \\ \ddot{\theta} \end{Bmatrix} + \begin{bmatrix} (k_1 + k_2) & k_2 l \\ k_2 l & k_2 l^2 \end{bmatrix} \begin{Bmatrix} x_1 \\ \theta \end{Bmatrix} = \begin{Bmatrix} 0 \\ 0 \end{Bmatrix}$$

and both static and dynamic coupling are now present.

Figure 5.2-4. Coordinates leading to static and dynamic coupling.

EXAMPLE 5.2-2

Determine the normal modes of vibration of an automobile simulated by the simplified two degrees of freedom system with the following numerical values (See Fig. 5.2-5):

$$W = 3220 \text{ lb} \qquad l_1 = 4.5 \text{ ft} \qquad k_1 = 2400 \text{ lb/ft}$$
$$J_c = \frac{W}{g} r^2 \qquad l_2 = 5.5 \text{ ft} \qquad k_2 = 2600 \text{ lb/ft}$$
$$r = 4 \text{ ft} \qquad l = 10 \text{ ft}$$

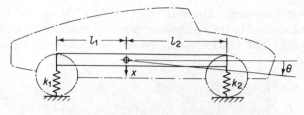

Figure 5.2-5.

The equations of motion indicate static coupling.

$$m\ddot{x} + k_1(x - l_1\theta) + k_2(x + l_2\theta) = 0$$

$$J_c\ddot{\theta} - k_1(x - l_1\theta)l_1 + k_2(x + l_2\theta)l_2 = 0$$

Assuming harmonic motion, we have

$$\begin{bmatrix} (k_1 + k_2 - \omega^2 m) & -(k_1 l_1 - k_2 l_2) \\ -(k_1 l_1 - k_2 l_2) & (k_1 l_1^2 + k_2 l_2^2 - \omega^2 J_c) \end{bmatrix} \begin{Bmatrix} x \\ \theta \end{Bmatrix} = \begin{Bmatrix} 0 \end{Bmatrix}$$

From the determinant of the matrix equation, the two natural frequencies are

$$\omega_1 = 6.90 \text{ rad/sec} = 1.10 \text{ cps}$$

$$\omega_2 = 9.06 \text{ rad/sec} = 1.44 \text{ cps}$$

The amplitude ratios for the two frequencies are

$$\left(\frac{x}{\theta}\right)_{\omega_1} = -14.6 \text{ ft/rad} = -3.06 \text{ in./deg}$$

$$\left(\frac{x}{\theta}\right)_{\omega_2} = 1.09 \text{ ft/rad} = 0.288 \text{ in./deg}$$

The mode shapes are illustrated by the diagrams of Fig. 5.2-6.

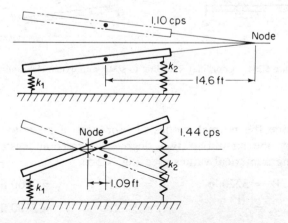

Figure 5.2-6.   Normal modes of system shown in Figure 5.2-5.

5.3 FORCED HARMONIC VIBRATION

We consider here a system excited by a harmonic force $F_1 \sin \omega t$. Assuming the motion to be represented by the matrix equation

$$\begin{bmatrix} m_1 & 0 \\ 0 & m_2 \end{bmatrix} \begin{Bmatrix} \ddot{x}_1 \\ \ddot{x}_2 \end{Bmatrix} + \begin{bmatrix} k_{11} & k_{12} \\ k_{21} & k_{22} \end{bmatrix} \begin{Bmatrix} x_1 \\ x_2 \end{Bmatrix} = \begin{Bmatrix} F_1 \\ 0 \end{Bmatrix} \sin \omega t \qquad (5.3\text{-}1)$$

we assume the solution to be

$$\begin{Bmatrix} x_1 \\ x_2 \end{Bmatrix} = \begin{Bmatrix} X_1 \\ X_2 \end{Bmatrix} \sin \omega t$$

Substituting this solution into the first equation, we obtain

$$\begin{bmatrix} (k_{11} - m_1\omega^2) & k_{12} \\ k_{21} & (k_{22} - m_2\omega^2) \end{bmatrix} \begin{Bmatrix} X_1 \\ X_2 \end{Bmatrix} = \begin{Bmatrix} F_1 \\ 0 \end{Bmatrix} \qquad (5.3\text{-}2)$$

or, in simpler notation

$$[Z(\omega)] \begin{Bmatrix} X_1 \\ X_2 \end{Bmatrix} = \begin{Bmatrix} F_1 \\ 0 \end{Bmatrix}$$

Premultiplying by $[Z(\omega)]^{-1}$ we obtain (See Appendix C)

$$\begin{Bmatrix} X_1 \\ X_2 \end{Bmatrix} = [Z(\omega)]^{-1} \begin{Bmatrix} F_1 \\ 0 \end{Bmatrix} = \frac{adj[Z(\omega)] \begin{Bmatrix} F_1 \\ 0 \end{Bmatrix}}{|Z(\omega)|} \qquad (5.3\text{-}3)$$

Referring to Eq. (5.3-2), the determinant $|Z(\omega)|$ can be expressed as

$$|Z(\omega)| = m_1 m_2 (\omega_1^2 - \omega^2)(\omega_2^2 - \omega^2) \qquad (5.3\text{-}4)$$

where ω_1 and ω_2 are the normal mode frequencies. Thus Eq. (5.3-3) becomes

$$\begin{Bmatrix} X_1 \\ X_2 \end{Bmatrix} = \frac{1}{|Z(\omega)|} \begin{bmatrix} (k_{22} - m_2\omega^2) & -k_{12} \\ -k_{21} & (k_{11} - m_1\omega^2) \end{bmatrix} \begin{Bmatrix} F \\ 0 \end{Bmatrix} \qquad (5.3\text{-}5)$$

or

$$X_1 = \frac{(k_{22} - m_2\omega^2)F}{m_1 m_2 (\omega_1^2 - \omega^2)(\omega_2^2 - \omega^2)}$$

$$X_2 = \frac{-k_{12}F}{m_1 m_2 (\omega_1^2 - \omega^2)(\omega_2^2 - \omega^2)} \qquad (5.3\text{-}6)$$

EXAMPLE 5.3-1

Apply Eqs. (5.3-6) of Sec. 5.3 to the system shown in Fig. 5.3-1 when m_1 is excited by the force $F_1 \sin \omega t$. Plot its frequency response curve.

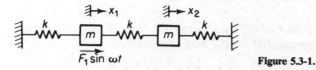

Figure 5.3-1.

Solution: The equation of motion in matrix form is

$$\begin{bmatrix} m & 0 \\ 0 & m \end{bmatrix}\begin{Bmatrix} \ddot{x}_1 \\ \ddot{x}_2 \end{Bmatrix} + \begin{bmatrix} 2k & -k \\ -k & 2k \end{bmatrix}\begin{Bmatrix} x_1 \\ x_2 \end{Bmatrix} = \begin{Bmatrix} F_1 \\ 0 \end{Bmatrix} \sin \omega t$$

Thus we have $k_{11} = k_{22} = 2k$; $k_{12} = k_{21} = -k$; $\omega_1^2 = k/m$; and $\omega_2^2 = 3k/m$. Eqs. (5.3-6) of Sec. 5.3 therefore become

$$X_1 = \frac{(2k - m\omega^2)F_1}{m^2(\omega_1^2 - \omega^2)(\omega_2^2 - \omega^2)}$$

$$X_2 = \frac{kF_1}{m^2(\omega_1^2 - \omega^2)(\omega_2^2 - \omega^2)}$$

It is convenient here to expand each of the above equations in partial fractions. For X_1 we obtain

$$\frac{(2k - m\omega^2)F_1}{m^2(\omega_1^2 - \omega^2)(\omega_2^2 - \omega^2)} = \frac{C_1}{(\omega_1^2 - \omega^2)} + \frac{C_2}{(\omega_2^2 - \omega^2)}$$

To solve for C_1, multiply by $(\omega_1^2 - \omega^2)$ and let $\omega = \omega_1$

$$C_1 = \frac{(2k - m\omega_1^2)F_1}{m^2(\omega_2^2 - \omega_1^2)} = \frac{F_1}{2m}$$

Similarly, C_2 is evaluated by multiplying by $(\omega_2^2 - \omega^2)$ and letting $\omega = \omega_2$

$$C_2 = \frac{(2k - m\omega_2^2)F_1}{m^2(\omega_1^2 - \omega_2^2)} = \frac{F_1}{2m}$$

An alternative form of X_1 is then

$$X_1 = \frac{F_1}{2m}\left[\frac{1}{\omega_1^2 - \omega^2} + \frac{1}{\omega_2^2 - \omega^2} \right]$$

$$= \frac{F_1}{2k}\left[\frac{1}{1 - (\omega/\omega_1)^2} + \frac{1}{3 - (\omega/\omega_1)^2} \right]$$

Treating X_2 in the same manner, its equation is

$$X_2 = \frac{F_1}{2k}\left[\frac{1}{1 - (\omega/\omega_1)^2} - \frac{1}{3 - (\omega/\omega_1)^2} \right]$$

The frequency response curve is shown in Fig. 5.3-2.

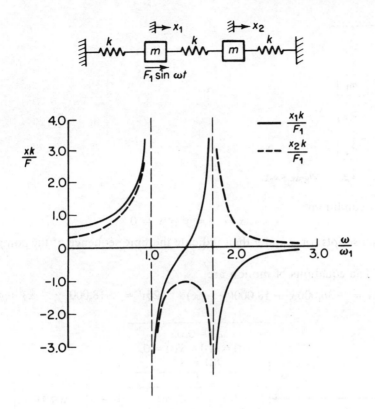

Figure 5.3-2. Forced response of the two degrees of freedom system.

5.4 DIGITAL COMPUTATION

The finite difference method of Sec. 4.5 can easily be extended to the solution of systems with two degrees of freedom. The procedure is illustrated by the following problem which is programmed and solved by the digital computer.

The system to be solved is shown in Fig. 5.4-1. To avoid confusion with subscripts, we let the displacements be x and y.

$$k_1 = 36 \text{ kN/m}$$
$$k_2 = 18 \text{ kN/m}$$
$$m_1 = 100 \text{ kg}$$
$$m_2 = 25 \text{ kg}$$
$$F \begin{cases} = 400 \text{ N} & t > 0 \\ = 0 & t < 0 \end{cases}$$

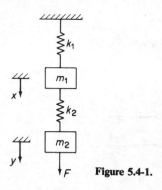

Figure 5.4-1.

Initial conditions:

$$x = \dot{x} = y = \dot{y} = 0$$

The subscripts for x and y then indicate the time sequence of the computation.

The equations of motion are

$$100\ddot{x} = -36{,}000x + 18{,}000(y - x) \qquad 25\ddot{y} = -18{,}000(y - x) + 400$$

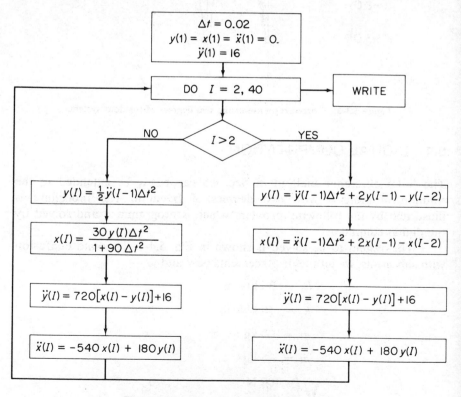

Figure 5.4-2.   Flow diagram for computation.

```
       $JOB
 1            DIMENSION X(40),Y(40),DX2(40),DY2(40),T(40),J(40),XC(40),YC(40)
 2            J(1)=1
 3            DT=0.02
 4            DT2=DT**2
 5            DX2(1)=0.0
 6            DY2(1)=16.0
 7            X(1)=0.0
 8            Y(1)=0.0
 9            XC(1)= 0.0
10            YC(1)= 0.0
11            T(1)=0.0
12            DO 100 I=2,40
13            J(I)=I
14            T(I)=DT*(I-1)
15            IF(I.GT.2) GO TO 200
16            Y(I)=DY2(I-1)*DT2/2
17            X(I)=30.0*Y(I)*DT2/(1+90.0*DT2)
18            DY2(I)=720.0*(X(I)-Y(I))+16.0
19            DX2(I)=-540.0*X(I)+180.0*Y(I)
20            YC(I)=Y(I)*100.0
21            XC(I)=X(I)*100.0
22            GO TO 100
23     200    Y(I)=DY2(I-1)*DT2+2*Y(I-1)-Y(I-2)
24            X(I)=DX2(I-1)*DT2+2*X(I-1)-X(I-2)
25            DY2(I)=720.0*(X(I)-Y(I))+16.0
26            DX2(I)=-540.0*X(I)+180.0*Y(I)
27            YC(I)=Y(I)*100.0
28            XC(I)=X(I)*100.0
29     100    CONTINUE
30            WRITE (6,300)
31     300    FORMAT('1','  J      TIME      DISPL-X,CM    DISPL-Y,CM',/)
32            WRITE(6,400) (J(I),T(I),XC(I),YC(I),I=1,40)
33     400    FORMAT(' ',I4,F10.4, 2F14.6)
34            STOP
35            END

       $ENTRY
```

Figure 5.4-3.   Computer program.

which can be rearranged to

$$\ddot{x} = -540x + 180y$$

$$\ddot{y} = 720(x - y) + 16$$

These equations are to be solved together with the recurrence equations of Sec. 4.5.

$$x_{i+1} = \ddot{x}_i \Delta t^2 + 2x_i - x_{i-1}$$

$$y_{i+1} = \ddot{y}_i \Delta t^2 + 2y_i - y_{i-1}$$

To establish a reasonable value for Δt, we note that

$$\sqrt{\frac{k_1}{m_1}} = 18.97 \qquad \tau_1 = 0.327$$

$$\sqrt{\frac{k_2}{m_2}} = 26.83 \qquad \tau_2 = 0.234$$

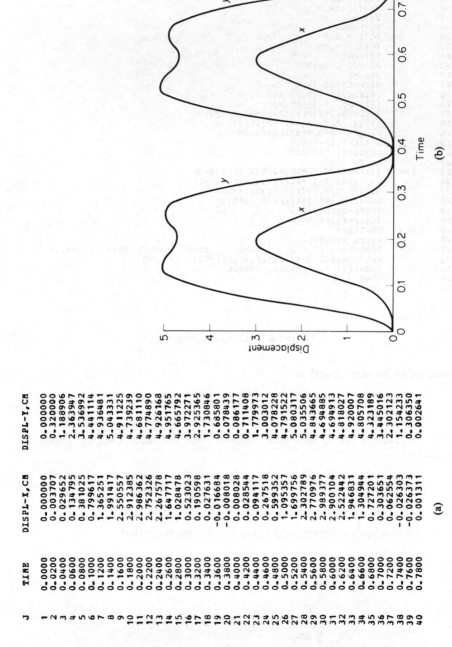

Figure 5.4-4.

J	TIME	DISPL-X,CM	DISPL-Y,CM
1	0.0000	0.000000	0.000000
2	0.0200	0.003707	0.320000
3	0.0400	0.029652	1.188906
4	0.0600	0.134795	2.363947
5	0.0800	0.381025	3.536992
6	0.1000	0.799617	4.441114
7	0.1200	1.365251	4.936481
8	0.1400	1.991417	5.043331
9	0.1600	2.550557	4.911225
10	0.1800	2.912385	4.739239
11	0.2000	2.986362	4.681110
12	0.2200	2.752326	4.774890
13	0.2400	2.267578	4.926168
14	0.2600	1.647717	4.951765
15	0.2800	1.028478	4.665792
16	0.3000	0.523023	3.972271
17	0.3200	0.190598	2.925365
18	0.3400	0.027631	1.730846
19	0.3600	-0.016684	0.685801
20	0.3800	-0.008018	0.078439
21	0.4000	-0.008028	0.086177
22	0.4200	0.028544	0.711408
23	0.4400	0.094117	1.779973
24	0.4600	0.267518	3.003012
25	0.4800	0.599352	4.078228
26	0.5000	1.095357	4.791522
27	0.5200	1.699756	5.080317
28	0.5400	2.302789	5.035506
29	0.5600	2.770976	4.843665
30	0.5800	2.989377	4.694885
31	0.6000	2.900104	4.694913
32	0.6200	2.522442	4.818027
33	0.6400	1.946831	4.920007
34	0.6600	1.304944	4.805708
35	0.6800	0.727201	4.323189
36	0.7000	0.303651	3.445016
37	0.7200	0.062554	2.302123
38	0.7400	-0.026303	1.154233
39	0.7600	-0.026373	0.306350
40	0.7800	0.001311	0.002641

(a)

(b)

We therefore arbitrarily choose a value $\Delta t = 0.020$ sec. It is also noted that the initial accelerations are $\ddot{x}_1 = 0$ and $\ddot{y}_1 = 16$, which requires us to use Eq. 4.5-8 for y and Eq. 4.5-10 for x. Using $\ddot{y}_1 = 16$ we have

$$y_2 = \tfrac{1}{2}(16)(\Delta t)^2 = 0.0032$$

The quantities x_2 and $\ddot{x}_2$ must be solved simultaneously from the equations

$$x_2 = \tfrac{1}{6}\ddot{x}_2 \Delta t^2$$
$$\ddot{x}_2 = -540x_2 + 180y_2$$

Eliminating $\ddot{x}_2$ the equation for x_2 becomes

$$x_2 = \frac{30y_2\Delta t^2}{1 + 90\Delta t^2}$$

The flow diagram for the computation is shown in Fig. 5.4-2 and the Fortran program is presented in Fig. 5.4-3. The computed results and the plot for x and y are shown in Fig. 5.4-4.

5.5 VIBRATION ABSORBER

A spring-mass system k_2, m_2, Fig. 5.5-1, tuned to the frequency of the exciting force such that $\omega^2 = k_2/m_2$, will act as a vibration absorber and reduce the motion of the main mass m_1 to zero. Making the substitution

$$\omega_{11}^2 = \frac{k_1}{m_1} \qquad \omega_{22}^2 = \frac{k_2}{m_2}$$

and assuming the motion to be harmonic, the equation for the amplitude X_1 can be shown to be equal to

$$\frac{X_1 k_1}{F_0} = \frac{\left[1 - \left(\dfrac{\omega}{\omega_{22}}\right)^2\right]}{\left[1 + \dfrac{k_2}{k_1} - \left(\dfrac{\omega}{\omega_{11}}\right)^2\right]\left[1 - \left(\dfrac{\omega}{\omega_{22}}\right)^2\right] - \dfrac{k_2}{k_1}} \qquad (5.5\text{-}1)$$

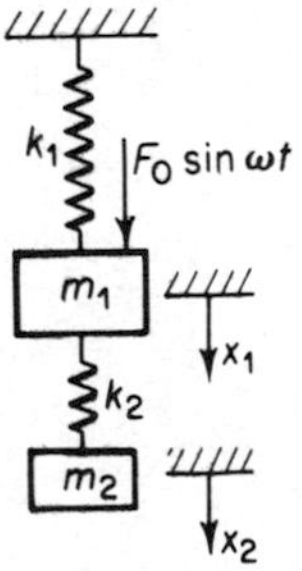

Figure 5.5-1. Vibration absorber.

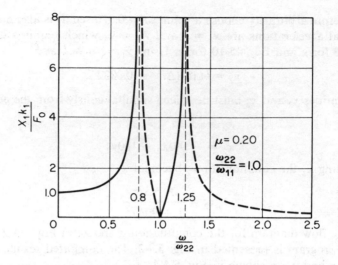

Figure 5.5-2.   Response vs. frequency.

Figure 5.5-2 shows a plot of this equation with $\mu = m_2/m_1$ as parameter. Note that $k_2/k_1 = \mu(\omega_{22}/\omega_{11})^2$. Since the system is one of two degrees of freedom, two natural frequencies exist. These are shown against μ in Fig. 5.5-3.

So far nothing has been said about the size of the absorber mass. At $\omega = \omega_{22}$, the amplitude $X_1 = 0$, but the absorber mass undergoes an

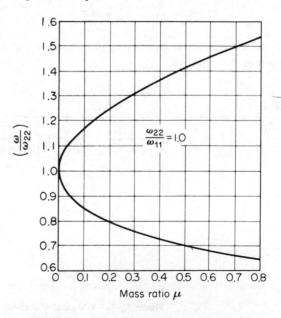

Figure 5.5-3.   Natural frequencies vs. $\mu = m_2/m_1$.

amplitude equal to

$$X_2 = -\frac{F_0}{k_2} \qquad (5.5\text{-}2)$$

Since the force acting on m_2 is

$$k_2 X_2 = \omega^2 m_2 X_2 = -F_0$$

the absorber system k_2, m_2 exerts a force equal and opposite to the disturbing force. Thus the size of k_2 and m_2 depends on the allowable value of X_2.

5.6   CENTRIFUGAL PENDULUM VIBRATION ABSORBER

The vibration absorber of Sec. 5.5 is only effective at one frequency, $\omega = \omega_{22}$. Also, with resonant frequencies on each side of ω_{22}, the usefulness of the spring-mass absorber is narrowly limited.

For a rotating system such as the automobile engine, the exciting torques are proportional to the rotational speed n, which may vary over a wide range. Thus for the absorber to be effective, its natural frequency must also be proportional to the speed. The characteristics of the centrifugal pendulum are ideally suited for this purpose.

Figure 5.6-1 shows the essentials of the centrifugal pendulum. It is a two degree of freedom nonlinear system; however, we will limit the oscillations to small angles, thereby reducing its complexity.

Placing the coordinates through point O' parallel and normal to r, the line r rotates with angular velocity $(\dot{\theta} + \dot{\phi})$. The acceleration of m is equal to the vector sum of the acceleration of O' and the acceleration of m relative to O'.

$$a_m = \left[R\ddot{\theta} \sin \phi - R\dot{\theta}^2 \cos \phi - r(\dot{\theta} + \dot{\phi})^2 \right] i$$
$$+ \left[R\ddot{\theta} \cos \phi + R\dot{\theta}^2 \sin \phi + r(\ddot{\theta} + \ddot{\phi}) \right] j \qquad (5.6\text{-}1)$$

Since the moment about O' is zero, we have

$$m \left[R\ddot{\theta} \cos \phi + R\dot{\theta}^2 \sin \phi + r(\ddot{\theta} + \ddot{\phi}) \right] r = 0 \qquad (5.6\text{-}2)$$

Assuming ϕ to be small, we let $\cos \phi = 1$ and $\sin \phi = \phi$, and arrive at the equation for the pendulum

$$\ddot{\phi} + \left(\frac{R}{r} \dot{\theta}^2 \right) \phi = - \left(\frac{R+r}{r} \right) \ddot{\theta} \qquad (5.6\text{-}3)$$

The above equation contains both θ and ϕ and hence a second equation is required, namely the torque equation for the wheel. Acting on the wheel are the exciting torque T and the pendulum torque which is

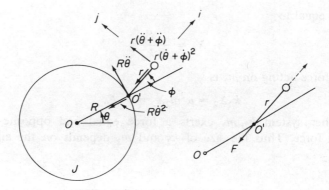

Figure 5.6-1. Centrifugal pendulum.

found from the cross product of the wheel radius R and the pendulum force ma_m. It is evident from Eqs. (5.6-1) and (5.6-3) that even for the small angle approximation we will end up with two nonlinear differential equations and that a simple solution is not possible without further approximation.

If we assume the motion of the wheel to be a steady rotation n plus a small sinusoidal oscillation, we can write

$$\theta = nt + \theta_0 \sin \omega t$$

$$\dot{\theta} = n + \omega \theta_0 \cos \omega t \cong n \qquad (5.6\text{-}4)$$

$$\ddot{\theta} = -\omega^2 \theta_0 \sin \omega t$$

Then Eq. (5.6-3) becomes

$$\ddot{\phi} + \left(\frac{R}{r} n^2\right)\phi = \left(\frac{R+r}{r}\right)\omega^2 \theta_0 \sin \omega t \qquad (5.6\text{-}3')$$

and we recognize the natural frequency of the pendulum to be

$$\omega_n = n\sqrt{\frac{R}{r}} \qquad (5.6\text{-}5)$$

Assuming a steady state solution $\phi = \phi_0 \sin(\omega t - \alpha)$, the amplitude ratio becomes

$$\frac{\theta_0}{\phi_0} = \frac{\dfrac{n^2 R}{r} - \omega^2}{\left(\dfrac{R+r}{r}\right)\omega^2} \qquad (5.6\text{-}6)$$

which clearly indicates that the oscillation θ_0 of the wheel becomes zero when $\omega = n\sqrt{R/r}$.

Also by recognizing that the largest term in Eq. (5.6-1) is due to $\dot{\theta}^2 = n^2$, the pendulum torque opposing the disturbing torque T becomes

$$M \cong m(R + r)n^2R\phi \qquad (5.6\text{-}7)$$

5.7 VIBRATION DAMPER

In contrast to the vibration absorber, where the exciting force is opposed by the absorber, energy is dissipated by the vibration damper. Figure 5.7-1 represents a friction type of vibration damper, commonly known as the Lanchester damper, which has found practical use in torsional systems such as gas and diesel engines in limiting the amplitudes of vibration at critical speeds. The damper consists of two flywheels a free to rotate on the shaft and driven only by means of the friction rings b when the normal pressure is maintained by the spring-loaded bolts c.

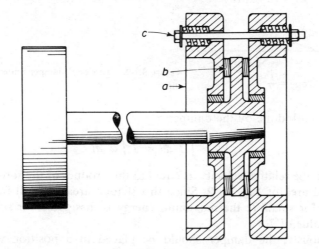

Figure 5.7-1. Torsional vibration damper.

When properly adjusted, the flywheels rotate with the shaft for small oscillations. However, when the torsional oscillations of the shaft tend to become large, the flywheels will not follow the shaft because of their large inertia, and energy is dissipated by friction due to the relative motion. The dissipation of energy thus limits the amplitude of oscillation, thereby preventing high torsional stresses in the shaft.

In spite of the simplicity of the torsional damper the mathematical analysis for its behavior is rather complicated. For instance, the flywheels may slip continuously, for part of the cycle, or not at all, depending on the pressure exerted by the spring bolts. If the pressure on the friction ring is either too great for slipping or zero, no energy is dissipated, and the damper becomes ineffective. Obviously, maximum energy dissipation takes place at some intermediate pressure, resulting in optimum damper effectiveness.

To obtain an insight to the problem, we consider briefly the case where the flywheels slip continuously. Assuming the shaft hub to be oscillating about its mean angular speed, as shown in Fig. 5.7-2, the flywheels will be acted upon by a constant frictional torque T while slipping. The acceleration of the flywheel, represented by the slope of the velocity curve, will hence be constant and equal to T/J, where J is the moment of inertia of the flywheels, and its velocity will be represented by a series of straight lines. The velocity of the flywheels will be increasing while the shaft speed is greater than that of the flywheels and decreasing when the shaft speed drops below that of the flywheels, as shown in the diagram.

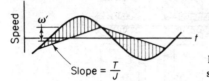

Figure 5.7-2. Torsional damper under continuous slip.

The work done by the damper,

$$W = \int T \, d\theta = T \int \omega' \, dt \tag{5.7-1}$$

where ω' is the relative velocity, is equal to the product of the torque T and the shaded area of Fig. 5.7-2. Since this shaded area is small for large T and large for small T, the maximum energy is dissipated for some intermediate value of T.*

Obviously, the damper should be placed in a position where the amplitude of oscillation is the greatest. This position generally is found on the side of the shaft away from the main flywheel, since the node is usually near the largest mass.

The Untuned Viscous Vibration Damper. In a rotating system such as an automobile engine, the disturbing frequencies for torsional oscillations are proportional to the rotational speed. However, there is generally more than one such frequency, and the centrifugal pendulum has the disadvantage that several pendulums tuned to the order number of the disturbance

---

*J. P. Den Hartog and J. Ormondroyd, "Torsional-Vibration Dampers," *Trans. ASME* APM-52-13 (September–December, 1930), pp. 133–152.

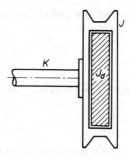

Figure 5.7-3.   Untuned viscous damper.

must be used. In contrast to the centrifugal pendulum, the untuned viscous torsional damper is effective over a wide operating range. It consists of a free rotational mass within a cylindrical cavity filled with viscous fluid, as shown in Fig. 5.7-3. Such a system is generally incorporated into the end pulley of a crankshaft which drives the fan belt, and is often referred to as the Houdaille damper.

   We can examine the untuned viscous damper as a two degrees of freedom system by considering the crankshaft, to which it is attached, as being fixed at one end with the damper at the other end. With the torsional stiffness of the shaft equal to K in. lb/rad the damper can be considered to be excited by a harmonic torque $M_0 e^{i\omega t}$. The damper torque results from the viscosity of the fluid within the pulley cavity, and we will assume it to be proportional to the relative rotational speed between the pulley and the free mass. Thus the two equations of motion for the pulley and the free mass are

$$J\ddot{\theta} + K\theta + c(\dot{\theta} - \dot{\varphi}) = M_0 e^{i\omega t}$$
$$J_d\dot{\varphi} - c(\dot{\theta} - \dot{\varphi}) = 0$$

(5.7-2)

Assuming the solution to be in the form

$$\theta = \theta_0 e^{i\omega t}$$
$$\varphi = \varphi_0 e^{i\omega t}$$

(5.7-3)

where θ_0 and φ_0 are complex amplitudes, their substitution into the differential equations results in

$$\left[\left(\frac{K}{J} - \omega^2\right) + i\frac{c\omega}{J}\right]\theta_0 - \frac{ic\omega}{J}\varphi_0 = \frac{M_0}{J}$$

and

$$\left(-\omega^2 + i\frac{c\omega}{J_d}\right)\varphi_0 = \frac{ic\omega}{J_d}\theta_0$$

(5.7-4)

Eliminating φ_0 between the two equations, the expression for the amplitude

θ_0 of the pulley becomes

$$\frac{\theta_0}{M_0} = \frac{(\omega^2 J_d - ic\omega)}{\left[\omega^2 J_d(K - J\omega^2)\right] + ic\omega\left[\omega^2 J_d - (K - J\omega^2)\right]} \qquad (5.7\text{-}5)$$

Letting $\omega_n^2 = K/J$ and $\mu = J_d/J$, the critical damping is

$$c_c = 2J\omega_n, \quad c = \frac{c}{c_c} 2J\omega_n = 2\zeta J\omega_n$$

The amplitude equation then becomes

$$\left|\frac{K\theta_0}{M_0}\right| = \sqrt{\frac{\mu^2(\omega/\omega_n)^2 + 4\zeta^2}{\mu^2(\omega/\omega_n)^2(1 - \omega^2/\omega_n^2)^2 + 4\zeta^2\left[\mu(\omega/\omega_n)^2 - (1 - \omega^2/\omega_n^2)\right]^2}}$$

$$(5.7\text{-}6)$$

which indicates that $|K\theta_0/M_0|$ is a function of three parameters, ζ, μ, and (ω/ω_n).

If μ is held constant and $|K\theta_0/M_0|$ plotted as a function of (ω/ω_n), the curve for any ζ will appear somewhat similar to that of a single degree of freedom system with a single peak. Of interest are the two extreme values of $\zeta = 0$ and $\zeta = \infty$. When $\zeta = 0$, we have an undamped system with resonant frequency $\omega_n = \sqrt{K/J}$, and the amplitude will be infinite at this frequency. If $\zeta = \infty$, the damper mass and the wheel will move together as a single mass, and again we have an undamped system but with natural frequency of $\sqrt{k/(J + J_d)}$.

Thus, like the Lanchester damper of the previous section, there is an optimum damping ζ_0 for which the peak amplitude is a minimum as shown in Fig. 5.7-4. The result can be presented as a plot of the peak values as a function of ζ for any given μ, as shown in Fig. 5.7-5.

$$\zeta_0 = \frac{\mu}{\sqrt{2(1 + \mu)(2 + \mu)}} \qquad (5.7\text{-}7)$$

and that the peak amplitude for optimum damping is found at a frequency equal to

$$\frac{\omega}{\omega_n} = \sqrt{2/(2 + \mu)} \qquad (5.7\text{-}8)$$

These conclusions can be arrived at by observing that the curves of Fig. 5.7-4 all pass through a common point P, regardless of the numerical values of ζ. Thus, by equating the equation for $|K\theta_0/M|$ for $\zeta = 0$ and $\zeta = \infty$, Eq. 5.7-8 is found. The curve for optimum damping then must pass through P with a zero slope, so that if we substitute $(\omega/\omega_n)^2 = 2/(2 + \mu)$ into the derivative of Eq. (5.7-6) equated to zero, the expression for ζ_0 is

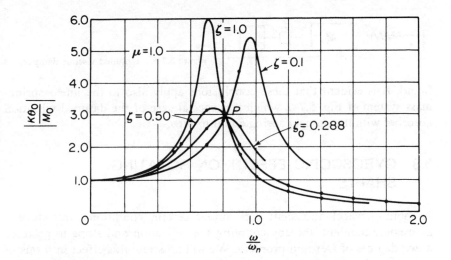

Figure 5.7-4. Response of an untuned viscous damper (all curves pass through P).

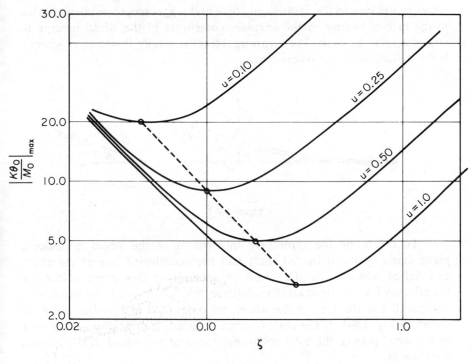

Figure 5.7-5.

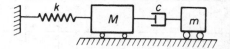

Figure 5.7-6. Untuned viscous damper.

found. It is evident that these conclusions apply also to the linear spring-mass system of Fig. 5.7-6, which is a special case of the damped vibration absorber with the damper spring equal to zero.

5.8 GYROSCOPIC EFFECT ON ROTATING SHAFTS

A rotating wheel and shaft can, under certain conditions, introduce a gyroscopic moment, thereby coupling the deflection and slope to produce a two degrees of freedom problem. We will illustrate this effect in terms of a wheel rotating on an overhanging shaft, as shown in Fig. 5.8-1.

If the wheel and shaft in the deflected position are spinning and whirling at the same time, there will be a force P acting outward at the wheel center due to the centrifugal force and a gyroscopic moment M due to the rate of change of the angular momentum of the wheel tending to straighten out the shaft. The centrifugal force is simply $P = m\omega_1^2 y$, where y is the deflection of the wheel center.

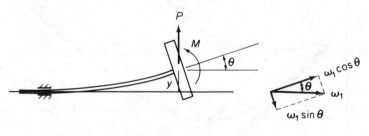

Figure 5.8-1.

To determine the gyroscopic moment M of the wheel, consider a plane made by the deflected shaft and the undeflected line of the shaft, and define whirl ω_1 as the speed of rotation of this plane about the undeflected line of the shaft. Resolving ω_1 into components perpendicular and parallel to the face of the wheel, we obtain $\omega_1 \sin \theta$ and $\omega_1 \cos \theta$, as shown in Fig. 5.8-1. If the shaft is given an additional rotation ω_2 (relative to the whirl plane), the total rotational speed of the wheel in the normal direction is

$$\omega = \omega_1 \cos \theta + \omega_2$$

which is actually the total shaft rotation or the rotational speed of the

Figure 5.8-2.

wheel. Thus the angular momentum H of the wheel is $J_p\omega$ perpendicular to the wheel and $J_d\omega_1 \sin\theta$ parallel to the face of the wheel, as shown in Fig. 5.8-2 where J_p and J_d are the polar and diametric moments of inertia of the wheel.

The gyroscopic moment M, which must be exerted on the wheel by the shaft, is the rate of change of this momentum vector due to ω_1. Assuming θ to be small, and resolving H into components parallel and perpendicular to ω_1, we obtain

$$M = (J_p\omega\theta - J_d\omega_1\theta)\omega_1 \qquad (5.8\text{-}1)$$

The moment exerted on the shaft by the wheel is the negative of the above

$$M_s = -\left(J_p\frac{\omega}{\omega_1} - J_d\right)\omega_1^2\theta$$

$$= -J_d\omega_1^2\theta\left(a\frac{\omega}{\omega_1} - 1\right) \qquad (5.8\text{-}2)$$

where $a = J_p/J_d$.

The deflection and slope at the wheel end of the shaft can be written in terms of the influence coefficients.

$$y = \alpha_{11}P + \alpha_{12}M_s$$
$$\theta = \alpha_{21}P + \alpha_{22}M_s \qquad (5.8\text{-}3)$$

Substituting for P and M_s, we obtain

$$y = (\alpha_{11}m\omega_1^2)y - \alpha_{12}J_d\omega_1^2\left(a\frac{\omega}{\omega_1} - 1\right)\theta$$

$$\theta = (\alpha_{21}m\omega_1^2)y - \alpha_{22}J_d\omega_1^2\left(a\frac{\omega}{\omega_1} - 1\right)\theta \qquad (5.8\text{-}4)$$

The influence coefficients for the case where the bearing is rigidly fixed are

$$\alpha_{11} = \frac{l^3}{3EI}, \qquad \alpha_{12} = \alpha_{21} = \frac{l^2}{2EI}, \qquad \alpha_{22} = \frac{l}{EI} \qquad (5.8\text{-}5)$$

Equations (5.8-4) also apply to the case where the bearing stiffness is finite so that translation and rotation of the bearing take place, as shown in Fig. 5.8-3.

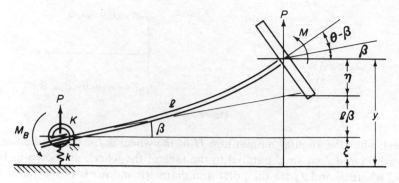

Figure 5.8-3.

The left side of the equation is now replaced by η and $(\theta - \beta)$

$$\eta = \alpha_{11}P + \alpha_{12}M_s$$
$$\theta - \beta = \alpha_{21}P + \alpha_{22}M_s \tag{5.8-6}$$

From the geometry of Fig. 5.8-3 we have

$$\eta = y - \xi - l\beta = y - \frac{P}{k} - \frac{lM_B}{K}$$

$$\beta = \frac{M_B}{K}, \qquad M_B = Pl + M_s$$

so that

$$y = \left(\alpha_{11} + \frac{1}{k} + \frac{l^2}{K}\right)P + \left(\alpha_{12} + \frac{l}{K}\right)M_s = \bar{\alpha}_{11}P + \bar{\alpha}_{12}M_s$$

$$\tag{5.8-7}$$

$$\theta = \left(\alpha_{12} + \frac{l}{k}\right)P + \left(\alpha_{22} + \frac{1}{K}\right)M_s = \bar{\alpha}_{21}P + \bar{\alpha}_{22}M_s$$

Thus the new equations for the flexible bearing differ only in the influence coefficients.

$$y = (\bar{\alpha}_{11}m\omega_1^2)y - \bar{\alpha}_{12}J_d\omega_1^2\left(a\frac{\omega}{\omega_1} - 1\right)\theta$$

$$\tag{5.8-8}$$

$$\theta = (\bar{\alpha}_{12}m\omega_1^2)y - \bar{\alpha}_{22}J_d\omega_1^2\left(a\frac{\omega}{\omega_1} - 1\right)\theta$$

Introducing the following nondimensional quantities

$$W = \omega_1\sqrt{\bar{\alpha}_{11}m} \qquad D = \frac{\bar{\alpha}_{22}}{\bar{\alpha}_{11}}\frac{J_d}{m}$$

$$\tag{5.8-9}$$

$$S = \omega\sqrt{\bar{\alpha}_{11}m} \qquad E = \frac{\bar{\alpha}_{12}^2}{\bar{\alpha}_{11}\bar{\alpha}_{22}}$$

the equations for y and θ take the form

$$(W^2 - 1)y - \frac{\bar{\alpha}_{11}}{\bar{\alpha}_{12}} EDW^2\left(a\frac{S}{W} - 1\right)\theta = 0$$

$$\frac{\bar{\alpha}_{12}}{\bar{\alpha}_{11}} W^2 y - \left[DW^2\left(a\frac{S}{W} - 1\right) + 1\right]\theta = 0$$

(5.8-10)

and by equating the determinant to zero, we obtain the relationship between whirl and wheel rotation which we will call spin.

$$S = \frac{W^4 + \dfrac{(D-1)W^2}{D(E-1)} - \dfrac{1}{D(E-1)}}{aW\left(W^2 + \dfrac{1}{E-1}\right)}$$

(5.8-11)

The mode shape is given by

$$\frac{y}{\theta} = \frac{\bar{\alpha}_{12}}{\bar{\alpha}_{22}} \frac{DW^2\left(a\dfrac{S}{W} - 1\right)}{(W^2 - 1)}$$

(5.8-12)

These equations can be solved by assuming a value for the nondimensional whirl W and solving for the nondimensional spin S. Fig. 5.8-4 shows a typical whirl–spin relationship. There are two retrograde and two forward whirl modes for a given spin speed.

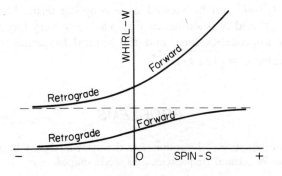

Figure 5.8-4.

Synchronous Whirl. For a wheel with an unbalance, we have found in Sec. 3.4 that the whirling speed ω_1 can be equal to the rotation speed ω. Thus the frequency equation, Eq. (5.8-4), for the synchronous whirl takes the form

$$(1 - \alpha_{11}m\omega^2)\left[1 + \alpha_{22}J_d\omega^2(a - 1)\right] + (\alpha_{21}m\omega_1^2)\left[\alpha_{12}J_d\omega^2(a - 1)\right] = 0$$

(5.8-13)

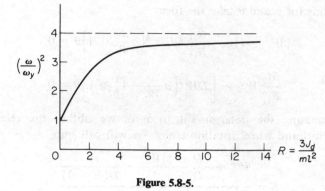

Figure 5.8-5.

For a wheel approaching a thin disk, $a = J_p/J_d = 2$, and the above equation reduces to

$$\omega^4 + \frac{12EI}{mJ_d l^3}\left(\frac{ml^2}{3} - J_d\right)\omega^2 - \frac{12}{mJ_d}\left(\frac{EI}{l^2}\right)^2 = 0 \qquad (5.8\text{-}14)$$

Since in the absence of the gyroscopic couple the natural frequency of the system is $\omega_y = \sqrt{3EI/ml^3}$, we can rewrite the frequency equation as

$$\left(\frac{\omega}{\omega_y}\right)^4 + 4\left(\frac{1}{R} - 1\right)\left(\frac{\omega}{\omega_y}\right)^2 - \frac{4}{R} = 0 \qquad (5.8\text{-}15)$$

where $R = 3J_d/ml^2$ can be viewed as a coupling term. The relationship between $(\omega/\omega_y)^2$ and R is shown in Fig. 5.8-5. For very large values of R, the ratio θ/y approaches zero, and the natural frequency of the system tends to the value $\omega = \sqrt{12EI/ml^2}$.

PROBLEMS

5-1   Write the equations of motion for the system shown in Fig. P5-1 and determine its natural frequencies and mode shapes.

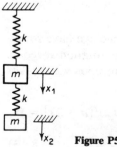

Figure P5-1.

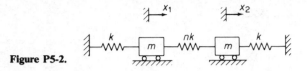

Figure P5-2.

5-2   Determine the normal modes and frequencies of the system shown in Fig. P5-2 when $n = 1$.

5-3   For the system of Problem 5-2, determine the natural frequencies as a function of n.

5-4   Determine the natural frequencies and mode shapes of the system shown in Fig. P5-4.

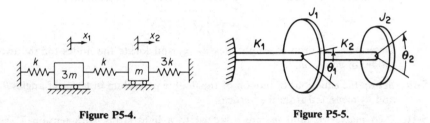

Figure P5-4.   Figure P5-5.

5-5   Determine the normal modes of the torsional system shown in Fig. P5-5 for $K_1 = K_2$ and $J_1 = 2J_2$.

5-6   If $K_1 = 0$ in the torsional system of Problem 5-5, the system becomes a degenerate two degrees of freedom system with only one natural frequency. Discuss the normal modes of this system as well as a linear spring-mass system equivalent to it. Show that the system can be treated as one of a single degree of freedom by using the coordinate $\phi = (\theta_1 - \theta_2)$.

5-7   Determine the natural frequency of the torsional system shown in Fig. P5-7, and draw the normal mode curve. $G = 11.5 \times 10^6$ psi.

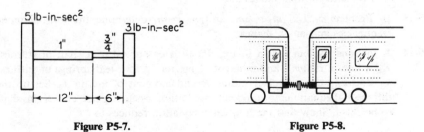

Figure P5-7.   Figure P5-8.

5-8   An electric train made up of two cars of weight 50,000 lb each is connected by couplings of stiffness equal to 16,000 lb/in. as shown in Fig. P5-8. Determine the natural frequency of the system.

5-9   Assuming small amplitudes, set up the differential equation of motion for the double pendulum using the coordinates shown in Fig. P5-9. Show that

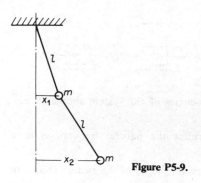

Figure P5-9.

the natural frequencies of the system are given by the equation

$$\omega = \sqrt{\frac{g}{l}(2 \pm \sqrt{2})}$$

Determine the ratio of amplitudes x_1/x_2 and locate the nodes for the two modes of vibration.

5-10 Set up the equations of motion of the double pendulum in terms of angles θ_1 and θ_2 measured from the vertical.

5-11 Two masses m_1 and m_2 are attached to a light string with tension T, as shown in Fig. P5-11. Assuming that T remains unchanged when the masses are displaced normal to the string, write the equations of motion expressed in matrix form.

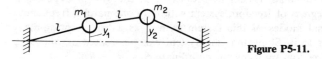

Figure P5-11.

5-12 In Problem 5-11 if the two masses are made equal, show that normal mode frequencies are $\omega = \sqrt{T/ml}$ and $\omega_2 = \sqrt{3T/ml}$. Establish the configuration for these normal modes.

5-13 In Problem 5-11 if $m_1 = 2m$, and $m_2 = m$, determine the normal mode frequencies and mode shapes.

5-14 A torsional system shown in Fig. P5-14 is composed of a shaft of stiffness K_1, a hub of radius r and moment of inertia J_1, four leaf springs of stiffness k_2, and an outer wheel of radius R and moment of inertia J_2. Set up the differential equations for torsional oscillation, assuming one end of the shaft to be fixed. Show that the frequency equation reduces to

$$\omega^4 - \left(\omega_{11}^2 + \omega_{22}^2 + \frac{J_2}{J_1}\omega_{22}^2\right)\omega^2 + \omega_{11}^2\omega_{22}^2 = 0$$

where ω_{11} and ω_{22} are uncoupled frequencies given by the expressions

$$\omega_{11}^2 = \frac{K_1}{J_1}, \qquad \omega_{22}^2 = \frac{4k_2R^2}{J_2}$$

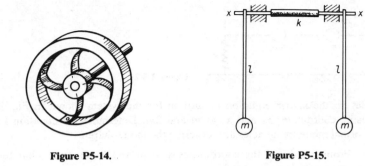

Figure P5-14.   Figure P5-15.

5-15 Two equal pendulums free to rotate about the x-x axis are coupled together by a rubber hose of torsional stiffness k lbin./rad, as shown in Fig. P5-15. Determine the natural frequencies for the normal modes of vibration, and describe how these motions may be started.

If $l = 19.3$ in., mg $= 3.86$ lb, and $k = 2.0$ lbin./rad, determine the beat period for a motion started with $\theta_1 = 0$ and $\theta_2 = \theta_0$. Examine carefully the phase of the motion as the amplitude approaches zero.

5-16 Determine the equations of motion for the system of Problem 5-4 when the initial conditions are $x_1(0) = A$, $\dot{x}_1(0) = x_2(0) = \dot{x}_2(0) = 0$.

5-17 The double pendulum of Problem 5-9 is started with the following initial conditions: $x_1(0) = x_2(0) = X$, $\dot{x}_1(0) = \dot{x}_2(0) = 0$. Determine the equations of motion.

5-18 The lower mass of Problem 5-1 is given a sharp blow, imparting to it an initial velocity $\dot{x}_2(0) = V$. Determine the equation of motion.

5-19 If the system of Problem 5-1 is started with initial conditions $x_1(0) = 0$, $x_2(0) = 1.0$, $\dot{x}_1(0) = \dot{x}_2(0) = 0$, show that the equations of motion are

$$x_1(t) = 0.447 \cos \omega_1 t - 0.447 \cos \omega_2 t$$

$$x_2(t) = 0.722 \cos \omega_1 t + 0.278 \cos \omega_2 t$$

$$\omega_1 = \sqrt{.382k/m} \qquad \omega_2 = \sqrt{2.618k/m}$$

5-20 Choose coordinates x for the displacement of c and θ clockwise for the rotation of the uniform bar shown in Fig. P5-20, and determine the natural frequencies and mode shapes.

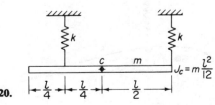

Figure P5-20.

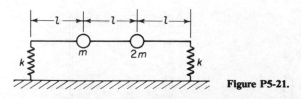

Figure P5-21.

5-21   Set up the matrix equation of motion for the system shown in Fig. P5-21, using coordinates x_1 and x_2 at m and $2m$. Determine the equation for the normal mode frequencies and describe the mode shapes.

5-22   In Problem 5-21, if the coordinates x at m and θ are used, what form of coupling will they result in?

5-23   Compare Problems 5-9 and 5-10 in matrix form and indicate the type of coupling present in each coordinate system.

5-24   The following information is given for a certain automobile shown in Fig. P5-24.

$$W = 3500 \text{ lb} \qquad k_1 = 2000 \text{ lb/ft}$$
$$l_1 = 4.4 \text{ ft} \qquad k_2 = 2400 \text{ lb/ft}$$
$$l_2 = 5.6 \text{ ft} \qquad r = 4 \text{ ft} = \text{radius of gyration about } cg.$$

Determine the normal modes of vibration and locate the node for each mode.

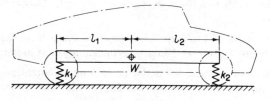

Figure P5-24.

5-25   An airfoil section to be tested in a wind tunnel is supported by a linear spring k and a torsional spring K, as shown in Fig. P5-25. If the center of gravity of the section is a distance e ahead of the point of support, determine the differential equations of motion of the system.

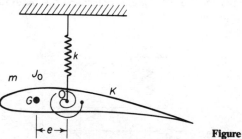

Figure P5-25.

5-26 Determine the natural frequencies and normal modes of the system shown in Fig. P5-26 when

$$gm_1 = 3.86 \text{ lb} \qquad k_1 = 20 \text{ lb/in.}$$
$$gm_2 = 1.93 \text{ lb} \qquad k_2 = 10 \text{ lb/in.}$$

When forced by $F_1 = F_0 \sin \omega t$, determine the equations for the amplitudes and plot them against ω/ω_{11}.

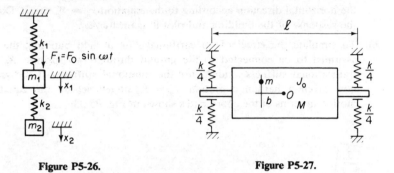

Figure P5-26.   **Figure P5-27.**

5-27 A rotor is mounted in bearings which are free to move in a single plane as shown in Fig. P5-27. The rotor is symmetrical about O with total mass M and moment of inertia J_0 about an axis perpendicular to the shaft. If a small unbalance mr acts at an axial distance b from its center O, determine the equations of motion for a rotational speed ω.

5-28 A two-story building is represented in Fig. P5-28 by a lumped mass system where $m_1 = \frac{1}{2}m_2$ and $k_1 = \frac{1}{2}k_2$. Show that its normal modes are

$$\left(\frac{x_1}{x_2}\right)^{(1)} = 2, \qquad \omega_1^2 = \frac{1}{2}\frac{k_1}{m_1}$$

$$\left(\frac{x_1}{x_2}\right)^{(2)} = -1, \qquad \omega_2^2 = 2\frac{k_1}{m_1}$$

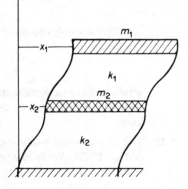

Figure P5-28.

5-29 In Problem 5-28, if a force is applied to m_1 to deflect it by unity, and the system is released from this position, determine the equation of motion of each mass by the normal mode summation method.

5-30 In Problem 5-29, determine the ratio of the maximum shear in the first and second stories.

5-31 Repeat Problem 5-29 if the load is applied to m_2, displacing it by unity.

5-32 Assume in Problem 5-28 that an earthquake causes the ground to oscillate in the horizontal direction according to the equation $x_g = X_g \sin \omega t$. Determine the response of the building and plot it against ω/ω_1.

5-33 To simulate the effect of an earthquake on a rigid building, the base is assumed to be connected to the ground through two springs; K_h for the translational stiffness, and K_r for the rotational stiffness. If the ground is now given a harmonic motion $Y_g = Y_G \sin \omega t$, set up the equations of motion in terms of the coordinates shown in Fig. P5-33.

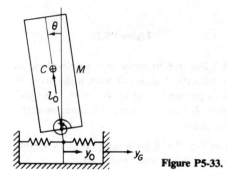

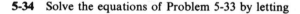

Figure P5-33.

5-34 Solve the equations of Problem 5-33 by letting

$$\omega_h^2 = \frac{K_h}{M}, \qquad \left(\frac{\rho_c}{l_0}\right)^2 = \frac{1}{3}$$

$$\omega_r^2 = \frac{K_r}{M\rho_c^2} \qquad \left(\frac{\omega_r}{\omega_h}\right)^2 = 4$$

The first natural frequency and mode shape are

$$\frac{\omega_1}{\omega_h} = 0.734 \text{ and } \frac{Y_0}{l_0\theta} = -1.14$$

which indicate a motion that is predominantly translational. Establish the second natural frequency and its mode ($Y_1 = Y_0 - 2l_0\theta_0$ = displacement of top).

5-35 The response and mode configuration for Problems 5-33 and 5-34 are shown in Fig. P5-35. Verify the mode shapes for several values of the frequency ratio.

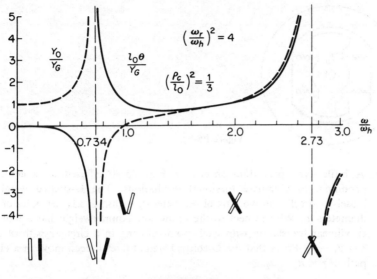

Figure P5-35.

5-36 The expansion joints of a concrete highway are 45 ft apart. These joints cause a series of impulses at equal intervals to affect cars traveling at a constant speed. Determine the speeds at which pitching motion and up-and-down motion are most apt to arise for the automobile of Problem 5-24.

5-37 For the system shown in Fig. P5-37, $W_1 = 200$ lb and the absorber weight $W_2 = 50$ lb. If W_1 is excited by a 2 lb in. unbalance rotating at 1800 rpm, determine the proper value of the absorber spring k_2. What will be the amplitude of W_2?

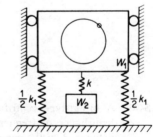

Figure P5-37.

5-38 In Problem 5-37, if a dashpot c is introduced between W_1 and W_2, determine the amplitude equations by the complex algebra method.

5-39 A flywheel of moment of inertia I has a torsional absorber of moment of inertia I_d free to rotate on the shaft and connected to the flywheel by four springs of stiffness k lb/in. as shown in Fig. P5-39. Set up the differential equations of motion for the system, and discuss the response of the system to an oscillatory torque.

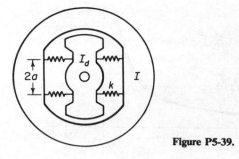

Figure P5-39.

5-40 A bifilar-type pendulum shown in Fig. P5-40 is used as a centrifugal pendulum to eliminate torsional oscillations. The U-shaped weight fits loosely and rolls on two pins of diameter d_2 within two larger holes of equal diameters d_1. With respect to the crank, the counterweight has a motion of curvilinear translation with each point moving in a circular path of radius $r = d_1 - d_2$. Prove that the U-shaped weight does indeed move in a circular path of $r = d_1 - d_2$.

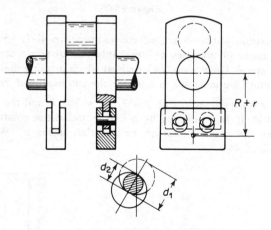

Figure P5-40.

5-41 A bifilar-type centrifugal pendulum is proposed to eliminate a torsional disturbance of frequency equal to four times the rotational speed. If the distance R to the center of gravity of the pendulum mass is made equal to 4.0 in. and $d_1 = \frac{3}{4}$ in., what must be the diameter d_2 of the pins?

5-42 A jig used to size coal contains a screen that reciprocates with a frequency of 600 cpm. The jig weighs 500 lb and has a fundamental frequency of 400 cpm. If an absorber weighing 125 lb is to be installed to eliminate the vibration of the jig frame, determine the absorber spring stiffness. What will be the resulting two natural frequencies of the system?

5-43 In a certain refrigeration plant a section of pipe carrying the refrigerant vibrated violently at a compressor speed of 232 rpm. To eliminate this difficulty it was proposed to clamp a spring-mass system to the pipe to act as an absorber. For a trial test a 2.0-lb absorber tuned to 232 cpm resulted in two natural frequencies of 198 and 272 cpm. If the absorber system is to be designed so that the natural frequencies lie outside the region 160 to 320 cpm, what must be the weight and spring stiffness?

5-44 A type of damper frequently used on automobile crankshafts is shown in Fig. P5-44. J represents a solid disk free to spin on the shaft, and the space between the disk and case is filled with a silicone oil of coefficient of viscosity μ. The damping action results from any relative motion between the two. Derive an equation for the damping torque exerted by the disk on the case due to a relative velocity of ω.

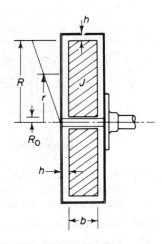

Figure P5-44.

5-45 For the Houdaille viscous damper with mass ratio $\mu = 0.25$, determine the optimum damping ζ_0 and the frequency at which the damper is most effective.

5-46 If the damping for the viscous damper of Problem 5-45 is equal to $\zeta = 0.10$, determine the peak amplitude as compared to the optimum.

5-47 Establish the relationships given by Eqs. (5.7-7) and (5.7-8) of Sec. 5.7.

5-48 A simply supported shaft of length l and stiffness EI has a thin but rigid disk keyed to it at the point $l/3$ as shown in Fig. P5-48. Establish the equations of motion for y and θ and plot $(\omega/\omega_y)^2$ vs. J_d/ml^2.

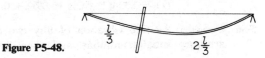

Figure P5-48.

5-49 Draw the flow diagram and develop the Fortran program for the computation of the response of the system shown in Prob. 5-4 when the mass $3m$ is excited by a rectangular pulse of magnitude 100 lb and duration $6\pi\sqrt{m/k}$ sec.

5-50 In Prob. 5-28 assume the following data, $k_1 = 4 \times 10^3$ lb/in, $k_2 = 6 \times 10^3$ lb/in, $m_1 = m_2 = 100$. Develop the flow diagram and the Fortran program for the case where the ground is given a displacement $y = 10''$ sin πt for 4 seconds.

5-51 Figure P5-51 shows a degenerate 3 DFS. Its characteristic equation yields one zero root and two elastic vibration frequencies. Discuss the physical significance of the fact that three coordinates are required but only two natural frequencies are obtained.

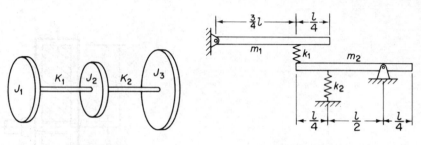

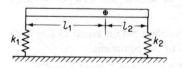

Figure P5-51.   **Figure P5-52.**

5-52 The two uniform rigid bars shown in Fig. P5-52 are of equal length but of different masses. Determine the equations of motion and the natural frequencies and mode shapes using matrix methods.

5-53 Show that the normal modes of the system of Prob. 5-51 are orthogonal.

5-54 For the system shown in Fig. P5-54 choose coordinates x_1 and x_2 at the ends of the bar and determine the type of coupling this introduces.

Figure P5-54.

5-55 Using the method of Laplace transforms, solve analytically the problem solved by the digital computer in Sec. 5.4 and show that the solution is

$$x(t) = 1.499(1 - \cos 16.09t) - 0.3875(1 - \cos 31.64t) \text{ cm}$$
$$y(t) = 2.334(1 - \cos 16.09t) + 0.993(1 - \cos 31.64t) \text{ cm}$$

5-56 Consider the free vibration of any two degrees of freedom system with arbitrary initial conditions, and show by examination of the subsidiary equations of Laplace transforms that the solution is the sum of normal modes.

5-57 Determine by the method of Laplace transformation the solution to the forced vibration problem shown in Fig. P5-57. Initial conditions are $x_1(0)$, $\dot{x}_1(0)$, $x_2(0)$ and $\dot{x}_2(0)$.

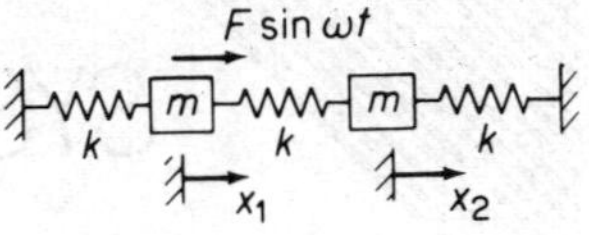

Figure P5-57.

6

PROPERTIES OF VIBRATING SYSTEMS

The vibration analysis of systems of many degrees of freedom requires a systematic approach for clarity of formulation and simplicity of computation. In this respect, matrix methods are ideally suited, not only for the above requirements but also in providing us with simpler discussions of some of the properties of vibrating systems.

In this chapter we will discuss the various properties of vibrating systems and the matrix techniques applicable to them. These concepts form the basis for the treatment and understanding of the behavior of large systems.

6.1 FLEXIBILITY AND STIFFNESS MATRIX

The flexibility influence coefficient a_{ij} is defined as the displacement at i due to a unit force applied at j. With forces f_1, f_2, and f_3 acting at stations 1, 2, and 3, the principle of superposition can be applied to determine the displacements in terms of the flexibility influence coefficients.

$$x_1 = a_{11} f_1 + a_{12} f_2 + a_{13} f_3$$
$$x_2 = a_{21} f_1 + a_{22} f_2 + a_{23} f_3 \qquad (6.1\text{-}1)$$
$$x_3 = a_{31} f_1 + a_{32} f_2 + a_{33} f_3$$

Expressed in matrix form, these equations become

$$\begin{Bmatrix} x_1 \\ x_2 \\ x_3 \end{Bmatrix} = \begin{bmatrix} a_{11} & a_{12} & a_{13} \\ a_{21} & a_{22} & a_{23} \\ a_{31} & a_{32} & a_{33} \end{bmatrix} \begin{Bmatrix} f_1 \\ f_2 \\ f_3 \end{Bmatrix} \qquad (6.1\text{-}2)$$

or

$$\{x\} = [a]\{f\} \qquad (6.1\text{-}3)$$

where $[a]$ is the flexibility matrix.

If Eq. (6.1-3) is premultiplied by the inverse of the flexibility matrix $[a]^{-1}$, we obtain

$$[a]^{-1}\{x\} = \{f\}$$

or

$$\{f\} = [k]\{x\} \qquad (6.1\text{-}4)$$

We thus recognize that the inverse of the flexibility is the stiffness matrix $[k]$.

$$[a]^{-1} = [k]$$
$$[a] = [k]^{-1} \qquad (6.1\text{-}5)$$

Writing out the terms of Eq. (6.1-4), we have

$$\begin{Bmatrix} f_1 \\ f_2 \\ f_3 \end{Bmatrix} = \begin{bmatrix} k_{11} & k_{12} & k_{13} \\ k_{21} & k_{22} & k_{23} \\ k_{31} & k_{32} & k_{33} \end{bmatrix} \begin{Bmatrix} x_1 \\ x_2 \\ x_3 \end{Bmatrix} \qquad (6.1\text{-}6)$$

The elements of the stiffness matrix have the following interpretation. If $x_1 = 1.0$ and $x_2 = x_3 = 0$, the forces at 1, 2, and 3 that are required to maintain this displacement according to Eq. (6.1-6) are k_{11}, k_{21}, and k_{31} in the first column. Similarly, the forces f_1, f_2, and f_3 required to maintain the displacement configuration $x_1 = 0$, $x_2 = 1.0$, and $x_3 = 0$ are k_{12}, k_{22}, and k_{32} in the second column. Thus the general rule for establishing the stiffness elements of any column is to set the displacement corresponding to that column to unity with all other displacements equal to zero and measure the forces required at each station.

Since flexible beam elements are often encountered, Table 6.1-1, "Stiffness for Beam Elements," will be helpful. These results can be determined by the area moment method or by superposition of different configurations.

TABLE 6.1-1
STIFFNESS FOR BEAM ELEMENTS

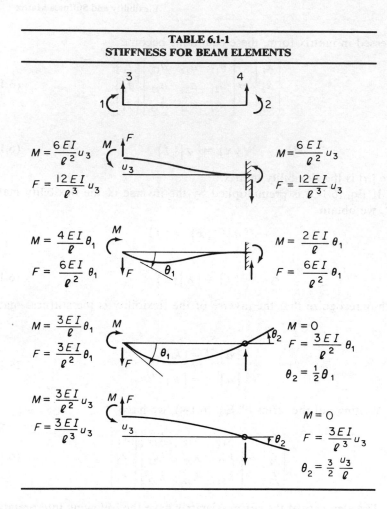

$$M = \frac{6EI}{\ell^2} u_3$$

$$F = \frac{12EI}{\ell^3} u_3$$

$$M = \frac{6EI}{\ell^2} u_3$$

$$F = \frac{12EI}{\ell^3} u_3$$

$$M = \frac{4EI}{\ell} \theta_1$$

$$F = \frac{6EI}{\ell^2} \theta_1$$

$$M = \frac{2EI}{\ell} \theta_1$$

$$F = \frac{6EI}{\ell^2} \theta_1$$

$$M = \frac{3EI}{\ell} \theta_1$$

$$F = \frac{3EI}{\ell^2} \theta_1$$

$$M = 0$$

$$F = \frac{3EI}{\ell^2} \theta_1$$

$$\theta_2 = \frac{1}{2}\theta_1$$

$$M = \frac{3EI}{\ell^2} u_3$$

$$F = \frac{3EI}{\ell^3} u_3$$

$$M = 0$$

$$F = \frac{3EI}{\ell^3} u_3$$

$$\theta_2 = \frac{3}{2}\frac{u_3}{\ell}$$

EXAMPLE 6.1-1

Determine the flexibility influence coefficients for the points (1), (2), and (3) of the uniform cantilever beam shown in Fig. 6.1-1.

Solution: The influence coefficients may be determined by placing unit loads at (1), (2) and (3) as shown, and calculating the deflections at these points. Using the area moment method,* the deflection at the various points is equal to the moment of the M/EI area about the point in question. For example, the value of $a_{21} = a_{12}$ is found from

*Egor P. Popov, *Introduction to Mechanics of Solids* (Englewood Cliffs, N.J.: Prentice-Hall, Inc., 1968), p. 411.

176

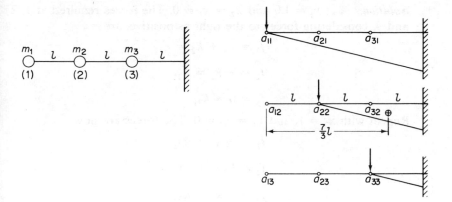

Figure 6.1-1.

Fig. 6.1-1 as follows

$$a_{12} = \frac{1}{EI}\left[\frac{1}{2}(2l)^2 \times \frac{7}{3}l\right] = \frac{14}{3}\frac{l^3}{EI}$$

The other values (determined as above) are

$$a_{11} = \frac{27}{3}\frac{l^3}{EI} \qquad a_{21} = a_{12} = \frac{14}{3}\frac{l^3}{EI}$$

$$a_{22} = \frac{8}{3}\frac{l^3}{EI} \qquad a_{23} = a_{32} = \frac{2.5}{3}\frac{l^3}{EI}$$

$$a_{33} = \frac{1}{3}\frac{l^3}{EI} \qquad a_{13} = a_{31} = \frac{4}{3}\frac{l^3}{EI}$$

The flexibility matrix can now be written as

$$a = \frac{l^3}{3EI}\begin{bmatrix} 27 & 14 & 4 \\ 14 & 8 & 2.5 \\ 4 & 2.5 & 1 \end{bmatrix}$$

and the symmetry about the diagonal should be noted.

EXAMPLE 6.1-2

Fig. 6.1-2 shows a three degrees of freedom system. Determine the stiffness matrix.

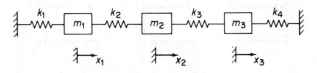

Figure 6.1-2

Solution: Let $x_1 = 1.0$ and $x_2 = x_3 = 0$. The forces required at 1, 2 and 3, considering forces to the right as positive, are

$$f_1 = k_1 + k_2 = k_{11}$$

$$f_2 = -k_2 = k_{21}$$

$$f_3 = 0 = k_{31}$$

Repeat with $x_2 = 1$, and $x_1 = x_3 = 0$. The forces are now

$$f_1 = -k_2 = k_{12}$$

$$f_2 = k_2 + k_3 = k_{22}$$

$$f_3 = -k_3 = k_{32}$$

For the last column of k's, let $x_3 = 1$ and $x_1 = x_2 = 0$. The forces are

$$f_1 = 0 = k_{13}$$

$$f_2 = -k_3 = k_{23}$$

$$f_3 = k_3 + k_4 = k_{33}$$

The stiffness matrix can now be written as

$$K = \begin{bmatrix} (k_1 + k_2) & -k_2 & 0 \\ -k_2 & (k_2 + k_3) & -k_3 \\ 0 & -k_3 & (k_3 + k_4) \end{bmatrix}$$

EXAMPLE 6.1-3

Consider the four-story building with rigid floors shown in Fig. 6.1-3. Show diagramatically the significance of the terms of the stiffness matrix.

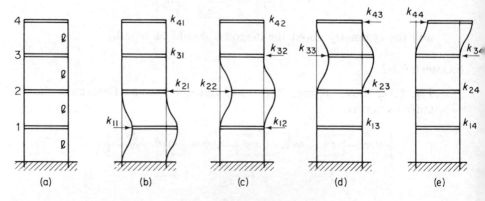

Figure 6.1-3.

Solution: The stiffness matrix for the problem is a 4×4. The elements of the first column are obtained by giving station 1 a unit displacement with the displacement of all other stations equal to zero, as shown in Fig. 6.1-3b. The forces required for this configuration are the elements of the first column. Similarly, the elements of the second column are the forces necessary to maintain the configuration shown in Fig. 6.1-3c, etc.

It is evident from these diagrams that $k_{11} = k_{22} = k_{33}$ and that they can be determined from the deflection of a fixed-fixed beam of length $2l$, which is

$$k_{11} = k_{22} = k_{33} = \frac{192EI}{(2l)^3} = 24\frac{EI}{l^3}$$

The stiffness matrix is then easily found as

$$[k] = \frac{EI}{l^3} \begin{bmatrix} 24 & -12 & 0 & 0 \\ -12 & 24 & -12 & 0 \\ 0 & -12 & 24 & -12 \\ 0 & 0 & -12 & 12 \end{bmatrix}$$

EXAMPLE 6.1-4

Determine the stiffness matrix for the square frame shown in Fig. 6.1-4. Assume the corners to remain at 90°.

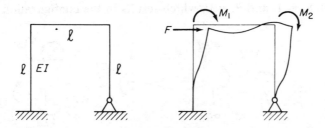

Figure 6.1-4.

Solution: With the applied forces equal to F, M_1, and M_2, the displacements of the corners are u, θ_1, and θ_2, and the stiffness matrix relating the force to the displacement is

$$\begin{Bmatrix} F \\ M_1 \\ M_2 \end{Bmatrix} = \begin{bmatrix} k_{11} & k_{12} & k_{13} \\ k_{21} & k_{22} & k_{23} \\ k_{31} & k_{32} & k_{33} \end{bmatrix} \begin{Bmatrix} u \\ \theta_1 \\ \theta_2 \end{Bmatrix}$$

The first column of the stiffness matrix can be determined by letting $u = 1$ and $\theta_1 = \theta_2 = 0$, which results in the configuration (a).

Considering each member and making use of Table 6.1-1, we have

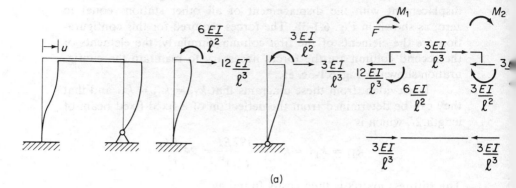

(a)

$$\left\{\begin{array}{c} F \\ M_1 \\ M_2 \end{array}\right\} = \left[\begin{array}{ccc} \dfrac{15EI}{l^3} & 0 & 0 \\ -\dfrac{6EI}{l^2} & 0 & 0 \\ -\dfrac{3EI}{l^2} & 0 & 0 \end{array}\right] \left\{\begin{array}{c} u \\ 0 \\ 0 \end{array}\right\}$$

The second column of the stiffness matrix is determined by letting $u = 0$, $\theta_1 = 1$, and $\theta_2 = 0$, which results in the configuration (b)

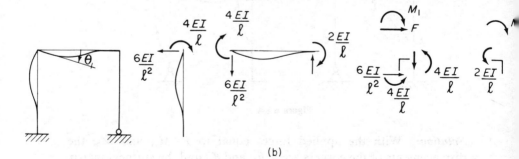

(b)

$$\left\{\begin{array}{c} F \\ M_1 \\ M_2 \end{array}\right\} = \left[\begin{array}{ccc} 0 & -\dfrac{6EI}{l^2} & 0 \\ 0 & \dfrac{8EI}{l} & 0 \\ 0 & \dfrac{2EI}{l} & 0 \end{array}\right] \left\{\begin{array}{c} 0 \\ \theta_1 \\ 0 \end{array}\right\}$$

In like manner, the determination of the third column is given as follows:

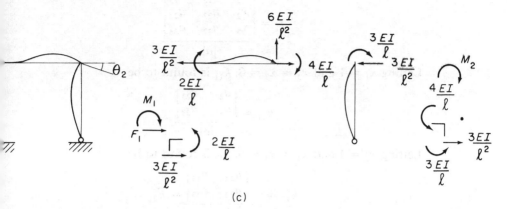

(c)

$$\left\{ \begin{array}{c} F \\ M_1 \\ M_2 \end{array} \right\} = \left[\begin{array}{ccc} 0 & 0 & -\dfrac{3EI}{l^2} \\ 0 & 0 & \dfrac{2EI}{l} \\ 0 & 0 & \dfrac{7EI}{l} \end{array} \right] \left\{ \begin{array}{c} 0 \\ 0 \\ \theta_2 \end{array} \right\}$$

The system stiffness is found from the superposition of the above three configurations, which becomes

$$\left\{ \begin{array}{c} F \\ M_1 \\ M_2 \end{array} \right\} = \dfrac{EI}{l} \left[\begin{array}{ccc} \dfrac{15}{l^2} & -\dfrac{6}{l} & -\dfrac{3}{l} \\ -\dfrac{6}{l} & 8 & 2 \\ -\dfrac{3}{l} & 2 & 7 \end{array} \right] \left\{ \begin{array}{c} u \\ \theta_1 \\ \theta_2 \end{array} \right\}$$

It should be noted that the matrix is symmetric about the diagonal.

EXAMPLE 6.1-5

Given the equation

$$\left\{ \begin{array}{c} x_1 \\ x_2 \\ x_3 \end{array} \right\} = \left[\begin{array}{ccc} a_{11} & a_{12} & a_{13} \\ a_{21} & a_{22} & a_{23} \\ a_{31} & a_{32} & a_{33} \end{array} \right] \left\{ \begin{array}{c} f_1 \\ f_2 \\ f_3 \end{array} \right\}$$

determine the stiffness matrix from Cramer's rule.

Solution: From Cramer's rule, f_1 can be written as

$$f_1 = \frac{\begin{vmatrix} x_1 & a_{12} & a_{13} \\ x_2 & a_{22} & a_{23} \\ x_3 & a_{32} & a_{33} \end{vmatrix}}{|a|}$$

Letting $x_1 = 1$ and $x_2 = x_3 = 0$, k_{11} is found to be

$$k_{11} = \frac{\begin{vmatrix} a_{22} & a_{23} \\ a_{32} & a_{33} \end{vmatrix}}{|a|}$$

Letting $x_2 = 1$ and $x_1 = x_3 = 0$, k_{12} is found to be

$$k_{12} = \frac{-\begin{vmatrix} a_{12} & a_{13} \\ a_{32} & a_{33} \end{vmatrix}}{|a|} = k_{21}$$

Similarly, all other terms can be found. It should be noted that the above procedure is simply that of inverting the matrix $[a]$.

6.2   RECIPROCITY THEOREM

The reciprocity theorem states that in a linear system $a_{ij} = a_{ji}$. For the proof of this theorem, we consider the work done by forces f_i and f_j where the order of loading is i followed by j and then by its reverse. Reciprocity results when we recognize that the work done is independent of the order of loading.

Applying f_i, the work done is $\frac{1}{2}f_i^2 a_{ii}$. Applying f_j, the work done by f_j is $\frac{1}{2}f_j^2 a_{jj}$. However, i undergoes further displacement $a_{ij}f_j$, and the additional work done by f_i becomes $a_{ij}f_j f_i$. Thus the total work done is

$$W = \tfrac{1}{2}f_i^2 a_{ii} + \tfrac{1}{2}f_j^2 a_{jj} + a_{ij}f_j f_i \tag{6.2-1}$$

We now reverse the order of loading, in which case the total work done is

$$W = \tfrac{1}{2}f_j^2 a_{jj} + \tfrac{1}{2}f_i^2 a_{ii} + a_{ji}f_i f_j \tag{6.2-2}$$

Since the work done in the two cases must be equal, we find that

$$a_{ij} = a_{ji} \tag{6.2-3}$$

EXAMPLE 6.2-1

Fig. 6.2-1 shows an overhanging beam with P first applied at 1 and then at 2. In (a) the deflection at 2 is

$$y_2 = a_{21}P$$

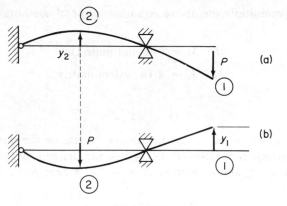

Figure 6.2-1.

In (b) the deflection at 1 is

$$y_1 = a_{12}P$$

Since $a_{12} = a_{21}$, y_1 will equal y_2.

6.3 EIGENVALUES AND EIGENVECTORS

For the free vibration of the undamped system of several degrees of freedom, the equations of motion expressed in matrix form become

$$[M]\{\ddot{x}\} + [k]\{x\} = \{0\} \qquad (6.3\text{-}1)$$

where

$$M = \begin{bmatrix} m_{11} & m_{12} \cdots \\ & \vdots \\ m_{n1} & m_{n2} \cdots m_{nn} \end{bmatrix} = \text{mass matrix (a square matrix)}$$

$$K = \begin{bmatrix} k_{11} & k_{12} \cdots \\ & \vdots \\ k_{n1} & k_{n2} \cdots k_{nn} \end{bmatrix} = \text{stiffness matrix (a square matrix)}$$

$$X = \begin{Bmatrix} x_1 \\ x_2 \\ \vdots \\ \vdots \\ x_n \end{Bmatrix} = \text{displacement vector (a column matrix)}$$

When there is no ambiguity, we will dispense with the brackets and braces and use capital letters and simply write the matrix equation as

$$M\ddot{X} + KX = 0 \qquad (6.3\text{-}2)$$

If we premultiply the above equation by M^{-1}, we obtain the following terms:

$$M^{-1}M = I \text{ (a unit matrix)}$$

$$M^{-1}K = A \text{ (a system matrix)}$$

and

$$I\ddot{X} + AX = 0 \tag{6.3-3}$$

The matrix A is referred to as the *system matrix*, or the dynamic matrix since the dynamic properties of the system are defined by this matrix.

Assuming harmonic motion $\ddot{X} = -\lambda X$, where $\lambda = \omega^2$, Eq. (6.3-3) becomes

$$[A - \lambda I]\{X\} = 0 \tag{6.3-4}$$

The *characteristic equation* of the system is the determinant equated to zero, or

$$|A - \lambda I| = 0 \tag{6.3-5}$$

The roots λ_i of the characteristic equation are called *eigenvalues*, and the natural frequencies of the system are determined from them by the relationship

$$\lambda_i = \omega_i^2 \tag{6.3-6}$$

By substituting λ_i into the matrix equation, Eq. (6.3-4), we obtain the corresponding mode shape X_i which is called the *eigenvector*. Thus for an n-degrees of freedom system, there will be n eigenvalues and n eigenvectors.

It is also possible to find the eigenvectors from the adjoint matrix (see Appendix C) of the system. If, for conciseness, we make the abbreviation $B = A - \lambda I$ and start with the definition of the inverse

$$B^{-1} = \frac{1}{|B|} \text{adj } B \tag{6.3-7}$$

we can premultiply by $|B|B$ to obtain

$$|B|I = B \text{ adj } B$$

or in terms of the original expression for B

$$|A - \lambda I|I = [A - \lambda I]\text{adj}[A - \lambda I] \tag{6.3-8}$$

If now we let $\lambda = \lambda_i$, an eigenvalue, then the determinant on the left side of the equation is zero and we obtain

$$[0] = [A - \lambda_i I]\text{adj}[A - \lambda_i I] \tag{6.3-9}$$

The above equation is valid for all λ_i and represents n equations for the n-degrees of freedom system. Comparing this equation with Eq. (6.3-4) for the ith mode

$$[A - \lambda_i I]\{X\}_i = 0$$

we recognize that the adjoint matrix, adj$[A - \lambda_i I]$, must consist of columns, each of which is the eigenvector X_i (multiplied by an arbitrary constant). Eigenvalues and eigenvectors can be calculated for any symmetric matrix by standard subroutine programs.

EXAMPLE 6.3-1

Consider the two-story building shown in Fig. 6.3-1. The equation of motion can be expressed in matrix notation as

$$\begin{bmatrix} 2m & 0 \\ 0 & m \end{bmatrix}\begin{Bmatrix} \ddot{x}_1 \\ \ddot{x}_2 \end{Bmatrix} + \begin{bmatrix} 3k & -k \\ -k & k \end{bmatrix}\begin{Bmatrix} x_1 \\ x_2 \end{Bmatrix} = \begin{Bmatrix} 0 \\ 0 \end{Bmatrix} \qquad (a)$$

Premultiplying by the inverse of the mass matrix

$$M^{-1} = \begin{bmatrix} \dfrac{1}{2m} & 0 \\ 0 & \dfrac{1}{m} \end{bmatrix}$$

and letting $\lambda = \omega^2$, Eq. (a) becomes

$$\begin{bmatrix} \left(\dfrac{3}{2}\dfrac{k}{m} - \lambda\right) & -\dfrac{k}{2m} \\ -\dfrac{k}{m} & \left(\dfrac{k}{m} - \lambda\right) \end{bmatrix}\begin{Bmatrix} x_1 \\ x_2 \end{Bmatrix} = \begin{Bmatrix} 0 \\ 0 \end{Bmatrix} \qquad (b)$$

The characteristic equation from the determinant of the above matrix is

$$\lambda^2 - \frac{5}{2}\frac{k}{m}\lambda + \left(\frac{k}{m}\right)^2 = 0 \qquad (c)$$

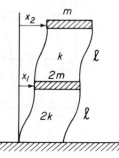

Figure 6.3-1.

from which the eigenvalues are found to be

$$\lambda_1 = \frac{1}{2}\frac{k}{m}$$

$$\lambda_2 = 2\frac{k}{m}$$

(d)

The eigenvectors can be found from Eq. (b) by substituting the above values of λ. We will, however, illustrate the use of the adjoint matrix in their evaluation.

The adjoint matrix from Eq. (b) is

$$\mathrm{adj}[A - \lambda I] = \begin{bmatrix} \left(\dfrac{k}{m} - \lambda_i\right) & \dfrac{k}{2m} \\ \dfrac{k}{m} & \left(\dfrac{3}{2}\dfrac{k}{m} - \lambda_i\right) \end{bmatrix}$$

(e)

Substituting $\lambda_i = \frac{1}{2}(k/m)$, we obtain from (e)

$$\begin{bmatrix} 0.50 & 0.50 \\ 1.00 & 1.00 \end{bmatrix}\frac{k}{m}$$

Here each column is already normalized to unity and the first eigenvector is

$$X_i = \left\{ \begin{matrix} 0.50 \\ 1.00 \end{matrix} \right\}$$

Similarly, when $\lambda_2 = 2(k/m)$ is used, the adjoint matrix gives

$$\begin{bmatrix} -1.0 & 0.5 \\ 1.0 & -0.5 \end{bmatrix}\frac{k}{m}$$

Normalizing to unity, the second eigenvector from either column is

$$X_2 = \left\{ \begin{matrix} -1.00 \\ 1.00 \end{matrix} \right\}$$

The two normal modes are shown in Fig. 6.3-2.

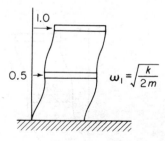
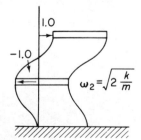

Figure 6.3-2.

6.4 EQUATIONS BASED ON FLEXIBILITY

In the previous section the characteristic equation was established from the equations of motion based on the stiffness matrix. It is also possible to arrive at the eigenvalues and eigenvectors starting from the flexibility matrix.

Rewriting Eq. (6.1-2)

$$\begin{Bmatrix} x_1 \\ x_2 \\ x_3 \end{Bmatrix} = \begin{bmatrix} a_{11} & a_{12} & a_{13} \\ a_{21} & a_{22} & a_{23} \\ a_{31} & a_{32} & a_{33} \end{bmatrix} \begin{Bmatrix} f_1 \\ f_2 \\ f_3 \end{Bmatrix} \qquad (6.1\text{-}2)$$

we assume harmonic motion and replace the forces by the inertia forces $f_i = -m_i \ddot{x}_i = \omega^2 m_i x_i$. The above equation then becomes

$$\begin{Bmatrix} x_1 \\ x_2 \\ x_3 \end{Bmatrix} = \omega^2 \begin{bmatrix} a_{11}m_1 & a_{12}m_2 & a_{13}m_3 \\ a_{21}m_1 & a_{22}m_2 & a_{23}m_3 \\ a_{31}m_1 & a_{32}m_2 & a_{33}m_3 \end{bmatrix} \begin{Bmatrix} x_1 \\ x_2 \\ x_3 \end{Bmatrix} \qquad (6.4\text{-}1)$$

which may be rearranged to

$$[am - \lambda I] X = 0, \qquad \lambda = \frac{1}{\omega^2}$$

or

$$\begin{bmatrix} \left(a_{11}m_1 - \dfrac{1}{\omega^2}\right) & a_{12}m_2 & a_{13}m_3 \\[2ex] a_{21}m_1 & \left(a_{22}m_2 - \dfrac{1}{\omega^2}\right) & a_{23}m_3 \\[2ex] a_{31}m_1 & a_{32}m_2 & \left(a_{33}m_3 - \dfrac{1}{\omega^2}\right) \end{bmatrix} \begin{Bmatrix} x_1 \\ x_2 \\ x_3 \end{Bmatrix} = \begin{Bmatrix} 0 \\ 0 \\ 0 \end{Bmatrix}$$

$$(6.4\text{-}2)$$

The characteristic equation is then the determinant of the square matrix above. The eigenvalues in this case are equal to $\lambda = 1/\omega^2$ instead of ω^2.

EXAMPLE 6.4-1

Using the flexibility influence coefficients of Example (6.1-1), determine the matrix equation for the normal modes of the system shown in Fig. 6.1-1.

Solution: The inverse of the system matrix A is

$$A^{-1} = K^{-1}M = [a]M$$

$$= \frac{l^3}{3EI} \begin{bmatrix} 27 & 14 & 4 \\ 14 & 8 & 2.5 \\ 4 & 2.5 & 1 \end{bmatrix} \begin{bmatrix} m_1 & 0 & 0 \\ 0 & m_2 & 0 \\ 0 & 0 & m_3 \end{bmatrix}$$

Equation (6.4-2) then becomes

$$\left[[a][m] - \frac{1}{\omega^2} \begin{bmatrix} 1 & 0 & 0 \\ 0 & 1 & 0 \\ 0 & 0 & 1 \end{bmatrix} \right] \{X\} = 0$$

or

$$\begin{Bmatrix} x_1 \\ x_2 \\ x_3 \end{Bmatrix} = \frac{\omega^2 l^3}{3EI} \begin{bmatrix} 27 & 14 & 4 \\ 14 & 8 & 2.5 \\ 4 & 2.5 & 1 \end{bmatrix} \begin{bmatrix} m_1 & 0 & 0 \\ 0 & m_2 & 0 \\ 0 & 0 & m_3 \end{bmatrix} \begin{Bmatrix} x_1 \\ x_2 \\ x_3 \end{Bmatrix}$$

6.5   ORTHOGONAL PROPERTIES OF THE EIGENVECTORS

The normal modes, or the eigenvectors of the system, can be shown to be *orthogonal* with respect to the mass and stiffness matrices as follows. Let the equation for the ith mode be

$$KX_i = \lambda_i MX_i \tag{6.5-1}$$

Premultiply by the transpose of mode j

$$X_j' KX_i = \lambda_i(X_j' MX_i) \tag{6.5-2}$$

Next, start with the equation for the jth mode and premultiply by X_i' to obtain

$$X_i' KX_j = \lambda_j(X_i' MX_j) \tag{6.5-3}$$

Since K and M are symmetric matrices, the following relationships hold*

$$X_j' MX_i = X_i' MX_j$$
$$X_j' KX_i = X_i' KX_j \tag{6.5-4}$$

Thus, subtracting Eq. (6.5-3) from Eq. (6.5-2), we obtain

$$0 = (\lambda_i - \lambda_j)X_i' MX_j \tag{6.5-5}$$

*See Appendix C.

If $\lambda_i \neq \lambda_j$, the above equation requires that

$$X_i' M X_j = 0 \qquad (6.5\text{-}6)$$

It is also evident from Eq. (6.5-2) or Eq. (6.5-3) that as a consequence of Eq. (6.5-6)

$$X_i' K X_j = 0 \qquad (6.5\text{-}7)$$

Equations (6.5-6) and (6.5-7) define the *orthogonal* character of the normal modes.

Finally, if $i = j$, Eq. (6.5-5) is satisfied for any finite value of the products given by Eqs. (6.5-6) or (6.5-7). We therefore let

$$
\begin{aligned}
X_i' M X_i &= M_i \\
X_i' K X_i &= K_i
\end{aligned}
\qquad (6.5\text{-}8)
$$

These are called the *generalized mass* and the *generalized stiffness* respectively.

EXAMPLE 6.5-1

Verify that the two normal modes of the system considered in Example 6.3-1 are orthogonal.

Solution: The mass matrix and the two normal modes are

$$M = \begin{bmatrix} 2m & 0 \\ 0 & m \end{bmatrix} \qquad X_1 = \begin{Bmatrix} 0.50 \\ 1.00 \end{Bmatrix} \qquad X_2 = \begin{Bmatrix} -1.00 \\ 1.00 \end{Bmatrix}$$

Substituting into Eq. (6.5-6), we have

$$X_1' M X_2 = (0.50 \quad 1.00) \begin{bmatrix} 2m & 0 \\ 0 & m \end{bmatrix} \begin{Bmatrix} -1.00 \\ 1.00 \end{Bmatrix}$$

$$= (0.50 \quad 1.00) \begin{Bmatrix} -2m \\ m \end{Bmatrix} = -m + m = 0$$

The student should verify that $X_1' K X_2$ also equals zero.

EXAMPLE 6.5-2

Consider the problem of initiating the free vibration of a system from a specified distribution of the displacement. As previously stated, free vibrations are the superposition of normal modes, and we wish to determine how much of each mode will be present.

Solution: We will first express the displacement at time zero by the equation

$$u(o) = c_1 X_1 + c_2 X_2 + \cdots c_i X_i + \cdots$$

where X_i are the normal modes and c_i are the coefficients indicating how much of each mode is present. Premultiplying the above equation by $X_i'M$ and taking note of the orthogonal property of X_i, we obtain

$$X_i'Mu(0) = 0 + 0 + \cdots c_i X_i'MX_i + 0 + \cdots$$

The coefficient c_i of any mode is then found as

$$c_i = \frac{X_i'\,Mu(0)}{X_i'MX_i}$$

6.6 REPEATED ROOTS

When repeated roots are found in the characteristic equation, the corresponding eigenvectors are not unique, and a linear combination of such eigenvectors may also satisfy the equation of motion. To illustrate this point, let X_1 and X_2 be eigenvectors belonging to a common eigenvalue λ_0, and X_3 be a third eigenvector belonging to λ_3 that is different from λ_0. We can then write

$$AX_1 = \lambda_0 X_1$$
$$AX_2 = \lambda_0 X_2 \qquad\qquad (6.6\text{-}1)$$
$$AX_3 = \lambda_3 X_3$$

By multiplying the second equation by a constant b and adding it to the first, we obtain another equation

$$A(X_1 + bX_2) = \lambda_0(X_1 + bX_2) \qquad\qquad (6.6\text{-}2)$$

Thus a new eigenvector, $X_{12} = X_1 + bX_2$, which is a linear combination of the first two, also satisfies the basic equation

$$AX_{12} = \lambda_0 X_{12} \qquad\qquad (6.6\text{-}3)$$

and hence no unique mode exists for λ_0.

Any of the modes corresponding to λ_0 must be orthogonal to X_3 if they are to be a normal mode. If all three modes are orthogonal, they are linearly independent and may be combined to describe the free vibration resulting from any initial condition.

EXAMPLE 6.6-1

Consider the system shown in Fig. 6.6-1 where the connecting bar is rigid and negligible in weight.

The two normal modes of vibration are shown to be translation and rotation, which are orthogonal. The natural frequencies for the

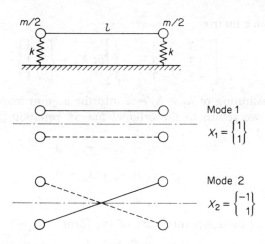

Figure 6.6-1.

two modes, however, are equal and can be calculated to be

$$\omega_n = \sqrt{\frac{2k}{m}}$$

The example illustrates that different eigenvectors may have equal eigenvalues.

EXAMPLE 6.6-2

Determine the eigenvalues and eigenvectors when

$$A = \begin{bmatrix} 0 & -1 & 1 \\ -1 & 0 & 1 \\ 1 & 1 & 0 \end{bmatrix}$$

Solution: The characteristic equation $|A - \lambda I| = 0$ yields

$$(\lambda - 1)^2(\lambda + 2) = 0$$

so the eigenvalues are $\lambda_1 = 1$, $\lambda_2 = 1$, and $\lambda_3 = -2$.
    Forming the adjoint matrix

$$\text{adj}[A - \lambda I] = \begin{bmatrix} (\lambda^2 - 1) & -(\lambda - 1) & (\lambda - 1) \\ -(\lambda - 1) & (\lambda^2 - 1) & (\lambda - 1) \\ (\lambda - 1) & (\lambda - 1) & (\lambda^2 - 1) \end{bmatrix}$$

the eigenvector corresponding to $\lambda_3 = -2$ is found from any column of the above matrix

$$\begin{bmatrix} 3 & 3 & -3 \\ 3 & 3 & -3 \\ -3 & -3 & 3 \end{bmatrix} \text{ or } X_3 = \left\{ \begin{array}{c} -1 \\ -1 \\ 1 \end{array} \right\}$$

Substitution of $\lambda_1 = \lambda_2 = 1$ into the adjoint matrix leads to all zeros, so we return to the original matrix equation $[A - \lambda I]X = 0$ with $\lambda = 1$

$$-x_1 - x_2 + x_3 = 0$$

$$-x_1 - x_2 + x_3 = 0$$

$$x_1 + x_2 - x_3 = 0$$

All three of these equations are of the form

$$x_1 = x_3 - x_2$$

and hence for the eigenvalue X_1 corresponding to $\lambda_1 = \lambda_2 = 1$ we can write

$$X_1 = \left\{ \begin{array}{c} x_3 - x_2 \\ x_2 \\ x_3 \end{array} \right\}$$

which is found to be orthogonal to X_3 for all values of x_2 and x_3, i.e.

$$(X_3)\{X_1\} = 0$$

Thus for $x_2 = x_3 = 1$, one could obtain

$$X_1 = \left\{ \begin{array}{c} 0 \\ 1 \\ 1 \end{array} \right\}$$

and for $x_3 = 1$ and $x_2 = -1$ the second eigenvector could be

$$X_2 = \left\{ \begin{array}{c} 2 \\ -1 \\ 1 \end{array} \right\}$$

As shown previously by Eq. (6.6-2), X_1 and X_2 are not unique, and any linear combination of X_1 and X_2 will also satisfy the original matrix equation.

6.7 MODAL MATRIX P

In Chapter 5 we found that static or dynamic coupling results from the choice of coordinates, and that for an undamped system, there exists a set of principal coordinates that will express the equations of motion in the uncoupled form. Such uncoupled coordinates are desirable since each equation can be solved independently of the others.

For a lumped mass multidegrees of freedom system, coordinates chosen at each mass point will result in a mass matrix that is diagonal, but the stiffness matrix will contain off-diagonal terms indicating static coupling. Coordinates chosen in another way may result in dynamic coupling or both dynamic and static coupling.

It is possible to uncouple the equations of motion of an n-degrees of freedom system, provided we know beforehand the normal modes of the system. When the n normal modes (or eigenvectors) are assembled into a square matrix with each normal mode represented by a column, we call it the *modal matrix P*. Thus the modal matrix for a three degrees of freedom system may appear as

$$P = \left[\left\{ \begin{matrix} x_1 \\ x_2 \\ x_3 \end{matrix} \right\}_1 \left\{ \begin{matrix} x_1 \\ x_2 \\ x_3 \end{matrix} \right\}_2 \left\{ \begin{matrix} x_1 \\ x_2 \\ x_3 \end{matrix} \right\}_3 \right] = [X_1 \, X_2 \, X_3] \qquad (6.7\text{-}1)$$

The modal matrix makes it possible to include all the orthogonality relations of Sec. 6.5 into one equation. For this operation we need also the transpose of P, which is

$$P' = \left[\begin{matrix} (x_1 \, x_2 \, x_3)_1 \\ (x_1 \, x_2 \, x_3)_2 \\ (x_1 \, x_2 \, x_3)_3 \end{matrix} \right] = [X_1 \, X_2 \, X_3]' \qquad (6.7\text{-}2)$$

with each row corresponding to a mode. If we now form the product $P'MP$ or $P'KP$, the result will be a diagonal matrix since the off-diagonal terms simply express the orthogonality relations which are zero.

As an example consider a two degrees of freedom system. Performing the indicated operation with the modal matrix, we have

$$P'MP = [X_1 \ X_2]'[M][X_1 \ X_2]$$

$$= \begin{bmatrix} X_1'MX_1 & X_1'MX_2 \\ X_2'MX_1 & X_2'MX_2 \end{bmatrix} \qquad (6.7\text{-}3)$$

$$= \begin{bmatrix} M_1 & 0 \\ 0 & M_2 \end{bmatrix}$$

In the above equation, the off-diagonal terms are zero because of orthogonality, and the diagonal terms are the generalized mass M_i.

It is evident that a similar formulation applies also to the stiffness matrix K that results in the following equation

$$P'KP = \begin{bmatrix} K_1 & 0 \\ 0 & K_2 \end{bmatrix} \tag{6.7-4}$$

The diagonal terms here are the generalized stiffness K_i.

If each of the columns of the modal matrix P is divided by the square root of the generalized mass M_i, the new matrix is called the *weighted modal matrix* and designated as $\tilde{P}$. It is easily seen that the diagonalization of the mass matrix by the weighted modal matrix results in the unit matrix

$$\tilde{P}'M\tilde{P} = I \tag{6.7-5}$$

Since $M_i^{-1}K_i = \lambda_i$, the stiffness matrix treated similarly by the weighted modal matrix becomes a diagonal matrix of the eigenvalues

$$\tilde{P}'K\tilde{P} = \begin{bmatrix} \lambda_1 & & & 0 \\ & \lambda_2 & & \\ & & \ddots & \\ 0 & & & \lambda_n \end{bmatrix} = \Lambda \tag{6.7-6}$$

EXAMPLE 6.7-1

Consider the symmetrical two degrees of freedom system shown in Fig. 6.7-1. The equation of motion in matrix form is

$$\begin{bmatrix} m & 0 \\ 0 & m \end{bmatrix} \begin{Bmatrix} \ddot{x}_1 \\ \ddot{x}_2 \end{Bmatrix} + \begin{bmatrix} 2k & -k \\ -k & 2k \end{bmatrix} \begin{Bmatrix} x_1 \\ x_2 \end{Bmatrix} = \{0\} \tag{a}$$

and the eigenvalues and eigenvectors can be shown to equal

$$\lambda_1 = \omega_1^2 = \frac{k}{m} \qquad \lambda_2 = \omega_2^2 = 3\frac{k}{m} \tag{b}$$

$$\begin{Bmatrix} x_1 \\ x_2 \end{Bmatrix}_{\lambda_1} = \begin{Bmatrix} 1 \\ 1 \end{Bmatrix}, \quad \begin{Bmatrix} x_1 \\ x_2 \end{Bmatrix}_{\lambda_2} = \begin{Bmatrix} -1 \\ 1 \end{Bmatrix}$$

The generalized mass for both modes is $2m$, and the modal matrix and the weighted modal matrix are

$$P = \begin{bmatrix} 1 & -1 \\ 1 & 1 \end{bmatrix} \qquad \tilde{P} = \frac{1}{\sqrt{2m}} \begin{bmatrix} 1 & -1 \\ 1 & 1 \end{bmatrix} \tag{c}$$

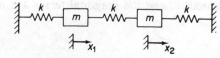

Figure 6.7-1.

To decouple the original equation we will use $\tilde{P}$ in the transformation

$$\left\{ \begin{matrix} x_1 \\ x_2 \end{matrix} \right\} = \frac{1}{\sqrt{2m}} \begin{bmatrix} 1 & -1 \\ 1 & 1 \end{bmatrix} \left\{ \begin{matrix} y_1 \\ y_2 \end{matrix} \right\} \tag{d}$$

and premultiply by $\tilde{P}'$ to obtain

$$\tilde{P}' M \tilde{P} \ddot{Y} + \tilde{P}' K \tilde{P} Y = 0$$

or

$$\begin{bmatrix} 1 & 0 \\ 0 & 1 \end{bmatrix} \left\{ \begin{matrix} \ddot{y}_1 \\ \ddot{y}_2 \end{matrix} \right\} + \frac{k}{m} \begin{bmatrix} 1 & 0 \\ 0 & 3 \end{bmatrix} \left\{ \begin{matrix} y_1 \\ y_2 \end{matrix} \right\} = 0 \tag{e}$$

Thus Eq. (a) has been transformed to the uncoupled Eq. (e) by the coordinate transformation of Eq. (d). The coordinates y_1 and y_2 are referred to as *principal or normal coordinates*.

The above equations in terms of the normal coordinates are similar to those of the single degree of freedom system and can be written as

$$\ddot{y}_i + \omega_i^2 y_i = 0 \tag{f}$$

Its general solution has been previously discussed and is

$$y_i(t) = y_i(0) \cos \omega_i t + \frac{1}{\omega_i} \dot{y}_i(0) \sin \omega_i t \tag{g}$$

The solution of the original two degrees of freedom system is then given by Eq. (d) to be

$$x_1(t) = \frac{1}{\sqrt{2m}} [y_1(t) - y_2(t)]$$

$$x_2(t) = \frac{1}{\sqrt{2m}} [y_1(t) + y_2(t)] \tag{h}$$

which is the sum of the normal mode solutions multiplied by appropriate constants. The initial conditions $y_i(0)$ and $\dot{y}_i(0)$ can be transformed in terms of $x_i(0)$ and $\dot{x}_i(0)$ by the inverse of Eq. (d), $y(0) = \tilde{P}^{-1} x(0)$. However, it is not necessary to carry out the inversion of $\tilde{P}$. Since $\tilde{P}' M \tilde{P} = I$, post multiplying this equation by $\tilde{P}^{-1}$, we obtain

$$\tilde{P}' M = \tilde{P}^{-1}$$

Thus the inverse of Eq. (d) becomes

$$y = \tilde{P}^{-1} x = \tilde{P}' M x$$

or

$$\left\{ \begin{array}{c} y_1(0) \\ y_2(0) \end{array} \right\} = \frac{\sqrt{2m}}{2} \left[\begin{array}{cc} 1 & 1 \\ -1 & 1 \end{array} \right] \left\{ \begin{array}{c} x_1(0) \\ x_2(0) \end{array} \right\}$$

Equation (g) can therefore be written in terms of the initial conditions $x(0)$ and $\dot{x}(0)$ as

$$y_1(t) = \frac{1}{2}\sqrt{2m} \left\{ \left[x_1(0) + x_2(0) \right] \cos \omega_1 t \right.$$

$$\left. + \frac{1}{\omega_1} \left[\dot{x}_1(0) + \dot{x}_2(0) \right] \sin \omega_1 t \right\}$$

$$y_2(t) = \frac{1}{2}\sqrt{2m} \left\{ \left[-x_1(0) + x_2(0) \right] \cos \omega_2 t \right.$$

$$\left. + \frac{1}{\omega_2} \left[-\dot{x}_1(0) + \dot{x}_2(0) \right] \sin \omega_2 t \right\}$$

Substituting into Eq. (h), the solution is entirely in terms of the original coordinates.

6.8   MODAL DAMPING IN FORCED VIBRATION

The equation of motion of an n-degree of freedom system with viscous damping and arbitrary excitation $F(t)$ can be presented in the matrix form

$$M\ddot{X} + C\dot{X} + KX = F \qquad (6.8\text{-}1)$$

It is generally a set of n coupled equations.

We have found that the solution of the homogeneous undamped equation

$$M\ddot{X} + KX = 0 \qquad (6.8\text{-}2)$$

leads to the eigenvalues and eigenvectors which describe the normal modes of the system and the modal matrix P or $\tilde{P}$. If we let $X = \tilde{P}Y$ and premultipy Eq. (6.8-1) by $\tilde{P}'$ as in Sec. 6.7, we obtain

$$\tilde{P}'M\tilde{P}\ddot{Y} + \tilde{P}'C\tilde{P}\dot{Y} + \tilde{P}'K\tilde{P}Y = \tilde{P}'F \qquad (6.8\text{-}3)$$

We have already shown that $\tilde{P}'M\tilde{P}$ and $\tilde{P}'K\tilde{P}$ are diagonal matrices. In general, $\tilde{P}'C\tilde{P}$ is not diagonal and the above equation is coupled by the damping matrix.

If C is proportional to M or K, it is evident that $\tilde{P}'C\tilde{P}$ becomes diagonal, in which case we can say that the system has *proportional damping*. Eq. (6.8-3) is then completely uncoupled and its ith equation will have the form

$$\ddot{y}_i + 2\zeta_i\omega_i\dot{y}_i + \omega_i^2 y_i = \tilde{f}_i(t) \tag{6.8-4}$$

Thus instead of n coupled equations we would have n uncoupled equations similar to that of a single degree of freedom system.

Rayleigh Damping. Rayleigh introduced proportional damping in the form

$$C = \alpha M + \beta K \tag{6.8-5}$$

where α and β are constants. The application of the weighted modal matrix $\tilde{P}$ here results in

$$\tilde{P}'C\tilde{P} = \alpha\tilde{P}'M\tilde{P} + \beta\tilde{P}'K\tilde{P}$$
$$= \alpha I + \beta\Lambda \tag{6.8-6}^*$$

where I is a unit matrix and Λ is a diagonal matrix of the eigenvalues [see Eq. (6.7-6)].

$$\Lambda = \begin{bmatrix} \omega_1^2 & & & \\ & \omega_2^2 & & \\ & & \ddots & \\ & & & \omega_n^2 \end{bmatrix} \tag{6.8-7}$$

Thus instead of Eq. (6.8-4), we obtain for the ith equation

$$\ddot{y}_i + (\alpha + \beta\omega_i^2)\dot{y}_i + \omega_i^2 y_i = \tilde{f}_i(t) \tag{6.8-8}$$

and the modal damping can be defined by the equation

$$2\zeta_i\omega_i = \alpha + \beta\omega_i^2 \tag{6.8-9}$$

6.9   NORMAL MODE SUMMATION

The forced vibration equation for the n-degree of freedom system

$$M\ddot{X} + C\dot{X} + KX = F \tag{6.9-1}$$

can be routinely solved by the digital computer. However, for systems of large numbers of degrees of freedom, the computation can be costly. It is possible, however, to cut down the size of the computation (or reduce the degrees of freedom of the system) by a procedure known as the *mode summation method*. Essentially, the displacement of the structure under

*It can be shown that $C = \alpha M^n + \beta K^n$ can also be diagonalized (see Problems 6-29 and 6-30).

forced excitation is approximated by the sum of a limited number of normal modes of the system multiplied by generalized coordinates.

For example, consider a 50-story building with 50 degrees of freedom. The solution of its undamped homogeneous equation will lead to 50 eigenvalues and 50 eigenvectors which describe the normal modes of the structure. If we know that the excitation of the building centers around the lower frequencies, the higher modes will not be excited and we would be justified in assuming the forced response to be the superposition of only a few of the lower frequency modes; perhaps $\phi_1(x)$, $\phi_2(x)$, and $\phi_3(x)$ may be sufficient. Then the deflection under forced excitation may be written as

$$x_i = \phi_1(x_i)q_1(t) + \phi_2(x_i)q_2(t) + \phi_3(x_i)q_3(t) \tag{6.9-2}$$

or in matrix notation the position of all n floors can be expressed in terms of the modal matrix P composed of only the three modes. (See Fig. 6.9-1).

$$\begin{Bmatrix} x_1 \\ x_2 \\ \cdot \\ \cdot \\ \cdot \\ x_n \end{Bmatrix} = \begin{bmatrix} \phi_1(x_1) & \phi_2(x_1) & \phi_3(x_1) \\ \cdot & \cdot & \cdot \\ \cdot & \cdot & \cdot \\ \cdot & \cdot & \cdot \\ \phi_1(x_n) & \phi_2(x_n) & \phi_3(x_n) \end{bmatrix} \begin{Bmatrix} q_1 \\ q_2 \\ \cdot \\ \cdot \\ \cdot \\ q_n \end{Bmatrix} \tag{6.9-3}$$

The use of the limited modal matrix then reduces the system to that equal to the number of modes used. For example, for the 50-story building, each of the matrices such as K is a 50×50 matrix; using three normal modes, P

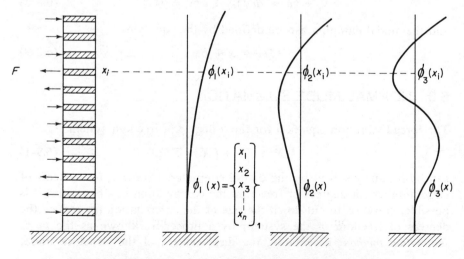

Figure 6.9-1. Building displacement represented by normal modes.

is a 50×3 matrix and the product $P'KP$ becomes

$$P'KP = (3 \times 50)(50 \times 50)(50 \times 3) = (3 \times 3) \text{ matrix}$$

Thus instead of solving the 50 coupled equations represented by Eq. (6.9-1), we need only solve the three by three equations represented by

$$P'M\ddot{q} + P'C\dot{q} + P'KPq = P'F$$

If the damping matrix is assumed to be proportional, the above equations become uncoupled, and if the force $F(x, t)$ is separable to $p(x)f(t)$, the three equations take the form

$$\ddot{q}_i + 2\zeta_i\omega_i\dot{q}_i + \omega_i^2 q_i = \Gamma_i f(t) \tag{6.9-4}$$

where the term

$$\Gamma_i = \frac{\sum_j \phi_i(x_j)p(x_j)}{\sum_j m_j\phi_i^2(x_j)} \tag{6.9-5}$$

is called the *mode participation factor*.

In many cases we are interested only in the maximum peak value of x_i, in which case the following procedure has been found to give acceptable results. We first find the maximum value of each $q_j(t)$ and combine them in the form

$$|x_i|_{\max} = |\phi_1(x_i)q_{1_{\max}}| + \sqrt{|\phi_2(x_i)q_{2_{\max}}|^2 + |\phi_3(x_i)q_{3_{\max}}|^2} \tag{6.9-6}*$$

Thus the first mode response is supplemented by the square root of the sum of the squares of the peaks for the higher modes. For the above computation, a shock spectrum for the particular excitation can be used to determine $q_{i_{\max}}$.

EXAMPLE 6.9-1

Consider the 10-story building of equal rigid floors and equal interstory stiffness. If the foundation of the building undergoes horizontal translation $u_0(t)$, determine the response of the building.

Solution: We will assume the normal modes of the building to be known. Given are the first three normal modes which have been computed from the undamped homogeneous equation and are as

*The method is used by the Shock and Vibration groups in various industries and military.

follows:

Floor $\omega_1 = 0.1495\sqrt{k/m}$ $\omega_2 = 0.4451\sqrt{k/m}$ $\omega_3 = 0.7307\sqrt{k/m}$

	$\varphi_1(x)$	$\varphi_2(x)$	$\varphi_3(x)$
10	1.0000	1.0000	1.0000
9	0.9777	0.8019	0.4662
8	0.9336	0.4451	−0.3165
7	0.8686	0.0000	−0.9303
6	0.7840	−0.4451	−1.0473
5	0.6822	−0.8019	−0.6052
4	0.5650	−1.0000	1.6010
3	0.4352	−1.0000	0.8398
2	0.2954	−0.8019	1.0711
1	0.1495	−0.4451	0.7307
0	0.0000	0.0000	0.0000

The equation of motion of the building due to ground motion $u_0(t)$ is

$$M\ddot{X} + C\dot{X} + KX = -M1\ddot{u}_0(t)$$

where 1 is a unit vector and X is a 10×1 vector. Using the three given modes, we make the transformation

$$X = Pq$$

where P is a 10×3 matrix and q is a 3×1 vector, i.e.,

$$P = \begin{bmatrix} \phi_1(x_1) & \phi_2(x_1) & \phi_3(x_1) \\ \phi_1(x_2) & \phi_2(x_2) & \phi_3(x_2) \\ \cdot & \cdot & \cdot \\ \cdot & \cdot & \cdot \\ \cdot & \cdot & \cdot \\ \phi_1(x_{10}) & \phi_2(x_{10}) & \phi_3(x_{10}) \end{bmatrix}, \quad q = \begin{Bmatrix} q_1 \\ q_2 \\ q_3 \end{Bmatrix}$$

Premultiplying by P', we obtain

$$P'MP\ddot{q} + P'CP\dot{q} + P'KPq = -P'M1u_0(t)$$

and by assuming C to be a proportional damping matrix, the above equation results in three uncoupled equations

$$m_{11}\ddot{q}_1 + c_{11}\dot{q}_1 + k_{11}q_1 = -\ddot{u}_0(t)\sum_{i=1}^{10} m_i\phi_1(x_i)$$

$$m_{22}\ddot{q}_2 + c_{22}\dot{q}_2 + k_{22}q_2 = -\ddot{u}_0(t)\sum_{i=1}^{10} m_i\phi_2(x_i)$$

$$m_{33}\ddot{q}_3 + c_{33}\dot{q}_3 + k_{33}q_3 = -\ddot{u}_0(t)\sum_{i=1}^{10} m_i\phi_3(x_i)$$

where m_{ii}, c_{ii}, and k_{ii} are generalized mass, generalized damping, and generalized stiffness. The $q_i(t)$ are then independently solved from each of the above equations. The displacement x_i of any floor must be found from the equation $X = Pq$ to be

$$x_i = \phi_1(x_i)q_1(t) + \phi_2(x_i)q_2(t) + \phi_3(x_i)q_3(t)$$

Thus the time solution for any floor is composed of the normal modes used.

From the numerical information supplied on the normal modes we will now determine the numerical values for the first equation which can be rewritten as

$$\ddot{q}_1 + 2\zeta_1\omega_1\dot{q}_1 + \omega_1^2 q_1 = -\frac{\sum m\varphi_1}{\sum m\varphi_1^2}\ddot{u}(t)$$

We have for the first mode

$$m_{11} = \sum m\varphi_1^2 = 5.2803m$$

$$\frac{c_{11}}{m_{11}} = 2\zeta_1\omega_1 = 0.299\sqrt{\frac{k}{m}}\,\zeta_1$$

$$\frac{k_{11}}{m_{11}} = \omega_1^2 = 0.02235\frac{k}{m}$$

$$\sum m\varphi_1 = 6.6912m$$

The equation for the first mode then becomes

$$\ddot{q}_1 + 0.299\sqrt{\frac{k}{m}}\,\zeta_1\dot{q}_1 + 0.02235\frac{k}{m}q_1 = -1.2672\ddot{u}_0(t)$$

Thus given the values for k/m and ζ_1, the above equation can be solved for any $\ddot{u}_0(t)$.

PROBLEMS

6-1 Determine the flexibility matrix for the spring-mass system shown in Fig. P6-1.

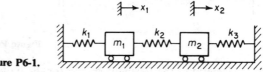

Figure P6-1.

6-2 Three equal springs of stiffness k lb/in. are joined at one end, the other ends being arranged symmetrically at 120° from each other, as shown in Fig. P6-2. Prove that the influence coefficients of the junction in a direction making an angle θ with any spring is independent of θ and equal to $1/1.5k$.

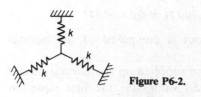

Figure P6-2.

6-3 A simply supported uniform beam of length l is loaded with weights at positions $0.25l$ and $0.6l$. Determine the flexibility influence coefficients for these positions.

6-4 Determine the flexibility matrix for the cantilever beam shown in Fig. P6-4 and calculate the stiffness matrix from its inverse.

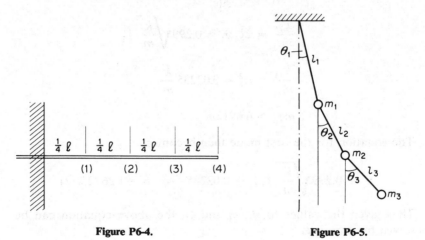

Figure P6-4.

Figure P6-5.

6-5 Determine the influence coefficients for the triple pendulum shown in Fig. P6-5.

6-6 Determine the stiffness matrix for the system shown in Fig. P6-6 and establish the flexibility matrix by its inverse.

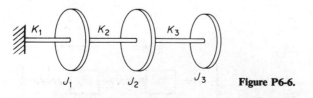

Figure P6-6.

6-7 Determine the flexibility matrix for the uniform beam of Fig. P6-7 by using the area-moment method.

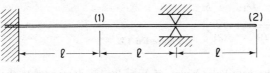

Figure P6-7.

6-8   Determine the flexibility matrix for the four-story building of Fig. 6.1-3 and invert it to arrive at the stiffness matrix given in the text.

6-9   Consider a system with *n* springs in series as presented in Fig. P6-9 and show that the stiffness matrix is a band matrix along the diagonal.

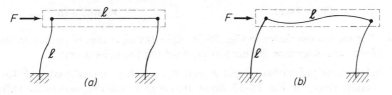

Figure P6-9.

6-10   Compare the stiffness of the framed building with rigid floor beam vs. that with flexible floor beam. Assume all length and EI to be equal. If the floor mass is pinned at the corners as shown in Fig. P6-10b, what is the ratio of the two natural frequencies?

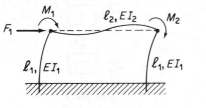

Figure P6-10.

6-11   The rectangular frame of Fig. P6-11 is fixed in the ground. Determine the stiffness matrix for the force system shown.

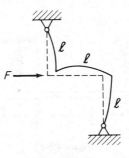

Figure P6-11.   **Figure P6-12.**

6-12   Determine the stiffness against the force F for the frame of Fig. P6-12, which is pinned at the top and bottom.

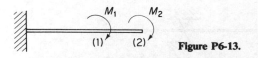

Figure P6-13.

6-13 Using the cantilever beam of Fig. P6-13, demonstrate that the reciprocity theorem holds for moment loads as well as fc rces.

6-14 Verify each of the results given in Table 6.1-1 by the area moment method and superposition.

6-15 Using the adjoint matrix, determine the normal modes of the spring-mass system shown in Fig. P6-15.

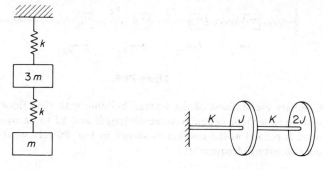

Figure P6-15.   Figure P6-16.

6-16 For the system shown in Fig. P6-16, write the equations of motion in matrix form and determine the normal modes from the adjoint matrix.

6-17 Determine the modal matrix P and the weighted modal matrix $\tilde{P}$ for the system shown in Fig. P6-17. Show that P or $\tilde{P}$ will diagonalize the stiffness matrix.

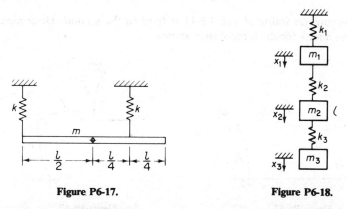

Figure P6-17.   Figure P6-18.

6-18 Determine the flexibility matrix for the spring-mass system of three degrees of freedom shown in Fig. P6-18 and write its equation of motion in matrix form.

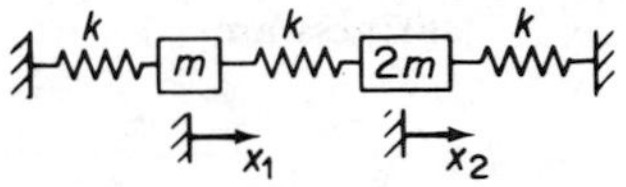

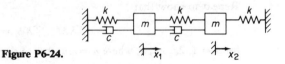

Figure P6-19.

6-19 Determine the modal matrix P and the weighted modal matrix $\tilde{P}$ for the system shown in Fig. P6-19 and diagonalize the stiffness matrix thereby decoupling the equations.

6-20 Determine $\tilde{P}$ for the double pendulum with coordinates θ_1 and θ_2. Show that $\tilde{P}$ decouples the equations of motion.

6-21 If in Prob. 6-11 masses and mass moment of inertia m_1, J_1 and m_2, J_2 are attached to the corners so that they rotate as well as translate, determine the equations of motion and find the natural frequencies and mode shapes.

6-22 Repeat the procedure of Prob. 6-21 with the frame of Fig. P6-12.

6-23 If the lower end of the frame of Prob. 6-12 is rigidly fixed to the ground, the rotation of the corners will differ. Determine its stiffness matrix and determine its matrix equation of motion for m_i, J_i at the corners.

6-24 Determine the damping matrix for the system presented in Fig. P6-24 and show that it is not proportional.

Figure P6-24.

6-25 Using the modal matrix $\tilde{P}$, reduce the system of Prob. 6-24 to one which is coupled only by damping and solve by the Laplace transform method.

6-26 Consider the viscoelastically damped system of Fig. P6-26. The system differs from the viscously damped system by the addition of the spring k_1 which introduces one more coordinate x_1 to the system. The equations of motion for the system in inertial coordinates x and x_1 are

$$m\ddot{x} = -kx - c(\dot{x} - \dot{x}_1) + F$$
$$0 = c(\dot{x} - \dot{x}_1) - k_1 x_1$$

Write the equation of motion in matrix form.

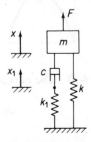

Figure P6-26.

6-27 Show, by comparing the viscoelastic system of Fig. P6-26 to the viscously damped system, that the equivalent viscous damping and the equivalent stiffness are

$$c_{eq} = \frac{c}{1 + \left(\dfrac{\omega c}{k_1}\right)^2}$$

$$k_{eq} = \frac{k + (k_1 + k)\left(\dfrac{\omega c}{k_1}\right)^2}{1 + \left(\dfrac{\omega c}{k_1}\right)^2}$$

6-28 Verify the relationship of Eq. (6.5-7)

$$X_i' K X_j = 0$$

by applying it to Prob. 6-16.

6-29 Starting with the matrix equation

$$K\phi_s = \omega_s^2 M \phi_s$$

premultiply first by KM^{-1} and, using the orthogonality relation $\phi_r' M \phi_s = 0$, show that

$$\phi_r' K M^{-1} K \phi_s = 0$$

Repeat to show that

$$\phi_r' [KM^{-1}]^h K \phi_s = 0$$

for $h = 1, 2, \ldots, n$, where $n = $ number of degrees of freedom of the system.

6-30 In a manner similar to Prob. 6-29, show that

$$\phi_r' [MK^{-1}]^h M \phi_s = 0, \qquad h = 1, 2, \ldots$$

6-31 Evaluate the numerical coefficients for the equations of motion for the second and third modes of Example 6.9-1.

6-32 If the acceleration $\ddot{u}(t)$ of the ground in Example 6.9-1 is a single sine pulse of amplitude a_0 and duration t_1 as shown in Fig. P6-32, determine the maximum q for each mode and the value of $x_{\max}$ as given in Sec. 6.9.

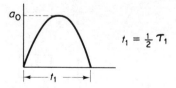

$$t_1 = \tfrac{1}{2} T_1$$

Figure P6-32.

6-33 The normal modes of the double pendulum of Prob. 5-9 are given as

$$\omega_1 = 0.764\sqrt{\frac{g}{l}} \qquad \omega_2 = 1.850\sqrt{\frac{g}{l}}$$

$$\phi_1 = \left\{ \begin{matrix} \theta_1 \\ \theta_2 \end{matrix} \right\}_{(1)} = \left\{ \begin{matrix} 0.707 \\ 1.00 \end{matrix} \right\} \qquad \phi_2 = \left\{ \begin{matrix} \theta_1 \\ \theta_2 \end{matrix} \right\}_{(2)} = \left\{ \begin{matrix} -0.707 \\ 1.00 \end{matrix} \right\}$$

If the lower mass is given an impulse $F_0\delta(t)$, determine the response in terms of the normal modes.

6-34   The normal modes of the three mass torsional system of Fig. P6-6 are given for $J_1 = J_2 = J_3$ and $K_1 = K_2 = K_3$.

$$\phi_1 = \begin{Bmatrix} 0.328 \\ 0.591 \\ 0.737 \end{Bmatrix} \qquad \lambda_1 = \frac{J\omega_1^2}{k} = 0.198 \qquad \phi_2 = \begin{Bmatrix} 0.737 \\ 0.328 \\ -0.591 \end{Bmatrix}$$

$$\lambda_2 = 1.555 \qquad \phi_3 = \begin{Bmatrix} 0.591 \\ -0.737 \\ 0.328 \end{Bmatrix} \qquad \lambda_3 = 3.247$$

Determine the equations of motion if a torque $M(t)$ is applied to the free end. If $M(t) = M_0 u(t)$ where $u(t)$ is a unit step function, determine the time solution and the maximum response of the end mass from the shock spectrum.

6-35   Using two normal modes, set up the equations of motion for the five-story building whose foundation stiffness in translation and rotation are k_t and $K_r = \infty$ (see Fig. P6-35).

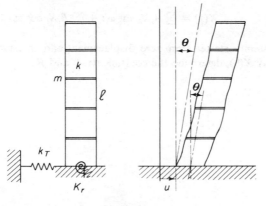

Figure P6-35.

6-36   The lateral and torsional oscillations of the system shown in Fig. P6-36 will have equal natural frequencies for a specific value of a/L. Determine this value and assuming that there is an eccentricity e of mass equal to me, determine the equations of motion.

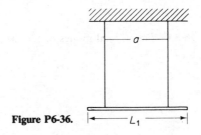

Figure P6-36.

6-37 Assume that a three-story building with rigid floor girders has Rayleigh damping. If the modal dampings for the first and second modes are 0.05% and 0.13% respectively, determine the modal damping for the third mode.

6-38 The normal modes of a three degree of freedom system with $m_1 = m_2 = m_3$ and $k_1 = k_2 = k_3$ are given as

$$X_1 = \begin{Bmatrix} 0.737 \\ 0.591 \\ 0.328 \end{Bmatrix} \qquad X_2 = \begin{Bmatrix} -0.591 \\ 0.328 \\ 0.737 \end{Bmatrix} \qquad X_3 = \begin{Bmatrix} 0.328 \\ -0.737 \\ 0.591 \end{Bmatrix}$$

Verify the orthogonal properties of these modes.

6-39 The system of Prob. 6-38 is given an initial displacement of

$$X = \begin{Bmatrix} 0.520 \\ -0.100 \\ 0.205 \end{Bmatrix}$$

and released. Determine how much of each mode will be present in the free vibration.

6-40 In general, the free vibration of an undamped system can be represented by the modal sum

$$X(t) = \sum_i A_i X_i \sin \omega_i t + \sum_i B_i X_i \cos \omega_i t$$

If the system is started from zero displacement and an arbitrary distribution of velocity $\dot{X}(0)$, determine the coefficients A_i and B_i.

7

NORMAL MODE VIBRATION OF CONTINUOUS SYSTEMS

The systems to be studied in this chapter have continuously distributed mass and elasticity. These bodies are assumed to be homogeneous and isotropic, obeying Hooke's law within the elastic limit. To specify the position of every particle in the elastic body, an infinite number of coordinates are necessary, and such bodies therefore possess an infinite number of degrees of freedom.

In general, the free vibration of these bodies is the sum of the principal modes as previously stated. For the principal mode of vibration every particle of the body performs simple harmonic motion at the frequency corresponding to the particular root of the frequency equation, each particle passing simultaneously through its respective equilibrium position. If the elastic curve of the body under which the motion is started coincides exactly with one of the principal modes, only that principal mode will be produced. However, the elastic curve resulting from a blow or a sudden removal of forces seldom corresponds to that of a principal mode, and thus all modes are excited. In many cases, however, a particular principal mode can be excited by proper initial conditions.

In this chapter some of the simpler problems of vibration of elastic bodies are taken up. The solutions to these problems are treated in terms of the principal modes of vibration.

7.1 VIBRATING STRING

A flexible string of mass ρ per unit length is stretched under tension T. By assuming the lateral deflection y of the string to be small, the change in tension with deflection is negligible and can be ignored.

In Fig. 7.1-1, a free-body diagram of an elementary length dx of the string is shown. Assuming small deflections and slopes, the equation of motion in the y-direction is

$$T\left(\theta + \frac{\partial \theta}{\partial x}\,dx\right) - T\theta = \rho\,dx\,\frac{\partial^2 y}{\partial t^2}$$

or

$$\frac{\partial \theta}{\partial x} = \frac{\rho}{T}\,\frac{\partial^2 y}{\partial t^2} \tag{7.1-1}$$

Since the slope of the string is $\theta = \partial y/\partial x$, the above equation reduces to

$$\frac{\partial^2 y}{\partial x^2} = \frac{1}{c^2}\,\frac{\partial^2 y}{\partial t^2} \tag{7.1-2}$$

where $c = \sqrt{T/\rho}$ can be shown to be the velocity of wave propagation along the string.

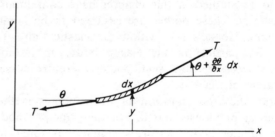

Figure 7.1-1. String element in lateral vibration.

The general solution of Eq. (7.1-2) can be expressed in the form

$$y = F_1(ct - x) + F_2(ct + x) \tag{7.1-3}$$

where F_1 and F_2 are arbitrary functions. Regardless of the type of function F, the argument $(ct \pm x)$ upon differentiation leads to the equation

$$\frac{\partial^2 F}{\partial x^2} = \frac{1}{c^2}\,\frac{\partial^2 F}{\partial t^2} \tag{7.1-4}$$

and hence the differential equation is satisfied.

Considering the component $y = F_1(ct - x)$, its value is determined by the argument $(ct - x)$ and hence by a range of values of t and x. For example, if $c = 10$, the equation for $y = F_1(100)$ is satisfied by $t = 0$, $x = -100$; $t = 1$, $x = -90$; $t = 2$, $x = -80$, etc. Therefore, the wave profile moves in the positive x-direction with speed c. In a similar manner we can show that $F_2(ct + x)$ represents a wave moving toward the negative x-direction with speed c. We therefore refer to c as the velocity of wave propagation.

One method of solving partial differential equations is that of separation of variables. In this method the solution is assumed in the form

$$y(x, t) = Y(x)G(t) \tag{7.1-5}$$

By substitution into Eq. (7.1-2) we obtain

$$\frac{1}{Y}\frac{d^2Y}{dx^2} = \frac{1}{c^2}\frac{1}{G}\frac{d^2G}{dt^2} \tag{7.1-6}$$

Since the left side of this equation is independent of t, whereas the right side is independent of x, it follows that each side must be a constant. Letting this constant be $-(\omega/c)^2$ we obtain two ordinary differential equations

$$\frac{d^2Y}{dx^2} + \left(\frac{\omega}{c}\right)^2 Y = 0 \tag{7.1-7}$$

$$\frac{d^2G}{dt^2} + \omega^2 G = 0 \tag{7.1-8}$$

with the general solutions

$$Y = A \sin\frac{\omega}{c}x + B \cos\frac{\omega}{c}x \tag{7.1-9}$$

$$G = C \sin \omega t + D \cos \omega t \tag{7.1-10}$$

The arbitrary constants A, B, C, D depend on the boundary conditions and the initial conditions. For example, if the string is stretched between two fixed points with distance l between them, the boundary conditions are $y(0, t) = y(l, t) = 0$. The condition that $y(0, t) = 0$ will require that $B = 0$ so that the solution will appear as

$$y = (C \sin \omega t + D \cos \omega t) \sin\frac{\omega}{c}x \tag{7.1-11}$$

The condition $y(l, t) = 0$ then leads to the equation

$$\sin\frac{\omega l}{c} = 0$$

or

$$\frac{\omega_n l}{c} = \frac{2\pi l}{\lambda} = n\pi, \qquad n = 1, 2, 3 \cdots$$

and $\lambda = c/f$ is the wavelength and f is the frequency of oscillation. Each n

represents a normal-mode vibration with natural frequency determined from the equation

$$f_n = \frac{n}{2l}c = \frac{n}{2l}\sqrt{\frac{T}{\rho}} \quad , \qquad n = 1, 2, 3, \cdots \qquad (7.1\text{-}12)$$

The mode shape is sinusoidal with the distribution

$$Y = \sin n\pi\frac{x}{l} \qquad (7.1\text{-}13)$$

In the more general case of free vibration initiated in any manner, the solution will contain many of the normal modes and the equation for the displacement may be written as

$$y(x, t) = \sum_{n=1}^{\infty} (C_n \sin \omega_n t + D_n \cos \omega_n t) \sin \frac{n\pi x}{l} \qquad (7.1\text{-}14)$$

$$\omega_n = \frac{n\pi c}{l}$$

Fitting this equation to the initial conditions of $y(x, 0)$ and $\dot{y}(x, 0)$, the C_n and D_n can be evaluated.

EXAMPLE 7.1-1

A uniform string of length l is fixed at the ends and stretched under tension T. If the string is displaced into an arbitrary shape $y(x, 0)$ and released, determine C_n and D_n of Eq. (7.1-14).

Solution: At $t = 0$, the displacement and velocity are

$$y(x, 0) = \sum_{n=1}^{\infty} D_n \sin \frac{n\pi x}{l}$$

$$\dot{y}(x, 0) = \sum_{n=1}^{\infty} \omega_n C_n \sin \frac{n\pi x}{l} = 0$$

Multiplying each equation by $\sin k\pi x/l$ and integrating from $x = 0$ to $x = l$ all of the terms on the right side will be zero, except the term $n = k$. Thus we arrive at the result

$$D_k = \frac{2}{l} \int_0^l y(x, 0) \sin \frac{k\pi x}{l}\, dx$$

$$C_k = 0 \qquad\qquad k = 1, 2, 3, \cdots$$

7.2   LONGITUDINAL VIBRATION OF RODS

The rod considered in this section is assumed to be thin and uniform along its length. Due to axial forces there will be displacements u along the rod which will be a function of both the position x and the time t. Since the

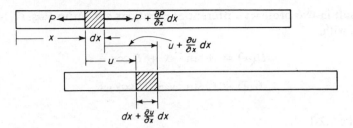

Figure 7.2-1.   Displacement of rod element.

rod has an infinite number of natural modes of vibration, the distribution of the displacement will differ with each mode.

Let us consider an element of this rod of length dx (Fig. 7.2-1). If u is the displacement at x, the displacement at $x + dx$ will be $u + (\partial u/\partial x)\, dx$. It is evident then that the element dx in the new position has changed in length by an amount $(\partial u/\partial x)\, dx$, and thus the unit strain is $\partial u/\partial x$. Since from Hooke's law the ratio of unit stress to unit strain is equal to the modulus of elasticity E, we can write

$$\frac{\partial u}{\partial x} = \frac{P}{AE} \tag{7.2-1}$$

where A is the cross-sectional area of the rod. Differentiating with respect to x

$$AE\frac{\partial^2 u}{\partial x^2} = \frac{\partial P}{\partial x} \tag{7.2-2}$$

We now apply Newton's law of motion for the element and equate the unbalanced force to the product of the mass and acceleration of the element

$$\frac{\partial P}{\partial x}\, dx = \rho A\, dx\,\frac{\partial^2 u}{\partial t^2} \tag{7.2-3}$$

where ρ is the density of the rod, mass per unit volume. Eliminating $\partial P/\partial x$ between Eqs. (7.2-2) and (7.2-3), we obtain the partial differential equation

$$\frac{\partial^2 u}{\partial t^2} = \left(\frac{E}{\rho}\right)\frac{\partial^2 u}{\partial x^2} \tag{7.2-4}$$

or

$$\frac{\partial^2 u}{\partial x^2} = \frac{1}{c^2}\frac{\partial^2 u}{\partial t^2} \tag{7.2-5}$$

which is similar to that of Eq. (7.1-2) for the string. The velocity of propagation of the displacement or stress wave in the rod is then equal to

$$c = \sqrt{E/\rho} \tag{7.2-6}$$

and a solution of the form

$$u(x, t) = U(x)G(t) \tag{7.2-7}$$

will result in two ordinary differential equations similar to Eqs. (7.1-7) and (7.1-8), with

$$U(x) = A \sin\frac{\omega}{c}x + B \cos\frac{\omega}{c}x \qquad (7.2\text{-}8)$$

$$G(t) = C \sin \omega t + D \cos \omega t \qquad (7.2\text{-}9)$$

EXAMPLE 7.2-1

Determine the natural frequencies and mode shapes of a free-free rod (a rod with both ends free).

Solution: For such a bar, the stress at the ends must be zero. Since the stress is given by the equation $E\, \partial u/\partial x$, the unit strain at the ends must also be zero; that is,

$$\frac{\partial u}{\partial x} = 0 \text{ at } x = 0, \text{ and } x = l$$

The two equations corresponding to the above boundary conditions are therefore

$$\left(\frac{\partial u}{\partial x}\right)_{x=0} = A\frac{\omega}{c}(C \sin \omega t + D \cos \omega t) = 0$$

$$\left(\frac{\partial u}{\partial x}\right)_{x=l} = \frac{\omega}{c}\left(A \cos\frac{\omega l}{c} - B \sin\frac{\omega l}{c}\right)(C \sin \omega t + D \cos \omega t) = 0$$

Since these equations must be true for any time t, A must be equal to zero from the first equation. Since B must be finite in order to have vibration, the second equation is satisfied when

$$\sin\frac{\omega l}{c} = 0$$

or

$$\frac{\omega_n l}{c} = \omega_n l\sqrt{\rho/E} = \pi, 2\pi, 3\pi, \ldots, n\pi$$

The frequency of vibration is thus given by

$$\omega_n = \frac{n\pi}{l}\sqrt{\frac{E}{\rho}}, \qquad f_n = \frac{n}{2l}\sqrt{\frac{E}{\rho}}$$

where n represents the order of the mode. The solution of the free-free rod with zero initial displacement can then be written as

$$u = u_0 \cos\frac{n\pi}{l}x \sin\frac{n\pi}{l}\sqrt{\frac{E}{\rho}}\,t$$

The amplitude of the longitudinal vibration along the rod is therefore a cosine wave having n nodes.

7.3 TORSIONAL VIBRATION OF RODS

The equation of motion of a rod in torsional vibration is similar to that of longitudinal vibration of rods discussed in the preceding section.

Letting x be measured along the length of the rod, the angle of twist in any length dx of the rod due to a torque T is

$$d\theta = \frac{T\,dx}{I_p G} \qquad (7.3-1)$$

where $I_p G$ is the torsional stiffness given by the product of the polar moment of inertia I_p of the cross-sectional area and the shear modulus of elasticity G. The torque on the two faces of the element being T and $T + (\partial T/\partial x)\,dx$, as shown in Fig. 7.3-1, the net torque from Eq. (7.3-1) becomes

$$\frac{\partial T}{\partial x}\,dx = I_p G \frac{\partial^2\theta}{\partial x^2}\,dx \qquad (7.3-2)$$

Equating this torque to the product of the mass moment of inertia $\rho I_p\,dx$ of the element and the angular acceleration $\partial^2\theta/\partial t^2$, where ρ is the density of the rod in mass per unit volume, the differential equation of motion becomes

$$\rho I_p\,dx\,\frac{\partial^2\theta}{\partial t^2} = I_p G \frac{\partial^2\theta}{\partial x^2}\,dx, \qquad \frac{\partial^2\theta}{\partial t^2} = \left(\frac{G}{\rho}\right)\frac{\partial^2\theta}{\partial x^2} \qquad (7.3-3)$$

This equation is of the same form as that of longitudinal vibration of rods where θ and G/ρ replace u and E/ρ, respectively. The general solution may hence be written immediately by comparison as

$$\theta = \left(A \sin \omega\sqrt{\frac{\rho}{G}}\,x + B \cos \omega\sqrt{\frac{\rho}{G}}\,x\right)(C \sin \omega t + D \cos \omega t) \qquad (7.3-4)$$

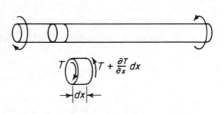

Figure 7.3-1. Torque acting on an element dx.

EXAMPLE 7.3-1

Determine the equation for the natural frequencies of a uniform rod in torsional oscillation with one end fixed and the other end free, as in Fig. 7.3-2.

215

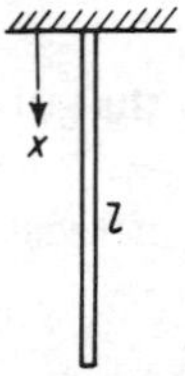

Figure 7.3-2.

Solution: Starting with equation

$$\theta = \left(A \sin \omega \sqrt{\rho/G}\, x + B \cos \omega \sqrt{\rho/G}\, x \right) \sin \omega t$$

apply the boundary conditions, which are
(1) when $x = 0$, $\theta = 0$,
(2) when $x = l$, torque = 0, or

$$\frac{\partial \theta}{\partial x} = 0$$

Boundary condition (1) results in $B = 0$.
Boundary condition (2) results in the equation

$$\cos \omega \sqrt{\rho/G}\, l = 0$$

which is satisfied by the following angles

$$\omega_n \sqrt{\frac{\rho}{G}}\, l = \frac{\pi}{2}, \frac{3\pi}{2}, \frac{5\pi}{2}, \ldots, \left(n + \frac{1}{2} \right) \pi$$

The natural frequencies of the bar are hence determined by the equation

$$\omega_n = \left(n + \frac{1}{2} \right) \frac{\pi}{l} \sqrt{\frac{G}{\rho}}$$

where $n = 0, 1, 2, 3 \cdots$.

EXAMPLE 7.3-2

The drill pipe of an oil well terminates at the lower end to a rod containing a cutting bit. Derive the expression for the natural

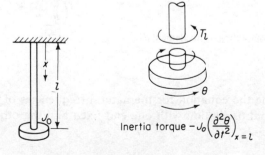

Inertia torque $-J_0 \left(\dfrac{\partial^2 \theta}{\partial t^2} \right)_{x=l}$

Figure 7.3-3.

frequencies, assuming the drill pipe to be uniform and fixed at the upper end and the rod and cutter to be represented by an end mass of moment of inertia J_0, as shown in Fig. 7.3-3.

Solution: The boundary condition at the upper end is $x = 0$, $\theta = 0$, which requires B to be zero in Eq. (7.3-4).

For the lower end, the torque on the shaft is due to the inertia torque of the end disk, as shown by the free-body diagram of Fig. 7.3-3. The inertia torque of the disk is $-J_0(\partial^2\theta/\partial^2 t)_{x=l} = J_0\omega^2(\theta)_{x=l}$, whereas the shaft torque from Eq. (7.3-1) is $T_l = GI_P(d\theta/dx)_{x=l}$. Equating the two, we have

$$GI_P\left(\frac{d\theta}{dx}\right)_{x=l} = J_0\omega^2\theta_{x=l}$$

Substituting from Eq. (7.3-4) with $B = 0$

$$GI_P\omega\sqrt{\frac{\rho}{G}}\ \cos\omega\sqrt{\frac{\rho}{G}}\ l = J_0\omega^2\sin\omega\sqrt{\frac{\rho}{G}}\ l$$

$$\tan\omega l\sqrt{\frac{\rho}{G}} = \frac{I_P}{\omega J_0}\sqrt{G\rho} = \frac{I_P\rho l}{J_0\omega l}\sqrt{\frac{G}{\rho}} = \frac{J_{\text{rod}}}{J_0\omega l}\sqrt{\frac{G}{\rho}}$$

This equation is of the form

$$\beta\tan\beta = \frac{J_{\text{rod}}}{J_0}, \qquad \beta = \omega l\sqrt{\frac{\rho}{G}}$$

which can be solved graphically or from tables.*

EXAMPLE 7.3-3

Using the frequency equation developed in the previous example, determine the first two natural frequencies of an oil-well drill pipe 5000 ft long, fixed at the upper end and terminating at the lower end to a drill collar 120 ft long. The average values for the drill pipe and drill collar are given as

$$\text{Drill pipe: outside diameter} = 4\tfrac{1}{2} \text{ in.}$$

$$\text{inside diameter} = 3.83 \text{ in.}$$

$$I_P = 0.00094 \text{ ft}^4 \qquad l = 5000 \text{ ft}$$

$$J_{\text{rod}} = I_P\rho l = 0.00094 \times \frac{490}{32.2} \times 5000 = 71.4 \text{ lb ft sec}^2$$

$$\text{Drill collar: outside diameter} = 7\tfrac{5}{8} \text{ in.}$$

$$\text{inside diameter} = 2.0 \text{ in.}$$

$$J_0 = 0.244 \times 120 \text{ ft} = 29.3 \text{ lb ft sec}^2$$

*See Jahnke and Emde, *Tables of Functions*, 4th Ed. (Dover Publications, Inc., 1945), Table V, p. 32.

Solution: The equation to be solved is

$$\beta \tan \beta = \frac{J_{\text{rod}}}{J_0} = 2.44$$

From Table V, p. 32, Jahnke and Emde, $\beta = 1.135, 3.722, \ldots$

$$\beta = \omega l \sqrt{\frac{\rho}{G}} = 5000\omega \sqrt{\frac{490}{12 \times 10^6 \times 12^2 \times 32.2}} = 0.470\omega$$

Solving for ω, the first two natural frequencies are found to be

$$\omega_1 = \frac{1.135}{0.470} = 2.41 \text{ rad/sec} = 0.384 \text{ cps}$$

$$\omega_2 = \frac{3.722}{0.470} = 7.93 \text{ rad/sec} = 1.26 \text{ cps}$$

7.4   EULER EQUATION FOR BEAMS

To determine the differential equation for the lateral vibration of beams, consider the forces and moments acting on an element of the beam shown in Fig. 7.4-1.

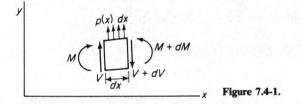

Figure 7.4-1.

V and M are shear and bending moments, respectively, and $p(x)$ represents the loading per unit length of the beam.

By summing forces in the y-direction

$$dV - p(x)\, dx = 0 \qquad\qquad (7.4\text{-}1)$$

By summing moments about any point on the right face of the element

$$dM - V\, dx - \tfrac{1}{2}p(x)(dx)^2 = 0 \qquad\qquad (7.4\text{-}2)$$

In the limiting process these equations result in the following important relationships

$$\frac{dV}{dx} = p(x), \qquad \frac{dM}{dx} = V \qquad\qquad (7.4\text{-}3)$$

The first part of Eq. (7.4-3) states that the rate of change of shear along the length of the beam is equal to the loading per unit length, and the second states that the rate of change of the moment along the beam is equal to the shear.

From Eq. (7.4-3) we obtain the following

$$\frac{d^2M}{dx^2} = \frac{dV}{dx} = p(x) \tag{7.4-4}$$

The bending moment is related to the curvature by the flexure equation, which, for the coordinates indicated in Fig. 7.4-1, is

$$M = EI\frac{d^2y}{dx^2} \tag{7.4-5}$$

Substituting this relation into Eq. (7.4-4), we obtain

$$\frac{d^2}{dx^2}\left(EI\frac{d^2y}{dx^2}\right) = p(x) \tag{7.4-6}$$

For a beam vibrating about its static equilibrium position under its own weight, the load per unit length is equal to the inertia load due to its mass and acceleration. Since the inertia force is in the same direction as $p(x)$ as shown in Fig. 7.4-1, we have, by assuming harmonic motion

$$p(x) = \rho\omega^2 y \tag{7.4-7}$$

where ρ is the mass per unit length of the beam. Using this relation, the equation for the lateral vibration of the beam reduces to

$$\frac{d^2}{dx^2}\left(EI\frac{d^2y}{dx^2}\right) - \rho\omega^2 y = 0 \tag{7.4-8}$$

In the special case where the flexural rigidity EI is a constant, the above equation may be written as

$$EI\frac{d^4y}{dx^4} - \rho\omega^2 y = 0 \tag{7.4-9}$$

On substituting

$$\beta^4 = \rho\frac{\omega^2}{EI} \tag{7.4-10}$$

we obtain the fourth-order differential equation

$$\frac{d^4y}{dx^4} - \beta^4 y = 0 \tag{7.4-11}$$

for the vibration of a uniform beam.

The general solution of Eq. (7.4-11) can be shown to be

$$y = A \cosh \beta x + B \sinh \beta x + C \cos \beta x + D \sin \beta x \tag{7.4-12}$$

To arrive at this result, we assume a solution of the form

$$y = e^{ax}$$

which will satisfy the differential equation when

$$a = \pm \beta, \text{ and } a = \pm i\beta$$

Since

$$e^{\pm \beta x} = \cosh \beta x \pm \sinh \beta x$$

$$e^{\pm i\beta x} = \cos \beta x \pm i \sin \beta x$$

the solution in the form of Eq. (7.4-12) is readily established.

The natural frequencies of vibration are found from Eq. (7.4-10) to be

$$\omega_n = \beta_n^2 \sqrt{\frac{EI}{\rho}} = (\beta_n l)^2 \sqrt{\frac{EI}{\rho l^4}}$$

where the number β_n depends on the boundary conditions of the problem. The following table lists numerical values of $(\beta_n l)^2$ for typical end conditions.

Beam configuration	$(\beta_1 l)^2$ Fundamental	$(\beta_2 l)^2$ Second Mode	$(\beta_3 l)^2$ Third Mode
Simply supported	9.87	39.5	88.9
Cantilever	3.52	22.0	61.7
Free-free	22.4	61.7	121.0
Clamped-clamped	22.4	61.7	121.0
Clamped-hinged	15.4	50.0	104.0
Hinged-free	0	15.4	50.0

EXAMPLE 7.4-1

Determine the natural frequencies of vibration of a uniform beam clamped at one end and free at the other.

Solution: The boundary conditions are

$$\text{at } x = 0 \begin{cases} y = 0 \\ \dfrac{dy}{dx} = 0 \end{cases}$$

$$\text{at } x = l \begin{cases} M = 0 \quad \text{or} \quad \dfrac{d^2 y}{dx^2} = 0 \\ V = 0 \quad \text{or} \quad \dfrac{d^3 y}{dx^3} = 0 \end{cases}$$

Substituting these boundary conditions in the general solution, we obtain

$$(y)_{x=0} = A + C = 0, \quad \therefore \ A = -C$$

$$\left(\frac{dy}{dx}\right)_{x=0} = \beta[A \sinh \beta x + B \cosh \beta x - C \sin \beta x$$

$$+ D \cos \beta x]_{x=0} = 0$$

$$\beta[B + D] = 0, \quad \therefore \quad B = -D$$

$$\left(\frac{d^2 y}{dx^2}\right)_{x=l} = \beta^2[A \cosh \beta l + B \sinh \beta l - C \cos \beta l - D \sin \beta l] = 0$$

$$A(\cosh \beta l + \cos \beta l) + B(\sinh \beta l + \sin \beta l) = 0$$

$$\left(\frac{d^3 y}{dx^3}\right)_{x=l} = \beta^3[A \sinh \beta l + B \cosh \beta l + C \sin \beta l - D \cos \beta l] = 0$$

$$A(\sinh \beta l - \sin \beta l) + B(\cosh \beta l + \cos \beta l) = 0$$

From the last two equations we obtain

$$\frac{\cosh \beta l + \cos \beta l}{\sinh \beta l - \sin \beta l} = \frac{\sinh \beta l + \sin \beta l}{\cosh \beta l + \cos \beta l}$$

which reduces to

$$\cosh \beta l \cos \beta l + 1 = 0$$

This last equation is satisfied by a number of values of βl, corresponding to each normal mode of oscillation, which for the first and second modes are 1.875 and 4.695, respectively. The natural frequency for the first mode is hence given by

$$\omega_1 = \frac{1.875^2}{l^2}\sqrt{\frac{EI}{\rho}} = \frac{3.515}{l^2}\sqrt{\frac{EI}{\rho}}$$

7.5 EFFECT OF ROTARY INERTIA AND SHEAR DEFORMATION

The Timoshenko theory accounts for both the rotary inertia and shear deformation of the beam. The free-body diagram and the geometry for the beam element are shown in Fig. 7.5-1. If the shear deformation is zero, the center line of the beam element will coincide with the perpendicular to the face of the cross section. Due to shear, the rectangular element tends to go into a diamond shape without rotation of the face and, the slope of the center line is diminished by the shear angle $(\psi - dy/dx)$. The following

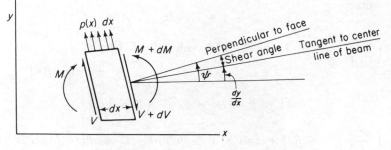

Figure 7.5-1. Effect of shear deformation.

quantities can then be defined

$$y = \text{deflection of the center line of the beam}$$

$$\frac{dy}{dx} = \text{slope of the center line of the beam}$$

$$\psi = \text{slope due to bending}$$

$$\psi - \frac{dy}{dx} = \text{loss of slope, equal to the shear angle}$$

There are two elastic equations for the beam, which are

$$\psi - \frac{dy}{dx} = \frac{V}{kAG} \tag{7.5-1}$$

$$\frac{d\psi}{dx} = \frac{M}{EI} \tag{7.5-2}$$

where A is the cross-sectional area, G the shear modulus, k a factor depending on the shape of the cross section, and EI the bending stiffness. For rectangular and circular cross sections, the values of k are $\frac{2}{3}$ and $\frac{3}{4}$ respectively. In addition, there are two dynamical equations

$$(\text{moment}) \ J\ddot{\psi} = \frac{dM}{dx} - V \tag{7.5-3}$$

$$(\text{force}) \ m\ddot{y} = -\frac{dV}{dx} + p(x, t) \tag{7.5-4}$$

where J and m are the rotary inertia and mass of the beam per unit length.

Substituting the elastic equations into the dynamical equations, we have

$$\frac{d}{dx}\left(EI\frac{d\psi}{dx}\right) + kAG\left(\frac{dy}{dx} - \psi\right) - J\ddot{\psi} = 0 \tag{7.5-5}$$

$$m\ddot{y} - \frac{d}{dx}\left[kAG\left(\frac{dy}{dx} - \psi\right)\right] - p(x, t) = 0 \tag{7.5-6}$$

which are the coupled equations of motion for the beam.

If ψ is eliminated and the cross section remains constant, these two equations can be reduced to a single equation

$$EI\frac{\partial^4 y}{\partial x^4} + m\frac{\partial^2 y}{\partial t^2} - \left(J + \frac{EIm}{kAG}\right)\frac{\partial^4 y}{\partial x^2 \partial t^2} + \frac{Jm}{kAG}\frac{\partial^4 y}{\partial t^4} = p(x, t)$$

$$+ \frac{J}{kAG}\frac{\partial^2 p}{\partial t^2} - \frac{EI}{kAG}\frac{\partial^2 p}{\partial x^2} \qquad (7.5\text{-}7)$$

It is evident then that the Euler equation

$$EI\frac{\partial^4 y}{\partial x^4} + m\frac{\partial^2 y}{\partial t^2} = p(x, t)$$

is a special case of the general beam equation including the rotary inertia and the shear deformation.

7.6   VIBRATION OF MEMBRANES

A membrane has no bending stiffness, and the lateral load on it is resisted only by the tension in the membrane itself. Its equation of motion can be derived by a procedure similar to that used in the string but applied in two dimensions.

Assume that the membrane is under uniform tension, T per unit length, which is large so that its variation due to lateral deflection is small. Defining the equilibrium position of the membrane in the xy plane, and letting w be the lateral deflection, we examine the forces on an element $dx\,dy$ as shown in Fig. 7.6-1. The resultant force in the w-direction due to

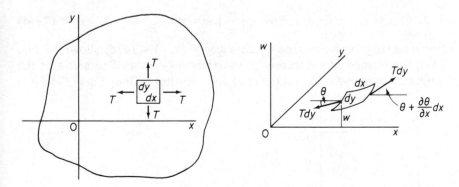

Figure 7.6-1.

the tension on the edges dy is

$$T \, dy \left(\theta + \frac{\partial \theta}{\partial x} dx \right) - T \, dy \, \theta = T \frac{\partial \theta}{\partial x} dy \, dx \qquad (7.6\text{-}1)$$

Similarly, the tension on the edges dx results in the component $T(\partial \phi / \partial y) \, dy \, dx$. Since the slopes in the x and y directions are $\theta = \partial w / \partial x$ and $\phi = \partial w / \partial y$, the total lateral force due to the tension T is

$$T \left(\frac{\partial^2 w}{\partial x^2} + \frac{\partial^2 w}{\partial y^2} \right) dx \, dy \qquad (7.6\text{-}2)$$

Letting ρ be the mass per unit area of the membrane and $p(x, y)$ the applied lateral pressure, the equation of motion becomes

$$\rho \, dx \, dy \frac{\partial^2 w}{\partial t^2} = T \left(\frac{\partial^2 w}{\partial x^2} + \frac{\partial^2 w}{\partial y^2} \right) dx \, dy + p(x, y) \, dx \, dy$$

or

$$\frac{\partial^2 w}{\partial t^2} = c^2 \nabla^2 w + \frac{1}{\rho} p(x, y) \qquad (7.6\text{-}3)$$

where

$$\nabla^2 = \frac{\partial^2}{\partial x^2} + \frac{\partial^2}{\partial y^2}$$

$$c = \sqrt{\frac{T}{\rho}}$$

This equation also applies in other coordinates with appropriate expression for ∇^2.

For the normal mode type of vibration, $p(x, y) = 0$ and $\partial^2 w / \partial t^2 = -\omega^2 w$, and the differential equation reduces to

$$\nabla^2 w + \left(\frac{\omega}{c} \right)^2 w = 0 \qquad (7.6\text{-}4)$$

For a rectangular membrane of dimensions $(x, y) = (a, b)$ shown in Fig. 7.6-2, the method of separation of variables may be used to arrive at the solution. Letting $w(x, y) = X(x) Y(y)$ and substituting into Eq. (7.6-4), it is

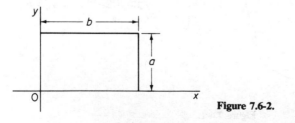

Figure 7.6-2.

easily shown that the solution is of the form

$$X(x) = C_1 \sin \alpha x + C_2 \cos \alpha x$$
$$Y(y) = C_3 \sin \beta y + C_4 \cos \beta y$$

(7.6-5)

where $\alpha^2 + \beta^2 = (\omega/c)^2$. The constants C_i in these equations must be determined from the boundary conditions.

7.7   DIGITAL COMPUTATION

When the motion of a structural member is represented by a partial differential equation, the method of separation of variables was found to eliminate the time variable and reduce the equation of motion to an ordinary differential equation in the spacial coordinate x. (See Sec. 7.1 for the string.) If the parameters of the system vary with the position, an analytical solution may not be possible. For such cases, the problem may be solved by the finite difference method. Special consideration must then be given to the boundary conditions.

Finite Difference.   In this method the differential equations and their boundary conditions are replaced by the corresponding finite difference equations. This then reduces the problem to a set of simultaneous algebraic equations which can be solved by the digital computer.

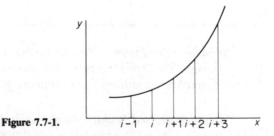

Figure 7.7-1.

Consider a function $y(x)$ which is shown in Fig. 7.7-1. At some point x_i the derivative is approximated by the equation

$$\left(\frac{dy}{dx}\right)_i \cong \frac{1}{h}(y_{i+1} - y_i) = \frac{1}{h}\Delta y$$

(7.7-1)

where $h = (x_{i+1} - x_i)$. The second derivative is

$$\left(\frac{d^2 y}{dx^2}\right)_i = \frac{d}{dx}\left(\frac{dy}{dx}\right)_i \cong \frac{1}{h}\left[\frac{1}{h}(y_{i+2} - y_{i+1}) - \frac{1}{h}(y_{i+1} - y_i)\right]$$
$$= \frac{1}{h^2}(y_{i+2} - 2y_{i+1} + y_i) = \frac{1}{h^2}\Delta^2 y$$

(7.7-2)

The above procedure can be repeated any number of times for higher order derivatives. The finite difference pattern up to the fourth derivative is shown in the following table.

FINITE DIFFERENCE TABLE

x	y	$\dfrac{\Delta y}{\Delta x}$	$\dfrac{\Delta^2 y}{\Delta x^2}$	$\dfrac{\Delta^3 y}{dx^3}$	$\dfrac{\Delta^4 y}{\Delta x^4}$
x_1	y_1				
		$\frac{1}{h}(y_2 - y_1)$			
x_2	y_2		$\frac{1}{h_2}(y_3 - 2y_2 + y_1)$		
		$\frac{1}{h}(y_3 - y_2)$		$\frac{1}{h^3}(y_4 - 3y_3 + 3y_2 - y_1)$	
x_3	y_3		$\frac{1}{h^2}(y_4 - 2y_3 + y_2)$		$\frac{1}{h^4}(y_5 - 4y_4 + 6y_3 - 4y_2 + y_1)$
		$\frac{1}{h}(y_4 - y_3)$		$\frac{1}{h^3}(y_5 - 3y_4 + 3y_3 - y_2)$	
x_4	y_4		$\frac{1}{h^2}(y_5 - 2y_4 + y_3)$		$\frac{1}{h^4}(y_6 - 4y_5 + 6y_4 - 4y_3 + y_2)$
		$\frac{1}{h}(y_5 - y_4)$		$\frac{1}{h^3}(y_6 - 3y_5 + 3y_4 - y_3)$	
x_5	y_5		$\frac{1}{h^2}(y_6 - 2y_5 + y_4)$		
		$\frac{1}{h}(y_6 - y_5)$			
x_6	y_6				

Boundary Conditions. To satisfy the boundary conditions, fictitious points outside the structure must be chosen. The following examples of typical boundary conditions for beams are given.

Simply Supported Beam. As shown in Fig. 7.7-2, let the point on the left of station 1 be p. The boundary conditions at the left end of the beam are

$$y_1 = 0, \qquad \left(\frac{d^2 y}{dx^2}\right)_1 = 0$$

Writing the difference equation for the second derivative at station 1, we have

$$\frac{1}{h^2}(y_2 - 2y_1 + y_p) = \frac{1}{h^2}(y_2 - 0 + y_p) = 0$$

Thus y_p must equal $-y_2$.

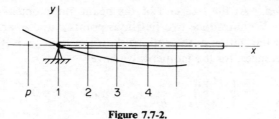

Figure 7.7-2.

Clamped End.   At the clamped end the deflection and slope are both zero as shown in Fig. 7.7-3. Again letting y_p be the deflection at the left of station 1, we have, using an interval of $2h$

$$\left(\frac{dy}{dx}\right)_1 = \frac{1}{2h}(y_2 - y_p) = 0$$

Thus $y_p = y_2$ and the deflection curve is symmetrical about the wall.

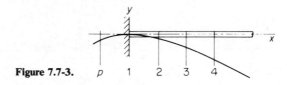

Figure 7.7-3.

Partially Restrained Beam.   Consider next the case where the left end of the beam is partially restrained. We can represent this condition by a torsional spring of stiffness K lb in./rad as shown in Fig. 7.7-4. The moment at the boundary is $M_1 = -K\theta_1$, but

$$\theta_1 = \left(\frac{dy}{dx}\right)_1 = \frac{1}{2h}(y_2 - y_p)$$

and

$$M_1 = EI\left(\frac{d^2y}{dx^2}\right)_1 = \frac{EI}{h^2}(y_2 - 0 + y_p)$$

Substituting into $M_1 = -K\theta_1$, and solving for y_p, we obtain

$$y_p = -y_2\left(\frac{2EI + Kh}{2EI - Kh}\right)$$

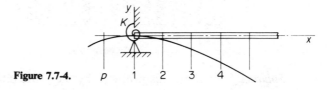

Figure 7.7-4.

Free End. At the free end of the beam, the moment and the shear must be zero. We introduce two fictitious points p and q, and an arbitrary number 4 for the station at the end, as shown in Fig. 7.7-5. Referring to the table of differences, for the moment, we have

$$\left(\frac{d^2 y}{dx^2}\right)_4 = \frac{1}{h^2}(y_p - 2y_4 + y_3) = 0$$

or

$$y_p = 2y_4 - y_3$$

For the shear we will average the third derivatives at the end as follows. Generally greater accuracy is obtained in this way.

$$\left(\frac{d^3 y}{dx^3}\right)_4 = \frac{1}{2}\left[\frac{1}{h^3}(y_q - 3y_p + 3y_4 - y_3) + \frac{1}{h^3}(y_p - 3y_4 + 3y_3 - y_2)\right]$$

$$= \frac{1}{2h^3}(y_q - 2y_p + 2y_3 - y_2) = 0$$

Thus

$$y_q = 4y_4 - 4y_3 + y_2$$

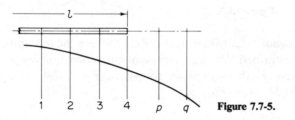

Figure 7.7-5.

EXAMPLE 7.7-1

A beam of non-uniform moment of inertia rests on an elastic foundation of stiffness k lb/in. as shown in Fig. 7.7-6. Its natural frequencies are to be found from its differential equation which is

$$\frac{d^2}{dx^2}\left(EI\frac{d^2 y}{dx^2}\right) + ky - \omega^2 my = 0 \qquad \text{(a)}$$

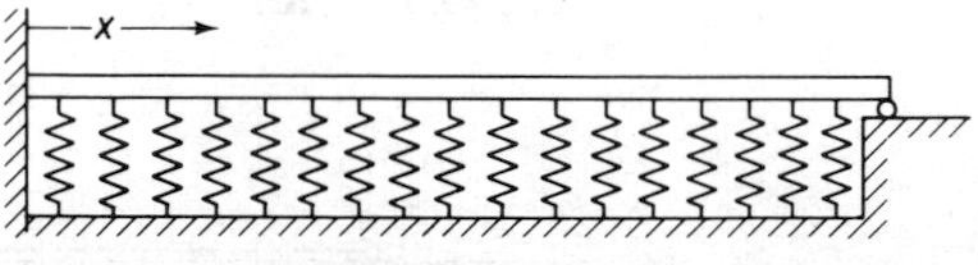

Figure 7.7-6.

Solution: To solve this problem by the finite difference method, we number the stations along the beam from 1 to n, and assign a new

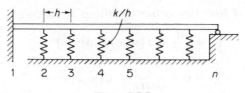

Figure 7.7-7.

foundation stiffness for each section, which is k/h as shown in Fig. 7.7-7. Equation (a) is also rewritten as

$$EI\frac{d^4 y}{dx^4} + 2E\frac{d^3 y}{dx^3}\frac{dI}{dx} + E\frac{d^2 y}{dx^2}\frac{d^2 I}{dx^2} + (k - m\omega^2)y = 0 \quad \text{(b)}$$

We will now write the finite difference equation for station 2, taking note of the boundary conditions at the left end. The derivatives encountered are

$$\left(\frac{d^2 y}{dx^2}\right)_2 = \frac{1}{h^2}(y_3 - 2y_2 + y_1) = \frac{1}{h^2}(y_3 - 2y_2)$$

$$\left(\frac{d^3 y}{dx^3}\right)_2 = \frac{1}{2}\left[\frac{1}{h^3}(y_4 - 3y_3 + 3y_2 - y_1) + \frac{1}{h^3}(y_3 - 3y_2 + 3y_1 - y_p)\right]$$

$$= \frac{1}{2h^3}(y_4 - 2y_3 - y_p) = \frac{1}{2h^3}(y_4 - 2y_3 - y_2)$$

$$\left(\frac{d^4 y}{dx^4}\right)_2 = \frac{1}{h^4}(y_4 - 4y_3 + 6y_2 - 0 + y_2)$$

$$= \frac{1}{h^4}(y_4 - 4y_3 + 7y_2)$$

With these derivatives, the finite difference equation for station 2 becomes

$$\frac{EI_2}{h^4}(y_4 - 4y_3 + 7y_2) + \frac{2E}{2h^3}(y_4 - 2y_3 - y_2)\frac{1}{2h}(I_3 - I_1)$$

$$+ \frac{E}{h^2}(y_3 - 2y_2)\frac{1}{h^2}(I_3 - 2I_2 + I_1) + \left(\frac{k}{h} - m\omega^2\right)y_2 = 0 \quad \text{(c)}$$

In a similar manner, equations for other stations can be written. The boundary conditions at the right end must also be considered, and the resulting set of algebraic equations can be programmed for digital computation.

Runge-Kutta Method. The Runge-Kutta method is another possibility for the structural problem. It is self-starting and results in good accuracy. The error is of order h^5.

To illustrate the procedure, we consider the beam with rotary inertia and shear terms, discussed in Sec. 7.5. The fourth order equation is first written in terms of four first order equations as follows

$$\frac{d\psi}{dx} = \frac{M}{EI} = F(x, \psi, y, M, V)$$

$$\frac{dy}{dx} = \psi - \frac{V}{kAG} = G(x, \psi, y, M, V)$$

$$\frac{dM}{dx} = V - \omega^2 J\psi = P(x, \psi, y, M, V)$$

$$\frac{dV}{dx} = \omega^2 my = K(x, \psi, y, M, V)$$

(7.7-3)

The Runge-Kutta procedure, discussed in Sec. 4.6 for a single coordinate, is now extended to the simultaneous solution of four variables listed below

$$\psi = \psi_1 + \frac{h}{6}(f_1 + 2f_2 + 2f_3 + f_4)$$

$$y = y_1 + \frac{h}{6}(g_1 + 2g_2 + 2g_3 + g_4)$$

$$M = M_1 + \frac{h}{6}(p_1 + 2p_2 + 2p_3 + p_4)$$

$$V = V_1 + \frac{h}{6}(k_1 + 2k_2 + 2k_3 + k_4)$$

(7.7-4)

where $h = \Delta x$.

Let f_i, g_i, p_i, k_i and F, G, P, and K be represented by vectors

$$l_i = \begin{Bmatrix} f_i \\ g_i \\ p_i \\ k_i \end{Bmatrix}, \qquad L = \begin{Bmatrix} F \\ G \\ P \\ K \end{Bmatrix}$$

Then the computation proceeds as follows.

$$l_1 = L(x_1, \psi_1, y_1, M_1, V_1)$$

$$l_2 = L\left(x_1 + \frac{h}{2}, \psi_1 + f_1\frac{h}{2}, y_1 + g_1\frac{h}{2}, M_1 + p_1\frac{h}{2}, V_1 + k_1\frac{h}{2}\right)$$

$$l_3 = L\left(x_1 + \frac{h}{2}, \psi_1 + f_2\frac{h}{2}, y_1 + g_2\frac{h}{2}, M_1 + p_2\frac{h}{2}, V_1 + k_2\frac{h}{2}\right)$$

$$l_4 = L(x_1 + h, \psi_1 + f_3 h, y_1 + g_3 h, M_1 + p_3 h, V_1 + k_3 h)$$

With these quantities substituted into Eq. (7.7-4), the dependent variables at the neighboring point x_2 are found, and the procedure is repeated for the point x_3, etc.

Returning to the beam equations, the boundary conditions at the beginning end x_1 provide a starting point. For example, in the cantilever beam with origin at the fixed end, the boundary conditions at the starting end are

$$\psi_1 = 0, \qquad M_1 = M_1$$
$$y_1 = 0, \qquad V_1 = V_1$$

These can be considered to be the linear combination of two boundary vectors as follows

$$\begin{Bmatrix} \psi_1 \\ y_1 \\ M_1 \\ V_1 \end{Bmatrix} = \begin{Bmatrix} 0 \\ 0 \\ 1 \\ 0 \end{Bmatrix} + \alpha \begin{Bmatrix} 0 \\ 0 \\ 0 \\ 1 \end{Bmatrix} = C_1 + \alpha D_1$$

Since the system is linear, we can start with each boundary vector separately. Starting with C_1, we obtain

$$C_N = \begin{Bmatrix} \psi_N \\ y_N \\ M_N \\ V_N \end{Bmatrix}_C$$

Starting with D_1, we obtain

$$\alpha D_N = \begin{Bmatrix} \psi_N \\ y_N \\ M_N \\ V_N \end{Bmatrix}_D$$

These must now add to satisfy the actual boundary conditions at the terminal end, which for a cantilever free end are

$$\begin{Bmatrix} \psi \\ y \\ M \\ V \end{Bmatrix}_N = \begin{Bmatrix} \psi \\ y \\ 0 \\ 0 \end{Bmatrix} = C_N + \alpha D_N$$

If the frequency chosen is correct, the above boundary equations lead to

$$M_{NC} + \alpha M_{ND} = 0$$
$$V_{NC} + \alpha V_{ND} = 0$$
$$\alpha = -\frac{M_{NC}}{M_{ND}} = -\frac{V_{NC}}{V_{ND}}$$

which is satisfied by the determinant

$$\begin{vmatrix} M_{NC} & V_{NC} \\ M_{ND} & V_{ND} \end{vmatrix} = 0$$

The iteration can be started with three different frequencies, which results in three values of the determinant. A parabola is passed through these three points and the zero of the curve is chosen for a new estimate of the frequency. When the frequency is close to the correct value, the new estimate may be made by a straight line between two values of the boundary determinant.

PROBLEMS

7-1   Find the wave velocity along a rope whose mass is 0.372 kg/m when stretched to a tension of 444 N.

7-2   Derive the equation for the natural frequencies of a uniform cord of length l fixed at the two ends. The cord is stretched to a tension T and its mass per unit length is ρ.

7-3   A cord of length l and mass per unit length ρ is under tension T with the left end fixed and the right end attached to a spring-mass system as shown in Fig. P7-3. Determine the equation for the natural frequencies.

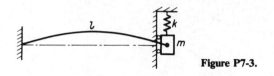

Figure P7-3.

7-4   A harmonic vibration has an amplitude that varies as a cosine function along the x-direction such that

$$y = a \cos kx \cdot \sin \omega t$$

Show that if another harmonic vibration of same frequency and equal amplitude displaced in space phase and time phase by a quarter wave length is added to the first vibration, the resultant vibration will represent a traveling wave with a propagation velocity equal to $c = \omega/k$.

7-5   Find the velocity of longitudinal waves along a thin steel bar. The modulus of elasticity and mass per unit volume of steel are 200×10^9 N/m^2 and 7810 kg/m^3.

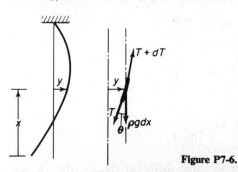

Figure P7-6.

7-6   Shown in Fig. P7-6 is a flexible cable supported at the upper end and free to oscillate under the influence of gravity. Show that the equation of lateral

motion is

$$\frac{\partial^2 y}{\partial t^2} = g\left(x\frac{\partial^2 y}{\partial x^2} + \frac{\partial y}{\partial x}\right)$$

7-7   In Prob. 7-6, assume a solution in the form $y = Y(x)\cos \omega t$ and show that $Y(x)$ can be reduced to a Bessel's differential equation

$$\frac{d^2Y(z)}{dz^2} + \frac{1}{z}\frac{dY(x)}{dz} + Y(z) = 0$$

with solution

$$Y(z) = J_0(z) \quad \text{or} \quad Y(x) = J_0\left(2\omega\sqrt{\frac{x}{g}}\right)$$

by a change in variable $z^2 = 4\omega^2 x/g$.

Figure P7-8.

7-8   A particular satellite consists of two equal masses m each, connected by a cable of length $2l$ and mass density ρ, as shown in Fig. P7-8. The assembly rotates in space with angular speed ω_0. Show that if the variation in the cable tension is neglected, the differential equation of lateral motion of the cable is

$$\frac{\partial^2 y}{\partial x^2} = \frac{\rho}{m\omega_0^2 l}\left(\frac{\partial^2 y}{\partial t^2} - \omega_0^2 y\right)$$

and that its fundamental frequency of oscillation is

$$\omega^2 = \left(\frac{\pi}{2l}\right)^2\left(\frac{m\omega_0 l}{\rho}\right) - \omega_0^2.$$

7-9   A uniform bar of length l is fixed at one end and free at the other end. Show that the frequencies of normal longitudinal vibrations are $f = (n + \frac{1}{2})c/2l$, where $c = \sqrt{E/\rho}$ is the velocity of longitudinal waves in the bar, and $n = 0, 1, 2, \cdots$.

7-10   A uniform rod of length l and cross-sectional area A is fixed at the upper end and is loaded with a weight W on the other end. Show that the natural frequencies are determined from the equation

$$\omega l\sqrt{\frac{\rho}{E}}\tan \omega l\sqrt{\frac{\rho}{E}} = \frac{A\rho lg}{W}$$

7-11   Show that the fundamental frequency for the system of Prob. 7-10 can be expressed in the form

$$\omega_1 = \beta_1\sqrt{k/rM}$$

where

$$n_1 l = \beta_1, \qquad r = \frac{M_{\text{rod}}}{M},$$

$$k = \frac{AE}{l}, \qquad M = \text{end mass}$$

Reducing the above system to a spring k and an end mass equal to $M + \frac{1}{3}M_{rod}$, determine an approximate equation for the fundamental frequency. Show that the ratio of the approximate to the exact frequency as found above is $(1/\beta_1)\sqrt{3r/(3 + r)}$.

7-12 The frequency of magnetostriction oscillators is determined by the length of the nickel alloy rod which generates an alternating voltage in the surrounding coils equal to the frequency of longitudinal vibration of the rod, as shown in Fig. P7-12. Determine the proper length of the rod clamped at the middle for a frequency of 20 kcps if the modulus of elasticity and density are given as $E = 30 \times 10^6$ lb/in.^{2} and $\rho = 0.31$ lb/in.^{3}

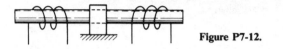

Figure P7-12.

7-13 The equation for the longitudinal oscillations of a slender rod with viscous damping is

$$m\frac{\partial^2 u}{\partial t^2} = AE\frac{\partial^2 u}{\partial x^2} - \alpha\frac{\partial u}{\partial t} + \frac{p_0}{l}p(x)f(t)$$

where the loading per unit length is assumed to be separable. Letting $u = \Sigma_i\phi_i(x)q_i(t)$ and $p(x) = \Sigma_i b_i\phi_i(x)$ show that

$$u = \frac{p_0}{ml\sqrt{1 - \zeta^2}}\sum_j \frac{b_j\phi_j}{\omega_j}\int_0^t f(t - \tau)e^{-\zeta\omega_j\tau}\sin\omega_j\sqrt{1 - \zeta^2}\,\tau\,d\tau$$

$$b_j = \frac{1}{l}\int_0^l p(x)\phi_j(x)\,dx$$

Derive the equation for the stress at any point x.

7-14 Show that $c = \sqrt{G/\rho}$ is the velocity of propagation of torsional strain along the rod. What is the numerical value of c for steel?

7-15 Determine the expression for the natural frequencies of torsional oscillations of a uniform rod of length l clamped at the middle and free at the two ends.

7-16 Determine the natural frequencies of a torsional system consisting of a uniform shaft of mass moment of inertia J_s with a disk of inertia J_0 attached to each end. Check the fundamental frequency by reducing the uniform shaft to a torsional spring with end masses.

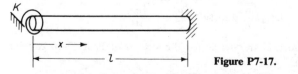

Figure P7-17.

7-17 A uniform bar has these specifications: length l, mass density per unit volume ρ, and torsional stiffness $I_p G$ where I_p is the polar moment of inertia of the cross section and G the shear modulus. The end $x = 0$ is fastened to a torsional spring of stiffness K lb in./rad, while the end l is fixed as shown in Fig. P7-17. Determine the transcendental equation from which natural frequencies can be established. Verify the correctness of this equation by considering special cases for $K = 0$ and $K = \infty$.

7-18 Determine the expression for the natural frequencies of a free-free bar in lateral vibration.

7-19 Determine the node position for the fundamental mode of the free-free beam by Rayleigh's method, assuming the curve to be $y = \sin(\pi x/l) - b$. By equating the momentum to zero, determine b. Substitute this value of b to find ω_1.

7-20 A concrete test beam $2 \times 2 \times 12$ in., supported at two points $0.224l$ from the ends, was found to resonate at 1690 cps. If the density of concrete is 153 lb/ft^3, determine the modulus of elasticity, assuming the beam to be slender.

7-21 Determine the natural frequencies of a uniform beam of length l clamped at both ends.

7-22 Determine the natural frequencies of a uniform beam of length l, clamped at one end and pinned at the other end.

7-23 A uniform beam of length l and weight W_b is clamped at one end and carries a concentrated weight W_0 at the other end. State the boundary conditions and determine the frequency equation.

7-24 The pinned end of a pinned-free beam is given a harmonic motion of amplitude y_0 perpendicular to the beam. Show that the boundary conditions result in the equation

$$\frac{y_0}{y_l} = \frac{\sinh \beta l \cos \beta l - \cosh \beta l \sin \beta l}{\sinh \beta l - \sin \beta l}$$

which for $y_0 \to 0$, reduces to

$$\tanh \beta l = \tan \beta l$$

7-25 A simply supported beam has an overhang of length l_2, as shown in Fig. P7-25. If the end of the overhang is free, show that boundary conditions require the deflection equation for each span to be

$$\phi_1 = C\left(\sin \beta x - \frac{\sin \beta l_1}{\sinh \beta l_1}\sinh \beta x\right)$$

$$\phi_2 = A\left\{\cos \beta x + \cosh \beta x - \left(\frac{\cos \beta l_2 + \cosh \beta l_2}{\sin \beta l_2 + \sinh \beta l_2}\right)(\sin \beta x + \sinh \beta x)\right\}$$

where x is measured from the left and right ends.

Figure P7-25.

7-26 Assume that the edges of the rectangular membrane of Fig. 7.6-2 are clamped and show that its solution is

$$w(x, y, t) = \sum_{m=1}^{\infty} \sum_{n=1}^{\infty} \sin\frac{m\pi x}{b}\sin\frac{n\pi y}{a}(A_{mn}\sin \omega_{mn}t + B_{mn}\cos \omega_{mn}t)$$

7-27   Show that the natural frequencies of the membrane of Prob. 7-26 are given by the equation

$$\omega_{m,n}^2 = c^2\pi^2\left(\frac{m^2}{b^2} + \frac{n^2}{a^2}\right)$$

where $m, n = 1, 2, 3, \ldots$.

7-28   Describe the natural mode shapes for the square membrane with clamped edges.

7-29   A membrane is stretched with large tension T lb/in., so that its lateral deflection y does not increase T appreciably. Using polar coordinates, show that the differential equation of lateral vibration is

$$\frac{\partial^2 y}{\partial t^2} = \frac{T}{\rho}\left(\frac{\partial^2 y}{\partial r^2} + \frac{1}{r}\frac{\partial y}{\partial r} + \frac{1}{r^2}\frac{\partial^2 y}{\partial \theta^2}\right)$$

7-30   Apply the results of Prob. 7-29 to a circular membrane of radius a with the boundary conditions $y(a) = 0$. The deflection of the symmetric modes without radial node lines can be shown to be given by $J_0(r\sqrt{\rho\omega^2/T})$. For the general case of radial and circumferential nodes, the natural frequencies are evaluated from the boundary conditions at $r = a$ and $r = 0$, which result in an equation of the form

$$\omega = \frac{\alpha_{n,m}}{a}\sqrt{\frac{T}{\rho}}$$

Values of $\alpha_n m$ are given in Fig. 7-30.

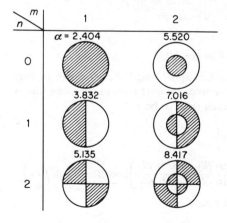

Figure P7-30.   Deflection of membranes.

7-31   When shear and rotary inertia are included, show that the differential equation of the beam may be expressed by the first order matrix equation

$$\frac{d}{dx}\begin{Bmatrix} \psi \\ y \\ M \\ V \end{Bmatrix} = \begin{bmatrix} 0 & 0 & \dfrac{1}{EI} & 0 \\ 1 & 0 & 0 & \dfrac{-1}{kAG} \\ -\omega^2 J & 0 & 0 & 1 \\ 0 & \omega^2 m & 0 & 0 \end{bmatrix}\begin{Bmatrix} \psi \\ y \\ M \\ V \end{Bmatrix}$$

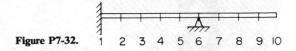

Figure P7-32. 1   2   3   4   5   6   7   8   9 10

7-32 For the beam configuration shown in Fig. P7-32, determine the finite difference equation for station 2.

7-33 For the beam of Prob. 7-32, establish the finite difference equations which apply to stations 5 and 7.

7-34 For the beam of Prob. 7-32, develop the finite difference equations for stations 9 and 10.

7-35 Using the tables of Appendix D, draw the normal mode deflection for each of the boundary conditions presented and give the corresponding natural frequencies.

8 LAGRANGE'S EQUATION

Lagrange* developed a general treatment of dynamical systems formulated from the scalar quantities of kinetic energy T, potential energy U, and work W. As the system becomes more complicated, the establishment of vector relationships required by Newton's laws becomes increasingly difficult, in which case the scalar approach based on energy and work offers considerable advantage. Furthermore, constraint forces of frictionless hinges and guides can be completely disregarded in Lagrange's formulation of the equations of motion.

8.1 GENERALIZED COORDINATES

The equations of motion of a system can be formulated in a number of different coordinate systems. However, n independent coordinates are necessary to describe the motion of a system of n degrees of freedom. Such independent coordinates are called *generalized coordinates* and are usually denoted by the letters q_i.

Constraints.   Motion of bodies are not always free motions and are often constrained to move in a predetermined manner. As an example, Fig. 8.1-1 shows a spherical pendulum of length l. Its position can be completely defined by the two independent coordinates ψ and ϕ. Hence ψ and

*Joseph L. C. Lagrange (1736–1813).

238

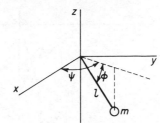

Figure 8.1-1.

ϕ are generalized coordinates, and the spherical pendulum represents a system of two degrees of freedom.

The position of the spherical pendulum can also be described by the three rectangular coordinates x, y, z, which exceed the degrees of freedom of the system by one. The coordinates x, y, z are, however, not independent, because they are related by the *constraint equation*

$$x^2 + y^2 + z^2 - l^2 = 0 \qquad (8.1\text{-}1)$$

One of the coordinates can be eliminated by the above equation, thereby reducing the number of necessary coordinates to two.

The excess coordinates exceeding the number of degrees of freedom of the system are called *superfluous coordinates*, and constraint equations equal in number to the superfluous coordinates are necessary for their elimination. Constraints are called *holonomic* if the excess coordinates can be eliminated through equations of constraint. Such constraints are in the form

$$C(q_1, q_2 \ldots q_n, t) = 0 \qquad (8.1\text{-}2)$$

We will deal only with holonomic systems in this text.

In the previous chapters we used the coordinate x for the position of each mass and since they were independent of each other and equal in number to the degrees of freedom of the system, they were also generalized coordinates.

Let us examine now the problem of defining the position of the double pendulum of Fig. 8.1-2. The double pendulum has only two degrees of freedom and the angles θ_1 and θ_2 completely define the position of m_1 and m_2. Thus θ_1 and θ_2 are generalized coordinates, i.e., $\theta_1 = q_1$ and $\theta_2 = q_2$.

The position of m_1 and m_2 can also be expressed in rectangular coordinates x, y. However they are related by the constraint equations

$$l_1^2 = x_1^2 + y_1^2$$

$$l_2^2 = (x_2 - x_1)^2 + (y_2 - y_1)^2$$

and hence are not independent.

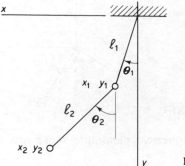

x

Figure 8.1-2.

We can express the rectangular coordinates x_i, y_i in terms of the generalized coordinates θ_1 and θ_2

$$x_1 = l_1 \sin \theta_1 \qquad x_2 = l_1 \sin \theta_1 + l_2 \sin \theta_2$$
$$y_1 = l_1 \cos \theta_1 \qquad y_2 = l_1 \cos \theta_1 + l_2 \cos \theta_2$$

and these can also be considered as constraint equations.

To determine the kinetic energy, the squares of the velocity can be written in terms of the generalized coordinates.

$$v_1^2 = \dot{x}_1^2 + \dot{y}_1^2 = \left(l_1 \dot{\theta}_1 \right)^2$$

$$v_2^2 = \dot{x}_2^2 + \dot{y}_2^2 = \left[l_1 \dot{\theta}_1 + l_2 \dot{\theta}_2 \cos(\theta_2 - \theta_1) \right]^2 + \left[l_2 \dot{\theta}_2 \sin(\theta_2 - \theta_1) \right]^2$$

The kinetic energy

$$T = \frac{1}{2} m_1 v_1^2 + \frac{1}{2} m_2 v_2^2$$

is then a function of both $q = \theta$ and $\dot{q} = \dot{\theta}$

$$T = T(q_1, q_2, \cdots \dot{q}_1, \dot{q}_2 \cdots) \tag{8.1-3}$$

For the potential energy, the reference can be chosen at the level of the support point.

$$U = -m_1(l_1 \cos \theta) - m_2(l_1 \cos \theta_1 + l_2 \cos \theta_2)$$

The potential energy is then seen to be a function only of the generalized coordinates

$$U = U(q_1, q_2 \cdots) \tag{8.1-4}$$

EXAMPLE 8.1-1

Consider the plane mechanism shown in Fig. 8.1-3 where the members are assumed to be rigid. Describe all possible motions in terms of generalized coordinates.

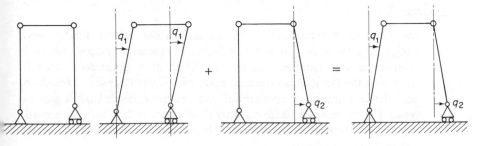

Figure 8.1-3.

Solution: As shown in Fig. 8.1-3, the displacements can be obtained by the superposition of two displacements q_1 and q_2. Since q_1 and q_2 are independent, they are generalized coordinates, and the system has two degrees of freedom.

EXAMPLE 8.1-2

The plane frame shown in Fig. 8.1-4 has flexible members. Determine a set of generalized coordinates of the system. Assume that the corners remain at 90°.

Solution: There are two translational modes, q_1 and q_2, and each of the four corners can rotate independently, making a total of six generalized coordinates, q_1, $q_2 \cdots q_6$. Allowing each of these displacements to take place with all others equal to zero, the displacement of the frame can be seen to be the superposition of the six generalized coordinates.

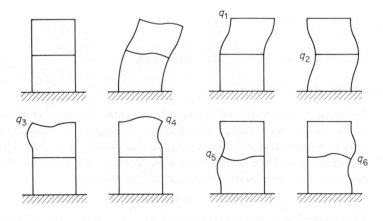

Figure 8.1-4.

EXAMPLE 8.1-3

In the lumped mass models we treated earlier, n coordinates were assigned to the n masses of the n degree of freedom system, and each coordinate was independent and qualified as a generalized coordinate. For the flexible continuous body of infinite degrees of freedom, an infinite number of coordinates are required. Such bodies can be treated as systems of finite number of degrees of freedom by considering its deflection to be the sum of its normal modes multiplied by generalized coordinates

$$y(x, t) = \phi_1(x)q_1(t) + \phi_2(x)q_2(t) + \phi_3(x)q_3(t) + \cdots$$

In many problems only a finite number of normal modes are sufficient, and the series can be terminated at n terms, thereby reducing the problem to that of a system of n degrees of freedom. For example, the motion of a slender free-free beam struck by a force P at point (a) can be described in terms of two rigid body motions of translation and rotation plus its normal modes of elastic vibration as shown in Fig. 8.1-5.

$$y(x, t) = \phi_T q_T + \phi_R q_R + \phi_1(x)q_1 + \phi_2(x)q_2 + \cdots$$

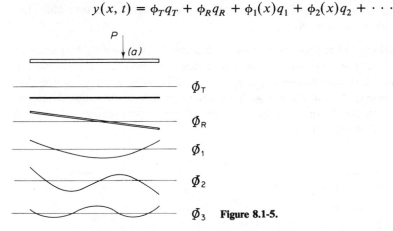

Figure 8.1-5.

EXAMPLE 8.1-4

In defining the motion of a framed structure, the number of coordinates chosen often exceed the number of degrees of freedom of the system so that constraint equations are involved. It is then desirable to express all of the coordinates u in terms of the fewer generalized coordinates q by a matrix equation of the form

$$u = Cq$$

The generalized coordinates q can be chosen arbitrarily from the coordinates u.

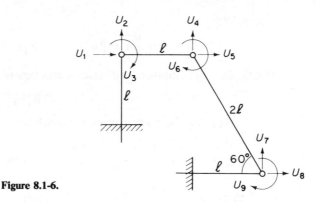

Figure 8.1-6.

As an illustration of this equation we will consider the framed structure of Fig. 8.1-6 consisting of four beam elements. We will be concerned only with the displacement of the nodes and not the stresses in the members, which would require an added consideration of the distribution of the masses.

In Fig. 8.1-6 we have four element members with three nodes which may undergo displacement. Two linear displacements and one rotation may be possible for each node. We can label them u_1 to u_9. For compatibility of displacement, the following constraints are observed

$$u_2 = u_8 = 0 \quad \text{(no axial extension)}$$
$$u_1 = u_5 \qquad \text{(axial length remains unchanged)}$$

$$(u_4 \cos 30° - u_5 \cos 60°) - (u_7 \cos 30° - u_8 \cos 60°) = 0$$

We will now disregard u_2 and u_8 which are zero and rewrite the above equations in matrix form.

$$\begin{bmatrix} 1 & 0 & -1 & 0 \\ 0 & 0.866 & -0.500 & -0.866 \end{bmatrix} \begin{Bmatrix} u_1 \\ u_4 \\ u_5 \\ u_7 \end{Bmatrix} = 0 \qquad \text{(a)}$$

Thus, the two constraint equations are in the form

$$[A]\{u\} = 0 \qquad \text{(b)}$$

We actually have seven coordinates (u_1, u_3, u_4, u_5, u_6, u_7, u_9) and two constraint equations. Thus, the degrees of freedom of the system are $7 - 2 = 5$ indicating that of the seven coordinates, five can be chosen as generalized coordinates q.

Of the four coordinates in the constraint equation, we will choose u_5 and u_7 as two of the generalized coordinates and partition

Eq. (a) as

$$[a \mid b]\left\{\frac{u}{q}\right\} = [a]\{u\} + [b]\{q\} = 0 \tag{c}$$

Thus, the superfluous coordinates u can be expressed in terms of q as

$$\{u\} = -[a]^{-1}[b]\{q\} \tag{d}$$

Applying the above procedure to Eq. (a), we have

$$\begin{bmatrix} 1 & 0 \\ 0 & 0.866 \end{bmatrix}\begin{Bmatrix} u_1 \\ u_4 \end{Bmatrix} + \begin{bmatrix} -1 & 0 \\ -0.5 & -0.866 \end{bmatrix}\begin{Bmatrix} u_5 \\ u_7 \end{Bmatrix} = \begin{Bmatrix} 0 \\ 0 \end{Bmatrix}$$

$$\begin{Bmatrix} u_1 \\ u_4 \end{Bmatrix} = \begin{bmatrix} 1 & 0 \\ 0 & \dfrac{1}{0.866} \end{bmatrix}\begin{bmatrix} 1 & 0 \\ 0.5 & 0.866 \end{bmatrix}\begin{Bmatrix} u_5 \\ u_7 \end{Bmatrix} = \begin{bmatrix} 1 & 0 \\ 0.578 & 1 \end{bmatrix}\begin{Bmatrix} u_5 \\ u_7 \end{Bmatrix}$$

By supplying the remaining q_i as indentities, all of the u's can be expressed in terms of the q's

$$\{u\} = [C]\{q\} \tag{e}$$

where the left side includes all the u's and the right column contains only the generalized coordinates. Thus, in our case, the seven u's expressed in terms of the five q's become

$$\begin{Bmatrix} u_1 \\ u_3 \\ u_4 \\ u_5 \\ u_6 \\ u_7 \\ u_9 \end{Bmatrix} = \begin{bmatrix} 0 & 1 & 0 & 0 & 0 \\ 1 & 0 & 0 & 0 & 0 \\ 0 & 0.578 & 0 & 1 & 0 \\ 0 & 1 & 0 & 0 & 0 \\ 0 & 0 & 1 & 0 & 0 \\ 0 & 0 & 0 & 1 & 0 \\ 0 & 0 & 0 & 0 & 1 \end{bmatrix}\begin{Bmatrix} u_3 \\ u_5 \\ u_6 \\ u_7 \\ u_9 \end{Bmatrix} \tag{f}$$

In Eq. (e) or (f) the matrix C is the constraint matrix relating u to q.

8.2   VIRTUAL WORK

A virtual displacement δx, $\delta \theta$, δr, etc., is an infinitesimal change in the coordinate which may be conceived in any manner irrespective of the time t, but without violating the constraints of the system.

Consider a system of particles acted upon by several forces. If the system is in static equilibrium, the resultant R_j of the forces acting on any particle j must be zero, and the work done by these forces in a virtual displacement δr_j is zero

$$\delta W = \sum_j \mathbf{R}_j \cdot \delta \mathbf{r}_j = 0 \tag{8.2-1}$$

If the force R_j is separated into an applied force F_j and a constraint force

f_i, then F_j is balanced by f_i, and neither force is zero. Limiting our discussion to constraint forces that do no work, such as the reaction of a smooth floor, the virtual work equation reduces to

$$\delta W = \sum_j \mathbf{F}_j \cdot \delta \mathbf{r}_j = 0 \qquad (8.2\text{-}2)$$

which expresses the principle of virtual work as presented by J. Bernoulli (1717). In summary, the above equation states that if a system is in static equilibrium, the work done by the *applied forces* in a virtual displacement compatible with the constraints is equal to zero.

EXAMPLE 8.2-1

To illustrate the method of virtual work, consider the problem of establishing the equilibrium position of the double pendulum of Fig. 8.2-1 when the lower mass is displaced by a horizontal force P.

The position of the double pendulum is completely established by the generalized coordinates θ_1 and θ_2. The position of each mass is written as

$$\mathbf{r}_1 = l(\sin \theta_1 \mathbf{i} + \cos \theta_1 \mathbf{j})$$
$$\mathbf{r}_2 = l(\sin \theta_1 + \sin \theta_2)\mathbf{i} + (\cos \theta_1 + \cos \theta_2)\mathbf{j}$$

Using the method of virtual work, we wish to satisfy Eq. (8.2-2)

$$\delta W = \sum_i \mathbf{F}_i \cdot \delta \mathbf{r}_i$$

There are three applied forces

$$\mathbf{F}_1 = m_1 g\, \mathbf{j}$$
$$\mathbf{F}_2 = m_2 g\, \mathbf{j}$$
$$\mathbf{F}_3 = P\, \mathbf{i}$$

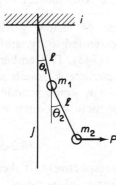

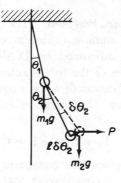

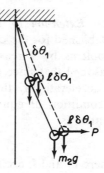

(a)   (b)   (c)

Figure 8.2-1.

The virtual displacements are

$$\delta \mathbf{r}_1 = \frac{\partial \mathbf{r}_1}{\partial \theta_1} \delta\theta_1 + \frac{\partial \mathbf{r}_1}{\partial \theta_2} \delta\theta_2$$

$$= l(\cos \theta_1 \mathbf{i} - \sin \theta_1 \mathbf{j}) \, \delta\theta_1$$

$$\delta \mathbf{r}_2 = l(\cos \theta_1 \, \delta\theta_1 + \cos \theta_2 \, \delta\theta_2)\mathbf{i} - l(\sin \theta_1 \, \delta\theta_1 + \sin \theta_2 \, \delta\theta_2)\mathbf{j}$$

$$\delta \mathbf{r}_3 = \delta \mathbf{r}_2$$

Substituting into δW and noting the dot product

$$\delta W = (Pl \cos \theta_1 - (m_1 + m_2)gl \sin \theta_1) \, \delta\theta_1$$

$$+ (Pl \cos \theta_2 - m_2 gl \sin \theta_2) \, \delta\theta_2 = 0$$

Since $\delta\theta_1$ and $\delta\theta_2$ are arbitrary, the above equation is satisfied by

$$Pl \cos \theta_1 - (m_1 + m_2)gl \sin \theta_2 = 0$$

$$Pl \cos \theta_2 - m_2 gl \sin \theta_2 = 0$$

or

$$\tan \theta_1 = \frac{P}{(m_1 + m_2)g}$$

$$\tan \theta_2 = \frac{P}{m_2 g}$$

These results can be visualized graphically by examining Fig. 8.2-1b and c which shows the two virtual displacements taken separately. In Fig. 8.2-1b the components of $l \, \delta\theta_2$ are $l \, \delta\theta_2(\cos \theta_2 i - \sin \theta_2 j)$, and the work done is

$$Pl \, \delta\theta_2 \cos \theta_2 - m_2 gl \, \delta\theta_2 \sin \theta_2 = 0$$

In Fig. 8.2-1c, θ_2 is unchanged while θ_1 is given a virtual displacement $\delta\theta_1$. Thus both m_1 and m_2 undergo the same displacement $l \, \delta\theta_1$, and all three forces do virtual work leading to the equation for $\delta\theta_1$.

Extension to Dynamic Problems. The principle of virtual work, established for the case of static equilibrium, can be extended to dynamic problems by a reasoning advanced by D'Alembert (1743). D'Alembert reasoned that since the sum of the forces acting on a particle m_i results in its acceleration $\ddot{\mathbf{r}}_i$, the application of a force equal to $-m_i\ddot{\mathbf{r}}_i$ would produce a condition of equilibrium. The equation for the particle can then be written as

$$\mathbf{F}_i + \mathbf{f}_i - m_i\ddot{\mathbf{r}}_i = 0 \qquad (8.2\text{-}3)$$

where $\mathbf{F}_i$ and $\mathbf{f}_i$ are the applied and constraint forces, respectively. It then follows from the principle of virtual work that for a system of particles

$$\delta W = \sum_i (\mathbf{F}_i - m_i\ddot{\mathbf{r}}_i) \cdot \delta \mathbf{r}_i = 0 \qquad (8.2\text{-}4)$$

where the work done by the constraint forces $\mathbf{f}_i$ is again zero.

EXAMPLE 8.2-2

We will illustrate the extension of the virtual work principle to dynamics by considering the simple pendulum of Fig. 8.2-2. The equation we are concerned with is Eq. (8.2-4).

$$\delta W = \sum_i (\mathbf{F}_i - m_i \ddot{\mathbf{r}}_i) \cdot \delta \mathbf{r}_i = 0$$

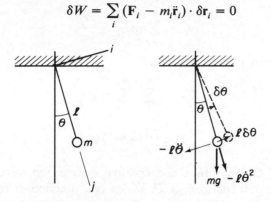

Figure 8.2-2.   Virtual work in dynamics.

In this simple problem we have only one generalized coordinate θ. The mass undergoes acceleration of

$$\ddot{\mathbf{r}} = l\ddot{\theta}\mathbf{i} - l\dot{\theta}^2 \mathbf{j}$$

and the force term including the D'Alembert force is

$$(\mathbf{F} - m\ddot{\mathbf{r}}) = (-mg \sin \theta - ml\ddot{\theta})\mathbf{i} + (mg \cos \theta + ml\dot{\theta}^2)\mathbf{j}$$

Its dot product with the virtual displacement $l\,\delta\theta\mathbf{i}$ then becomes

$$\delta W = -(mg \sin \theta + ml\ddot{\theta})l\,\delta\theta = 0$$

and we obtain the well-known equation of motion for the pendulum

$$\ddot{\theta} + \frac{g}{l} \sin \theta = 0$$

8.3   KINETIC ENERGY, POTENTIAL ENERGY, AND GENERALIZED FORCE

Kinetic Energy.   Representing the system by N particles, the instantaneous position of each particle may be expressed in terms of the N generalized coordinates

$$\mathbf{r}_j = \mathbf{r}_j(q_1 q_2 \cdots q_N) \qquad (8.3\text{-}1)$$

The velocity of the jth particle is

$$\mathbf{v}_j = \sum_{i=1}^{N} \frac{\partial \mathbf{r}_j}{\partial q_i} \dot{q}_i \qquad (8.3\text{-}2)$$

and the kinetic energy of the system becomes

$$T = \frac{1}{2} \sum_{j=1}^{N} m_j \mathbf{v}_j \cdot \mathbf{v}_j = \frac{1}{2} \sum_{i=1}^{N} \sum_{j=1}^{N} \left(\sum_{i=1}^{N} m_i \frac{\partial \mathbf{r}_i}{\partial q_j} \cdot \frac{\partial \mathbf{r}_i}{\partial q_i} \right) \dot{q}_i \dot{q}_j \qquad (8.3\text{-}3)$$

Defining the *generalized mass* as

$$m_{ij} = \sum_{i=1}^{N} m_i \frac{\partial \mathbf{r}_i}{\partial q_j} \cdot \frac{\partial \mathbf{r}_i}{\partial q_i} \qquad (8.3\text{-}4)$$

the kinetic energy may be written as

$$T = \frac{1}{2} \sum_{i=1}^{N} \sum_{j=1}^{N} m_{ij} \dot{q}_i \dot{q}_j$$

$$= \frac{1}{2} \{\dot{q}\}'[m]\{\dot{q}\} \qquad (8.3\text{-}5)$$

Potential Energy. In a conservative system, the forces can be derived from the potential energy U, which is a function of the generalized coordinates q_j. Expanding U in a Taylor series about the equilibrium position, we have for a system of n degrees of freedom

$$U = U_0 + \sum_{j=1}^{n} \left(\frac{\partial U}{\partial q_j} \right)_0 q_j + \frac{1}{2} \sum_{j=1}^{n} \sum_{l=1}^{n} \left(\frac{\partial^2 U}{\partial q_j \partial q_l} \right)_0 q_j q_l + \cdots \qquad (8.3\text{-}6)$$

In this expression U_0 is an arbitrary constant which we can set equal to zero. The derivatives of U are evaluated at the equilibrium position 0 and are constants when the q_j's are small quantities equal to zero at the equilibrium position. Since U is a minimum in the equilibrium position, the first derivative $(\partial U/\partial q_j)_0$ is zero, which leaves only $(\partial^2 U/\partial q_j \partial q_l)_0$ and higher order terms.

In the theory of small oscillations about the equilibrium position, terms beyond the second order are ignored and the equation for the potential energy reduces to

$$U = \frac{1}{2} \sum_{j=1}^{n} \sum_{l=1}^{n} \left(\frac{\partial^2 U}{\partial q_j \partial q_l} \right)_0 q_j q_l \qquad (8.3\text{-}7)$$

The second derivative evaluated at 0 is a constant associated with the *generalized stiffness*, which is

$$k_{jl} = \left(\frac{\partial^2 U}{\partial q_j \partial q_l} \right)_0 \qquad (8.3\text{-}8)$$

and the potential energy is written as

$$U = \frac{1}{2} \sum_{j=1}^{n} \sum_{l=1}^{n} k_{jl} q_j q_l$$

$$= \frac{1}{2} \{q\}'[k]\{q\} \qquad (8.3\text{-}9)$$

Generalized Force.   For the development of the generalized force, we start from Eq. (8.3-1). The virtual displacement of the coordinate $\mathbf{r}_j$ is

$$\delta\mathbf{r}_j = \sum_i \frac{\delta\mathbf{r}_j}{\delta q_i} \delta q_i \qquad (8.3\text{-}10)$$

and the time t is not involved.

When the system is in equilibrium, the virtual work can now be expressed in terms of the generalized coordinates q_i

$$\delta W = \sum_j \mathbf{F}_j \cdot \delta\mathbf{r}_j = \sum_j \sum_i \mathbf{F}_j \cdot \frac{\delta\mathbf{r}_j}{\delta q_i} \delta q_i \qquad (8.3\text{-}11)$$

Interchanging the order of summation and letting

$$Q_i = \sum_j \mathbf{F}_j \cdot \frac{\delta\mathbf{r}_j}{\delta q_i} \qquad (8.3\text{-}12)$$

be defined as the *generalized force*, the virtual work for the system, expressed in terms of the generalized coordinates, becomes

$$\delta W = \sum_i Q_i \delta q_i \qquad (8.3\text{-}13)$$

EXAMPLE 8.3-1

Determine the generalized mass when the displacement at position x is represented by the equation

$$r(x, t) = \phi_1(x)q_1(t) + \phi_2(x)q_2(t) + \cdots \phi_N(x)q_N(t)$$

$$= \sum_{i=1}^{N} \phi_i(x)q_i(t) \qquad (a)$$

where $\phi_i(x)$ are shape functions of only x.

Solution: The velocity is

$$v(x) = \sum_{i\text{-}1}^{N} \phi_i(x)\dot{q}_i(t) \qquad (b)$$

and the kinetic energy becomes

$$T = \frac{1}{2} \sum_{i=1}^{N} \sum_{j=1}^{N} \dot{q}_i\dot{q}_j \int \phi_i(x)\phi_j(x)\, dm$$

$$= \frac{1}{2} \sum_{i=1}^{N} \sum_{j=1}^{N} m_{ij}\dot{q}_i\dot{q}_j \qquad (c)$$

Thus, the generalized mass is

$$m_{ij} = \int \phi_i(x)\phi_j(x)\, dm \qquad (d)$$

where the integration is carried out over the entire system. In case the system consists of discrete masses, m_{ij} becomes

$$m_{ij} = \sum_{p=1}^{N} m_p \phi_i(x_p) \phi_j(x_p) \qquad (e)$$

EXAMPLE 8.3-2

Determine the generalized stiffness for a beam of constant cross section EI when the displacement $y(x, t)$ is represented by the sum

$$y(x, t) = \sum_{i=1}^{n} \varphi_i(x) q_i(t) \qquad (a)$$

The potential energy of a beam in bending is

$$U = \frac{1}{2} \int EI \left(\frac{d^2 y}{dx^2} \right)^2 dx \qquad (b)$$

Substituting for

$$\frac{d^2 y}{dx^2} = \sum_{i=1}^{n} \varphi_i''(x) q_i(t)$$

we obtain

$$U = \frac{1}{2} \sum_i \sum_j q_i q_j \int EI \varphi_i'' \varphi_j'' \, dx$$

$$= \frac{1}{2} \sum_i \sum_j k_{ij} q_i q_j \qquad (c)$$

and the generalized stiffness is

$$k_{ij} = \int EI \varphi_i'' \varphi_j'' \, dx \qquad (d)$$

EXAMPLE 8.3-3

The frame of Fig. 8.1-3 with rigid members is acted upon by the forces and moments shown in Fig. 8.3-1. Determine the generalized forces.

Solution: We will let δq_1 be the virtual displacement of the upper left corner and δq_2 the translation of the right support hinge. Due to δq_1 the virtual work done is

$$Q_1 \, \delta q_1 = F_1 \, \delta q_1 - F_2 \frac{a}{l} \delta q_1 + (M_1 - M_2) \frac{1}{l} \delta q_1$$

$$\therefore Q_1 = F_1 - \frac{a}{l} F_2 + \frac{1}{l} (M_1 - M_2)$$

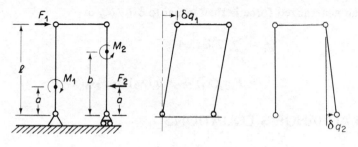

Figure 8.3-1.

The virtual work done due to δq_2 is

$$Q_2 \, \delta q_2 = -F_2(l-a)\frac{\delta q_2}{l} + M_2\frac{\delta q_2}{l}$$

$$\therefore Q_2 = [-F_2(l-a) + M_2]\frac{1}{l}$$

It should be noted that the dimension of Q_1 and Q_2 is that of a force.

EXAMPLE 8.3-4

Three forces F_1, F_2, and F_3 act at discrete points x_1, x_2, and x_3 of a structure whose displacement is expressed by the equation

$$y(x, t) = \sum_{i=1}^{n} \varphi_i(x) q_i(t)$$

Determine the generalized force Q_i.

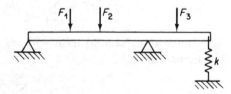

Figure 8.3-2.

Solution: The virtual displacement δy is

$$\sum_{i=1}^{n} \varphi_i(x) \, \delta q_i$$

and the virtual work due to this displacement is

$$\delta W = \sum_{j=1}^{3} F_j \cdot \left(\sum_{i=1}^{n} \varphi_i(x_j) \, \delta q_i \right)$$

$$= \sum_{i=1}^{n} \delta q_i \left(\sum_{j=1}^{3} F_j \varphi_i(x_j) \right) = \sum_{i=1}^{n} Q_i \, \delta q_i$$

The generalized force is then equal to $\delta W/\delta q_i$ or

$$Q_i = \sum_{j=1}^{3} F_j \varphi_i(x_j)$$

$$= F_1 \varphi_i(x_1) + F_2 \varphi_i(x_2) + F_3 \varphi_i(x_3)$$

8.4 LAGRANGE'S EQUATIONS

Lagrange's equations are differential equations of motion expressed in terms of generalized coordinates. We present here a brief development for the general form of these equations in terms of the kinetic and potential energies.

Consider first a conservative system where the sum of the kinetic and potential energies is a constant. The differential of the total energy is then zero.

$$d(T + U) = 0 \tag{8.4-1}$$

The kinetic energy T is a function of the generalized coordinates q_i and the generalized velocities $\dot{q}_i$, whereas the potential energy U is a function only of q_i.

$$T = T(q_1, q_2, \cdots q_N, \dot{q}_1, \dot{q}_2 \cdots \dot{q}_N)$$
$$U = U(q_1, q_2 \cdots q_N) \tag{8.4-2}$$

The differential of T is

$$dT = \sum_{i=1}^{N} \frac{\partial T}{\partial q_i} dq_i + \sum_{i=1}^{N} \frac{\partial T}{\partial \dot{q}_i} d\dot{q}_i \tag{8.4-3}$$

To eliminate the second term with $d\dot{q}_i$, we start with the equation for the kinetic energy

$$T = \frac{1}{2} \sum_{i=1}^{N} \sum_{j=1}^{N} m_{ij} \dot{q}_i \dot{q}_j \tag{8.4-4}$$

Differentiating this equation with respect to $\dot{q}_i$, multiplying by $\dot{q}_i$, and summing over i from 1 to N, we obtain a result which is equal to

$$\sum_{i=1}^{N} \frac{\partial T}{\partial \dot{q}_i} \dot{q}_i = \sum_{i=1}^{N} \sum_{j=1}^{N} m_{ij} \dot{q}_j \dot{q}_i = 2T$$

or

$$2T = \sum_{i=1}^{N} \frac{\partial T}{\partial \dot{q}_i} \dot{q}_i \tag{8.4-5}$$

We now form the differential of $2T$ from the above equation by using the product rule in calculus

$$2 \, dT = \sum_{i=1}^{N} d\left(\frac{\partial T}{\partial \dot{q}_i}\right) \dot{q}_i + \sum_{i=1}^{N} \frac{\partial T}{\partial \dot{q}_i} d\dot{q}_i \qquad (8.4\text{-}6)$$

Subtracting Eq. (8.4-3) from this equation, the second term with $d\dot{q}_i$ is eliminated. By shifting the scalar quantity dt, the term $d(\partial T/\partial \dot{q}_i)\dot{q}_i$ becomes $d/dt(\partial T/\partial \dot{q}_i) \, dq_i$ and the result is

$$dT = \sum_{i=1}^{N} \left[\frac{d}{dt}\left(\frac{\partial T}{\partial \dot{q}_i}\right) - \frac{\partial T}{\partial q_i} \right] dq_i \qquad (8.4\text{-}7)$$

From Eq. (8.4-2) the differential of U is

$$dU = \sum_{i=1}^{N} \frac{\partial U}{\partial q_i} dq_i \qquad (8.4\text{-}8)$$

Thus, Eq. (8.4-1) for the invariance of the total energy becomes

$$d(T + U) = \sum_{i=1}^{N} \left[\frac{d}{dt}\left(\frac{\partial T}{\partial \dot{q}_i}\right) - \frac{\partial T}{\partial q_i} + \frac{\partial U}{\partial q_i} \right] dq_i = 0 \qquad (8.4\text{-}9)$$

Since the N generalized coordinates are independent of one another, the dq_i may assume arbitrary values. Therefore, the above equation is satisfied only if

$$\frac{d}{dt}\left(\frac{\partial T}{\partial \dot{q}_i}\right) - \frac{\partial T}{\partial q_i} + \frac{\partial U}{\partial q_i} = 0 \qquad i = 1, 2, \cdots N \qquad (8.4\text{-}10)$$

This is Lagrange's equation for the case in which all forces have a potential U. They can be somewhat modified by introducing the Lagrangian $L = (T - U)$. Since $\partial U/\partial \dot{q}_i = 0$, Eq. (8.4-10) can be written in terms of L as

$$\frac{d}{dt}\left(\frac{\partial L}{\partial \dot{q}_i}\right) - \frac{\partial L}{\partial q_i} = 0 \qquad i = 1, 2, \cdots N \qquad (8.4\text{-}11)$$

When the system is also subjected to given forces that do not have a potential, we have instead of Eq. (8.4-1)

$$d(T + U) = dW \qquad (8.4\text{-}12)$$

where dW is the work of the nonpotential forces when the system is subjected to an arbitrary infinitesimal displacement. From Eq. (8.3-13) dW is expressed in terms of the generalized coordinates q_i.

$$dW = \sum_{i=1}^{N} Q_i \, dq_i \qquad (8.4\text{-}13)$$

where the quantities Q_i are known as the generalized forces associated with the generalized coordinate q_i. Lagrange's equation including nonconservative forces then becomes

$$\frac{d}{dt}\left(\frac{\partial T}{\partial \dot{q}_i}\right) - \frac{\partial T}{\partial q_i} + \frac{\partial U}{\partial q_i} = Q_i \qquad i = 1, 2, \cdots N \qquad (8.4\text{-}14)$$

EXAMPLE 8.4-1

Using Lagrange's method, set up the equations of motion for the system shown in Fig. 8.4-1.

Solution: The kinetic and potential energies are

$$T = \tfrac{1}{2}m\dot{q}_1^2 + \tfrac{1}{2}J\dot{q}_2^2$$

$$U = \tfrac{1}{2}kq_1^2 + \tfrac{1}{2}k(rq_2 - q_1)^2$$

and from the work done by the external moment, the generalized force is

$$\delta W = \mathcal{M}(t)\,\delta q_2 \qquad \therefore Q_2 = \mathcal{M}(t)$$

Substituting into Lagrange's equation, the equations of motion are

$$m\ddot{q}_1 + 2kq_1 - krq_2 = 0$$

$$J\ddot{q}_2 - krq_1 + kr^2q_2 = \mathcal{M}(t)$$

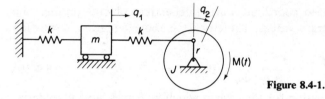

Figure 8.4-1.

EXAMPLE 8.4-2

Write the Lagrangian for the system shown in Fig. 8.4-2.

Solution:

$$T = \frac{1}{2}m(\dot{q}_1 \; \dot{q}_2)\begin{bmatrix} 3 & 0 \\ 0 & 1 \end{bmatrix}\begin{Bmatrix} \dot{q}_1 \\ \dot{q}_2 \end{Bmatrix}$$

$$U = \frac{1}{2}k(q_1 \; q_2)\begin{bmatrix} 2 & -1 \\ -1 & 2 \end{bmatrix}\begin{Bmatrix} q_1 \\ q_2 \end{Bmatrix}$$

$$L = T - U$$

$$= \frac{1}{2}(3m)\dot{q}_1^2 + \frac{1}{2}(m)\dot{q}_2^2 - kq_1^2 - kq_2^2 + kq_1q_2$$

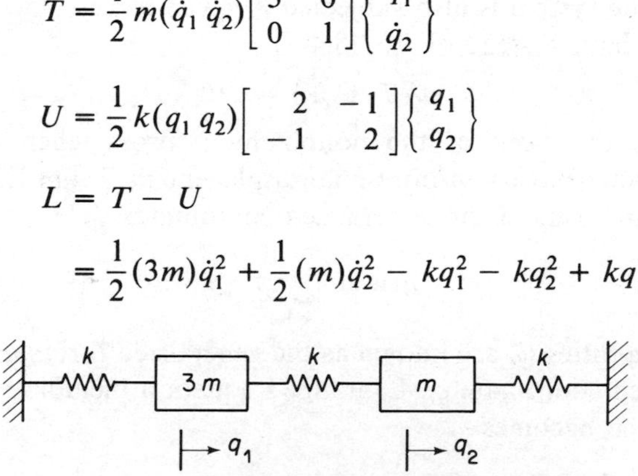

Figure 8.4-2.

EXAMPLE 8.4-3

Fig. 8.4-3 shows a simplified model of a two-story building whose foundation is subject to translation and rotation. Determine T and U and the equations of motion.

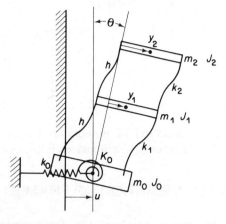

Figure 8.4-3.

Solution: We choose u and θ for the translation and rotation of the foundation and y for the elastic displacement of the floors. The equations for T and U become

$$T = \frac{1}{2} m_0 \dot{u}^2 + \frac{1}{2} J_0 \dot{\theta}^2 + \frac{1}{2} m_1 (\dot{u} + h\dot{\theta} + \dot{y}_1)^2 + \frac{1}{2} J_1 \dot{\theta}^2$$

$$+ \frac{1}{2} m_2 (\dot{u} + 2h\dot{\theta} + \dot{y}_2)^2 + \frac{1}{2} J_2 \dot{\theta}^2$$

$$U = \frac{1}{2} k_0 u^2 + \frac{1}{2} K_0 \theta^2 + \frac{1}{2} k_1 y_1^2 + \frac{1}{2} k_2 (y_2 - y_1)^2$$

where u, θ, y_1, and y_2 are the generalized coordinates. Substituting into Lagrange's equation, we obtain, for example,

$$\frac{\partial T}{\partial \dot{\theta}} = (J_0 + J_1 + J_2)\dot{\theta} + m_1 h (\dot{u} + h\dot{\theta} + \dot{y}_1) + m_2 2h (\dot{u} + 2h\dot{\theta} + \dot{y}_2)$$

$$\frac{\partial U}{\partial \theta} = K_0$$

The four equations in matrix form become

$$
\begin{bmatrix}
(m_0 + m_1 + m_2) & (m_1 + 2m_2)h & \vdots & m_1 & m_2 \\
(m_1 + 2m_2)h & \left(\sum J + m_1 h^2 + 2m_2 h^2\right) & \vdots & m_1 h & 2m_2 h \\
\hline
m_1 & m_1 h & \vdots & m_1 & 0 \\
m_2 & 2m_2 h & \vdots & 0 & m_2
\end{bmatrix}
\begin{Bmatrix}
\ddot{u} \\
\ddot{\theta} \\
\hline
\ddot{y}_1 \\
\ddot{y}_2
\end{Bmatrix}
$$

$$
+
\begin{bmatrix}
k_0 & 0 & \vdots & 0 & 0 \\
0 & K_0 & \vdots & 0 & 0 \\
\hline
0 & 0 & \vdots & (k_1 + k_2) & -k_2 \\
0 & 0 & \vdots & -k_2 & k_2
\end{bmatrix}
\begin{Bmatrix}
u \\
\theta \\
\hline
y_1 \\
y_2
\end{Bmatrix} = \{0\}
$$

It should be noted that the equation represented by the upper left corner of the matrices is that of rigid body translation and rotation.

8.5   VIBRATION OF FRAMED STRUCTURES

In framed structures the displacement and rotation of corners or joints, called *nodes*, can often serve as generalized coordinates. We can also assign mass at these nodes so that the equations of motion can be written in terms of generalized coordinates.

The determination of the stiffness matrix for the structure, however, requires the use of beam element stiffnesses, a collection of which was presented in Chapter 6.

The structure must be cut at the nodes to form beam elements and the force displacement relation of each node can be determined from the free-body diagram of the node joint.

EXAMPLE 8.5-1

Determine the generalized coordinates for the system shown in Fig. 8.5-1 and evaluate the stiffness and the mass matrices for the equations of motion.

Solution: It appears that three generalized coordinates are required, as shown in Fig. 8.5-1; however, the right end of the horizontal member is pinned and free to slide horizontally and cannot sustain a moment. The configuration of case (3) Table 6.1-1 fits the boundary conditions of the horizontal member, and since $\theta_2 = \frac{1}{2}\theta_1$ in this table, $q_3 = \frac{1}{2}q_2$ and the problem can be solved in terms of q_1 and q_2 as a two degree of freedom system.

The stiffness matrix can be determined by considering the superposition of the two configurations broken down into beam

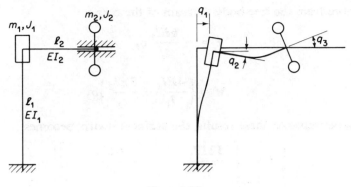

Figure 8.5-1.

elements shown in Fig. 8.5-2. Configuration (a) corresponds to case (1) in Table 6.1-1 and we can write the equations

$$F = \frac{12EI_1}{l_1^3} q_1$$

$$M = -\frac{6EI_1}{l_1^2} q_1$$

Configuration (b) is broken down into case (2) and case (3), and we

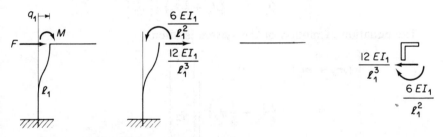

(a)

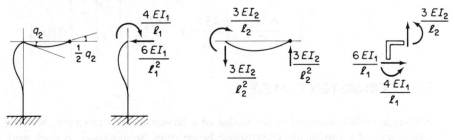

(b)

Figure 8.5-2.

have from the free-body diagram of the corner

$$F = -\frac{6EI_1}{l_1^2}q_2$$

$$M = \left(\frac{4EI_1}{l_1} + \frac{3EI_2}{l_2}\right)q_2$$

Superimposing these results, the stiffness matrix becomes

$$\left\{\begin{matrix}F\\M\end{matrix}\right\} = \begin{bmatrix} \dfrac{12EI_1}{l_1^3} & -\dfrac{6EI_1}{l_1^2} \\[2mm] -\dfrac{6EI_1}{l_1^2} & \left(\dfrac{4EI_1}{l_1} + \dfrac{3EI_2}{l_2}\right) \end{bmatrix}\left\{\begin{matrix}q_1\\q_2\end{matrix}\right\}$$

The kinetic energy equation can be written from inspection of Fig. 8.5-1.

$$T = \frac{1}{2}(m_1 + m_2)\dot{q}_1^2 + \frac{1}{2}J_1\dot{q}_2^2 + \frac{1}{2}J_2\left(\frac{\dot{q}_2}{2}\right)^2$$

and the dynamic term of Lagrange's equation becomes

$$\begin{bmatrix} (m_1 + m_2) & 0 \\ 0 & (J_1 + \tfrac{1}{4}J_2) \end{bmatrix}\left\{\begin{matrix}\ddot{q}_1\\\ddot{q}_2\end{matrix}\right\}$$

The equation of motion of the system is, then,

$$\begin{bmatrix} (m_1 + m_2) & 0 \\ 0 & \left(J_1 + \dfrac{1}{4}J_2\right) \end{bmatrix}\left\{\begin{matrix}\ddot{q}_1\\\ddot{q}_2\end{matrix}\right\}$$

$$+ \begin{bmatrix} \dfrac{12EI_1}{l_1^3} & -\dfrac{6EI_1}{l_1^2} \\[2mm] -\dfrac{6EI_1}{l_1^2} & \left(\dfrac{4EI_1}{l_1} + \dfrac{3EI_2}{l_2}\right) \end{bmatrix}\left\{\begin{matrix}q_1\\q_2\end{matrix}\right\} = \left\{\begin{matrix}F(t)\\M(t)\end{matrix}\right\}$$

8.6 CONSISTENT MASS

Masses are often lumped at the nodes of a structure. For example, half the total mass of a uniformly distributed beam may be assigned to each end. The advantage of this simple procedure is that the mass matrix becomes diagonal. However, a more accurate representation of mass is obtained by

Identification
of forces and moments from Table 6.1-1.

expressing a uniformly distributed mass in terms of the end deflections and slope according to the beam convention of Sec. 6.1, which is repeated here. This will result in a full matrix which is called the *consistent mass matrix*.

To derive the equation for the consistent mass matrix, we recognize that the general shape of a beam can be represented by a cubic equation

$$y = p_1 + \xi p_2 + \xi^2 p_3 + \xi^3 p_4, \quad \xi = \frac{x}{l} \qquad (8.6\text{-}1)$$

The mass is identified by the kinetic energy equation

$$T = \frac{l}{2} \int_0^1 m\dot{y}^2 \frac{dx}{l} \qquad (8.6\text{-}2)$$

We will now rewrite the deflection in matrix form as

$$y = \begin{bmatrix} 1 & \xi & \xi^2 & \xi^3 \\ 0 & 0 & 0 & 0 \\ 0 & 0 & 0 & 0 \\ 0 & 0 & 0 & 0 \end{bmatrix} \begin{Bmatrix} p_1 \\ p_2 \\ p_3 \\ p_4 \end{Bmatrix} = Lp \qquad (8.6\text{-}3)$$

The square of the velocity then becomes

$$\dot{y}^2 = (L\dot{p})(L\dot{p}) = \dot{p}'(L'L)\dot{p} \qquad (8.6\text{-}4)$$

We have for $L'L$

$$L'L = \begin{bmatrix} 1 \\ \xi \\ \xi^2 \\ \xi^3 \end{bmatrix} \begin{bmatrix} 1 & \xi & \xi^2 & \xi^3 \end{bmatrix} = \begin{bmatrix} 1 & \xi & \xi^2 & \xi^3 \\ \xi & \xi^2 & \xi^3 & \xi^4 \\ \xi^2 & \xi^3 & \xi^4 & \xi^5 \\ \xi^3 & \xi^4 & \xi^5 & \xi^6 \end{bmatrix}$$

and integrating from $\xi = 0$ to 1, the kinetic energy becomes

$$T = \frac{l}{2}\dot{p}' \int_0^1 mL'L \, d\xi \, \dot{p}$$

$$= \frac{1}{2}(\dot{p}_1 \dot{p}_2 \dot{p}_3 \dot{p}_4)ml \begin{bmatrix} 1 & \dfrac{1}{2} & \dfrac{1}{3} & \dfrac{1}{4} \\ \dfrac{1}{2} & \dfrac{1}{3} & \dfrac{1}{4} & \dfrac{1}{5} \\ \dfrac{1}{3} & \dfrac{1}{4} & \dfrac{1}{5} & \dfrac{1}{6} \\ \dfrac{1}{4} & \dfrac{1}{5} & \dfrac{1}{6} & \dfrac{1}{7} \end{bmatrix} \begin{Bmatrix} \dot{p}_1 \\ \dot{p}_2 \\ \dot{p}_3 \\ \dot{p}_4 \end{Bmatrix} = \frac{1}{2}\dot{p}' B\dot{p} \qquad (8.6\text{-}5)$$

The above square matrix represents the uniformly distributed mass in the p

coordinates. To express the mass in terms of the end displacements, we need the equation relating p to the rotation and translation of the ends. From the equations

$$y = p_1 + \xi p_2 + \xi^2 p_3 + \xi^3 p_4$$
$$\theta l = \quad\quad p_2 + 2\xi p_3 + 3\xi^2 p_4$$

this relationship is available by letting $\xi = 0$ and 1, which is

$$\begin{Bmatrix} p_1 \\ p_2 \\ p_3 \\ p_4 \end{Bmatrix} = \begin{bmatrix} 0 & 0 & 1 & 0 \\ 1l & 0 & 0 & 0 \\ -2l & -1l & -3 & 3 \\ 1l & 1l & 2 & -2 \end{bmatrix} \begin{Bmatrix} \theta_1 \\ \theta_2 \\ y_3 \\ y_4 \end{Bmatrix} = C\delta \qquad (8.6\text{-}6)$$

Substituting for p in terms of δ, we obtain

$$T = \frac{1}{2}\delta'[\,C'BC\,]\,\dot{\delta} = \frac{1}{2}\dot{\delta}'\mathfrak{M}\dot{\delta} \qquad (8.6\text{-}7)$$

where

$$\mathfrak{M} = \frac{ml}{420}\begin{bmatrix} 4l^2 & -3l^2 & 22l & 13l \\ -3l^2 & 4l^2 & -13l & -22l \\ 22l & -13l & 156 & 54 \\ 13l & -22l & 54 & 156 \end{bmatrix} \qquad (8.6\text{-}8)$$

is the consistent mass matrix* in terms of the end deflections.

EXAMPLE 8.6-1

Determine the consistent mass matrix for the rectangular frame of Fig. 8.6-1 in terms of coordinates q_1, q_2, and q_3.

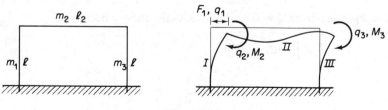

Figure 8.6-1.

Solution: The consistent mass matrix given by Eq. (8.6-8) must be applied to each member. Since Eq. (8.6-7) is that of kinetic energy, we will first determine the forces or moments associated with

*J. S. Archer, "Consistent Mass Matrix for Distributed Mass Systems," *Jour. Struct. Div. ASCE*, Vol. 89, No.STA4, (August 1963), pp. 161–178.

Lagrange's term $d/dt(\partial T/\partial \dot{\delta}_j)$. Since the mass matrix $\mathfrak{M}$ is symmetric, the term $(d/dt)(\partial T/\partial \dot{\delta}_j)$ is simply the jth row multiplied by $\ddot{\delta}$, or

$$\frac{d}{dt}\frac{\partial T}{\partial \dot{\delta}_j} = [\,j^{th} \text{ row of } \mathfrak{M}\,]\begin{Bmatrix}\ddot{\theta}_1 \\ \ddot{\theta}_2 \\ \ddot{y}_3 \\ \ddot{y}_4\end{Bmatrix}$$

This results in either a force or moment according to whether we have $d/dt(\partial T/\partial \dot{y})$ or $d/dt(\partial T/\partial \dot{\theta})$.

Let us now consider member I. (See Fig. 8.6-2.) The upper end has deflection $\theta_1 = q_2$ and $y_3 = q_1$ and the lower end has $\theta_2 = y_4 = 0$. Thus from the third row (or column) of $\mathfrak{M}$, we have

$$\frac{d}{dt}\frac{\partial T}{\partial \dot{y}_3} = F_1 = \frac{m_1 l}{420}\left[22 l\ddot{\theta}_1 + 156\ddot{y}_3\right] = \frac{m_1 l}{420}\left[22 l\ddot{q}_2 + 156\ddot{q}_1\right]$$

$$= (1, 2), (1, 1)$$

Similarly, from the first row of $\mathfrak{M}$ we have

$$\frac{d}{dt}\left(\frac{\partial T}{\partial \dot{\theta}_1}\right) = M_2 = \frac{m_1 l}{420}\left[4l^2\ddot{q}_2 + 22l\ddot{q}_1\right] = (2, 2), (2, 1)$$

where the notation (i, j) indicates the position in the new consistent mass matrix.

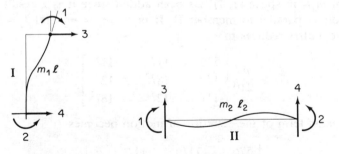

Figure 8.6-2.   **Figure 8.6-3.**

We next consider member II, with $\theta_1 = q_2$, $\theta_2 = -q_3$, and $y_3 = y_4 = 0$ (see Fig. 8.6-3).

$$\frac{d}{dt}\left(\frac{\partial T}{\partial \dot{\theta}_1}\right) = M_2 = \frac{m_2 l_2}{420}\left[4l_2^2\ddot{q}_2 + 3l_2^2\ddot{q}_3\right] = (2, 2), (2, 3)$$

$$\frac{d}{dt}\left(\frac{\partial T}{\partial \dot{\theta}_2}\right) = -M_3 = \frac{m_2 l_2}{420}\left[-3l_2^2\ddot{q}_2 - 4l_2^2\ddot{q}_3\right] = (3, 2), (3, 3)$$

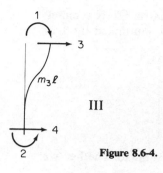

III

Figure 8.6-4.

For the third member, we have $\theta_1 = q_3$, $y_3 = q_1$, $\theta_2 = y_4 = 0$ (see Fig. 8.6-4).

$$\frac{d}{dt}\left(\frac{\partial T}{\partial \dot{y}_3}\right) = F_1 = \frac{m_3 l}{420}\left[22l\ddot{q}_3 + 156\ddot{q}_1\right] = (1, 3), (1, 1)$$

$$\frac{d}{dt}\left(\frac{\partial T}{\partial \theta_1}\right) = M_3 = \frac{m_3 l}{420}\left[4l^2\ddot{q}_3 + 22l\ddot{q}_1\right] = (3, 3), (3, 1)$$

Thus adding terms in corresponding $\mathfrak{M}$ space (i, j), we obtain for the consistent mass the following:

$$\mathfrak{M} = \frac{1}{420}\begin{bmatrix} [156(m_1l + m_3l) + 420(m_2l_2)] & 22l(m_1l) & 22l(m_3l) \\ 22l(m_1l) & [4l^2(m_1l) + 4l_2^2(m_2l_2)] & 3l_2^2(m_2l_2) \\ 22l(m_3l) & 3l_2^2(m_2l_2) & [4l^2(m_3l) + 4l_2^2(m_2l_2)] \end{bmatrix}$$

The term m_2l_2 in space $(1, 1)$ has been added since it is a result of translation q_1 parallel to member II. If $m_1 = m_2 = m_3$ and $l_2 = 2l$, the above matrix reduces to

$$\mathfrak{M} = \frac{ml}{210}\begin{bmatrix} 576 & 11l & 11l \\ 11l & 18l^2 & 12l^2 \\ 11l & 12l^2 & 18l^2 \end{bmatrix}$$

and the mass term of the equation of motion becomes

$$\frac{ml}{210}\begin{bmatrix} 576 & 11l & 11l \\ 11l & 18l^2 & 12l^2 \\ 11l & 12l^2 & 18l^2 \end{bmatrix}\begin{Bmatrix} \ddot{q}_1 \\ \ddot{q}_2 \\ \ddot{q}_3 \end{Bmatrix}$$

PROBLEMS

8-1   Using the method of virtual work, determine the equilibrium position of a carpenter's square hooked over a peg as shown in Fig. P8-1.

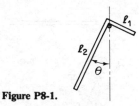

Figure P8-1.

8-2 Determine the equilibrium position of the two uniform bars shown in Fig. P-82 when a force P is applied as shown. All surfaces are friction free.

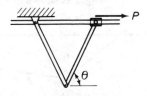

Figure P8-2.

8-3 Determine the equilibrium position of two point masses m_1 and m_2 connected by a massless rod and placed in a smooth hemispherical bowl of radius R as shown in Fig. P8-3.

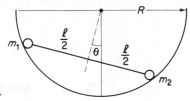

Figure P8-3.

8-4 The four masses on the string in Fig. P8-4 are displaced by a horizontal force F. Determine its equilibrium position by using virtual work.

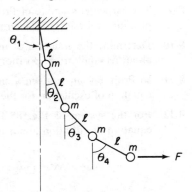

Figure P8-4.

8-5 A mass m is supported by two springs of unstretched length r_0 attached to a pin and a slider as shown in Fig. P8-5. There is coulomb friction with coefficient μ between the massless slider and the rod. Determine its equilibrium position by virtual work.

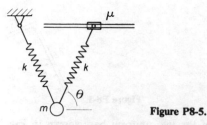

Figure P8-5.

8-6   Determine the equilibrium position of m_1 and m_2 attached to strings of equal length as shown in Fig. P8-6.

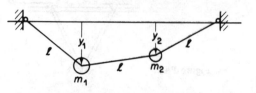

Figure P8-6.

8-7   A rigid uniform rod of length l is supported by a spring and a smooth floor as shown in Fig. P8-7. Determine its equilibrium position by virtual work. The unstretched length of the spring is $h/4$.

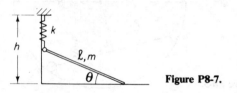

Figure P8-7.

8-8   Determine the equation of motion for small oscillation about the equilibrium position in Prob. 8-7.

8-9   The carpenter's square of Prob. 8-1 is displaced slightly from its equilibrium position and released. Determine its equation of oscillation.

8-10   Determine the equation of motion and the natural frequency of oscillation about its equilibrium position for the system in Prob. 8-3.

8-11   In Prob. 8-6 m_1 is given a small displacement and released. Determine the equation of oscillation for the system.

8-12   For the system of Fig. P8-12, determine the equilibrium position and its equation of vibration about it. Spring force $= 0$ when $\theta = 0$.

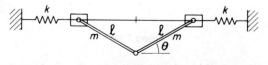

Figure P8-12.

8-13   Write Lagrange's equations of motion for the system shown in Fig. P8-13.

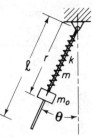

Figure P8-13.

8-14 The following constants are given for the beam of Fig. P8-14:

$$k = \frac{EI}{l^3} \qquad \frac{EI}{ml^4} = N \qquad \frac{k}{ml} = N$$

$$K = 5\frac{EI}{l} \qquad \frac{K}{ml^3} = 5\,N$$

Using the modes $\phi_1 = x/l$ and $\phi_2 = \sin(\pi x/l)$, determine the equation of motion by Lagrange's method, and determine the first two natural frequencies and mode shapes.

Figure P8-14.

8-15 Using Lagrange's method, determine the equations for the small oscillation of the bars shown in Fig. P8-15.

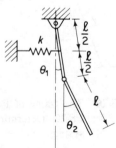

Figure P8-15.

8-16 Starting with the equation $T = \frac{1}{2}\dot{q}'M\dot{q}$, show that

$$\frac{\partial T}{\partial \dot{q}_i} = \frac{1}{2}\left(\dot{q}'M\frac{\partial \dot{q}}{\partial \dot{q}_i} + \frac{\partial \dot{q}'}{\partial q_i}M\dot{q}\right) = (i^{th} \text{ row of } M)\{\dot{q}\}$$

8-17 The rigid bar linkages of Example 8.1-1 are loaded by springs and masses as shown in Fig. P8-17. Write Lagrange's equations of motion.

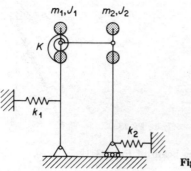

Figure P8-17.

8-18 Equal masses are placed at the nodes of the frame of Example 8.1-2 as shown in Fig. P8-18. Determine the stiffness matrix and the matrix equation of motion. (Let $l_2 = l_1$.)

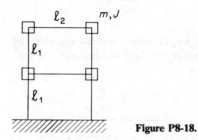

Figure P8-18.

8-19 Determine the stiffness matrix for the frame shown in Fig. P8-19.

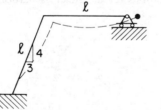

Figure P8-19.

8-20 The frame of Prob. 8-19 is loaded by springs and masses as shown in Fig. P8-20. Determine the equations of motion and the normal modes of the system.

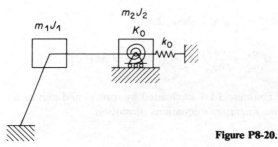

Figure P8-20.

8-21   Using area moment and superposition, determine M_1 and R_2 for the beam shown in Fig. P8-21. Let $EI_1 = 2EI_2$.

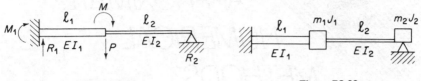

Figure P8-21.   Figure P8-22.

8-21   With loads m, J placed as shown in Fig. P8-22, set up the equations of motion.

8-23   Determine the consistent mass for the beam of Fig. P8-21 with m_1 and m_2 for mass per unit length of each section.

8-24   Determine the consistent mass matrix for the framed structure of Fig. P8-24 where the columns are pinned at the floor.

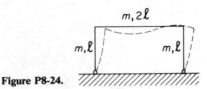

Figure P8-24.

8-25   Using the superposition of the beam elements shown in Table 6.1-1, show that the consistent stiffness matrix for the uniform beam element is

$$K = \frac{EI}{l^3}\begin{bmatrix} 4l^2 & 2l^2 & 6l & -6l \\ 2l^2 & 4l^2 & 6l & -6l \\ 6l & 6l & 12 & -12 \\ -6l & -6l & -12 & 12 \end{bmatrix}$$

8-26   For the extension of the double pendulum to the dynamic problem, the actual algebra can become long and tedious. Instead, draw in the components of $-\ddot{r}$ as shown. By taking each $\delta\theta$ separately, the virtual work equation can be easily determined visually. Complete the equations of motion for the system in Fig. P8-26. Compare with Lagrange's derivation.

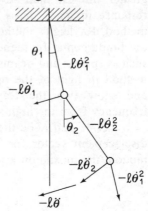

Figure P8-26.

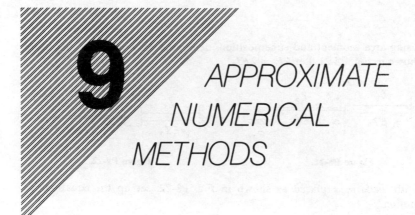

9 APPROXIMATE NUMERICAL METHODS

The exact analysis for the vibration of systems of many degrees of freedom is generally difficult, and its associated calculations are laborious. In many cases, all of the normal modes of the system are not required, and an estimate of the fundamental and a few of the lower modes are sufficient. In this chapter some of the approximate methods for determining the natural frequencies and mode shapes of the first few vibration modes will be presented.

9.1 RAYLEIGH METHOD

The fundamental frequency of multidegree of freedom systems is often of greater interest than its higher natural frequencies because its forced response in many cases is the largest. In Chapter 2, under the energy method, Rayleigh's method was introduced to obtain a better estimate to the fundamental frequency of systems which contained flexible elements such as springs and beams. In this section we wish to examine the Rayleigh method in light of the matrix techniques presented in previous chapters and show that the Rayleigh frequency approaches the fundamental frequency from the high side.

Let M and K be the mass and stiffness matrices and X the assumed displacement vector for the amplitude of vibration. Then for harmonic motion, the maximum kinetic and potential energies can be written as

$$T_{\max} = \tfrac{1}{2}\omega^2 X' M X \qquad (9.1\text{-}1)$$

and

$$U_{\max} = \tfrac{1}{2}X'KX \qquad (9.1\text{-}2)$$

Equating the two and solving for ω^2, we obtain the Rayleigh quotient

$$\omega^2 = \frac{X'KX}{X'MX} \qquad (9.1\text{-}3)$$

This quotient approaches the lowest natural frequency (or fundamental frequency) from the high side, and its value is somewhat insensitive to the choice of the assumed amplitudes. To show these qualities, we will express the assumed displacement curve in terms of the normal modes X_i as follows

$$X = X_1 + C_2 X_2 + C_3 X_3 + \cdots \qquad (9.1\text{-}4)$$

Then

$$X'KX = X_1'KX_1 + C_2^2 X_2'KX_2 + C_3^2 X_3'KX_3 + \cdots$$

and

$$X'MX = X_1'MX_1 + C_2^2 X_2'MX_2 + C_3^2 X_3'MX_3 + \cdots$$

where cross terms of the form $X_i'KX_j$ and $X_i'MX_j$ have been eliminated by the orthogonality conditions.

Noting that

$$X_i'KX_i = \omega_i^2 X_i'MX_i \qquad (9.1\text{-}5)$$

the Rayleigh quotient becomes

$$\omega^2 = \omega_1^2 \left\{ 1 + C_2^2 \left(\frac{\omega_2^2}{\omega_1^2} - 1 \right) \frac{X_2'MX_2}{X_1'MX_1} + \cdots \right\} \qquad (9.1\text{-}6)$$

If $X_i'MX_i$ is normalized to the same number, the above equation reduces to

$$\omega^2 = \omega_1^2 \left\{ 1 + C_2^2 \left(\frac{\omega_2^2}{\omega_1^2} - 1 \right) + \cdots \right\} \qquad (9.1\text{-}7)$$

It is evident, then, that ω^2 is greater than ω_1^2 because $\omega_2^2/\omega_1^2 > 1$. Since C_2 represents the deviation of the assumed amplitudes from the exact amplitudes X_1, the error in the computed frequency is only proportional to the square of the deviation of the assumed amplitudes from their exact values.

This analysis shows that if the exact fundamental deflection (or mode) X_1 is assumed, the fundamental frequency found by this method will be the correct frequency, since C_2, C_3, etc., will then be zero. For any other curve, the frequency determined will be higher than the fundamental. This fact can be explained on the basis that any deviation from the natural curve requires additional constraints, a condition that implies greater stiffness and higher frequency. In general, the use of the static deflection curve of the elastic body results in a fairly accurate value of the fundamental frequency. If greater accuracy is desired, the approximate curve can be repeatedly improved.

In our previous discussion of the Rayleigh method the potential energy was determined by the work done by the static weights in the assumed deformation. This work is, of course, stored in the flexible member as strain energy. For beams, the elastic strain energy can be calculated in terms of its flexural rigidity EI.

Letting M be the bending moment and θ the slope of the elastic curve, the strain energy stored in an infinitesimal beam element is

$$dU = \frac{1}{2} M \, d\theta \qquad (9.1\text{-}8)$$

Since the deflection in beams is generally small, the following geometric relations are assumed to hold (see Fig. 9.1-1).

$$\theta = \frac{dy}{dx} \qquad \frac{1}{R} = \frac{d\theta}{dx} = \frac{d^2y}{dx^2}$$

In addition, we have from the theory of beams, the flexure equation

$$\frac{1}{R} = \frac{M}{EI} \qquad (9.1\text{-}9)$$

where R is the radius of curvature. Substituting for $d\theta$ and $1/R$, U may be written as

$$U_{max} = \frac{1}{2} \int \frac{M^2}{EI} \, dx = \frac{1}{2} \int EI \left(\frac{d^2y}{dx^2} \right)^2 dx \qquad (9.1\text{-}10)$$

where the integration is carried out over the entire beam.

The kinetic energy is simply

$$T_{max} = \frac{1}{2} \int \dot{y}^2 \, dm = \frac{1}{2} \omega^2 \int y^2 \, dm \qquad (9.1\text{-}11)$$

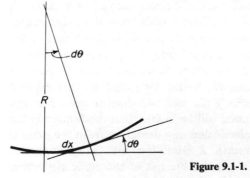

Figure 9.1-1.

where y is the assumed deflection curve. Thus, by equating the kinetic and potential energies, an alternative equation for the fundamental frequency of the beam is

$$\omega^2 = \frac{\int EI \left(\frac{d^2y}{dx^2} \right)^2 dx}{\int y^2 \, dm} \tag{9.1-12}$$

EXAMPLE 9.1-1

In applying this procedure to a simply supported beam of uniform cross section, shown in Fig. 9.1-2, we assume the deflection to be represented by a sine curve as follows

$$y = \left(y_0 \sin \frac{\pi x}{l} \right) \sin \omega t$$

where y_0 is the maximum deflection at mid-span. The second derivative then becomes

$$\frac{d^2y}{dx^2} = - \left(\frac{\pi}{l} \right)^2 y_0 \sin \frac{\pi x}{l} \sin \omega t$$

Substituting into Eq. (9.1-12) we obtain

$$\omega^2 = \frac{EI \left(\frac{\pi}{l} \right)^4 \int_0^l \sin^2 \frac{\pi x}{l} \, dx}{m \int_0^l \sin^2 \frac{\pi x}{l} \, dx} = \pi^4 \frac{EI}{ml^4}$$

The fundamental frequency is therefore found to be

$$\omega_1 = \pi^2 \sqrt{EI/ml^4}$$

In this case, the assumed curve happened to be the natural vibration curve, and the exact frequency is obtained by Rayleigh's method. Any other curve assumed for the case can be considered to be the result of additional constraints, or stiffness which will result in a constant greater then π^2 in the frequency equation.

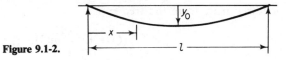

Figure 9.1-2.

EXAMPLE 9.1-2

If the distance between the ends of the beam of Fig. 9.1-2 is rigidly fixed, a tensile stress σ will be developed by the lateral deflection. Account for this additional strain energy in the frequency equation.

Solution: Due to the lateral deflection, the length dx of the beam is increased by an amount

$$\left[\sqrt{1 + (dy/dx)^2} - 1\right]dx \cong \frac{1}{2}\left(\frac{dy}{dx}\right)^2 dx$$

The additional strain energy in the element dx is

$$dU = \tfrac{1}{2}\sigma A\varepsilon \, dx = \tfrac{1}{2} EA\varepsilon^2 \, dx$$

where A is the cross-sectional area, σ the stress due to tension, and $\varepsilon = \tfrac{1}{2}(dy/dx)^2$ is the unit strain.

Equating the kinetic energy to the total strain energy of bending and tension, we obtain

$$\frac{1}{2}\omega^2 \int y^2 \, dm = \frac{1}{2}\int EI\left(\frac{d^2y}{dx^2}\right)^2 dx + \frac{1}{2}\int \frac{EA}{4}\left(\frac{dy}{dx}\right)^4 dx$$

The above equation then leads to the frequency equation

$$\omega_1^2 = \frac{\displaystyle\int EI\left(\frac{d^2y}{dx^2}\right)^2 dx + \int \frac{EA}{4}\left(\frac{dy}{dx}\right)^4 dx}{\displaystyle\int y^2 \, dm}$$

which contains an additional term due to the tension.

Accuracy of the Integral Method Over Differentiation. In using Rayleigh's method of determining the fundamental frequency, we must choose an assumed curve. Although the deviation of this assumed deflection curve compared to the exact curve may be slight, its derivative could be in error by a large amount and hence the strain energy computed from the equation

$$U = \frac{1}{2}\int EI\left(\frac{d^2y}{dx^2}\right)^2 dx$$

may be inaccurate. To avoid this difficulty, the following integral method for the evaluation of U is recommended for some beam problems.

We first recognize that the shear V is the integral of the inertia loading $m\omega^2 y$ from the free end

$$V(\xi) = \omega^2 \int_{\xi}^{l} m(\xi)y(\xi) \, d\xi \tag{9.1-13}$$

Since the bending moment is related to the shear by the equation

$$\frac{dM}{dx} = V \tag{9.1-14}$$

the moment at x is found from the integral

$$M(x) = \int_x^l V(\xi)\, d\xi \qquad (9.1\text{-}15)$$

The strain energy of the beam is then found from

$$U = \frac{1}{2}\int_0^l \frac{M(x)^2}{EI}\, dx \qquad (9.1\text{-}16)$$

which avoids any differentiation of the assumed deflection curve.

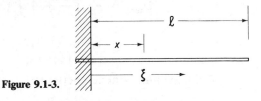

Figure 9.1-3.

EXAMPLE 9.1-3

Determine the fundamental frequency of the uniform cantilever beam shown in Fig. 9.1-5, using the simple curve $y = cx^2$.

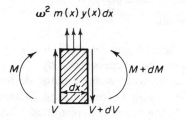

$\omega^2 m(x) y(x) dx$

Figure 9.1-4. Free-body diagram of beam element.

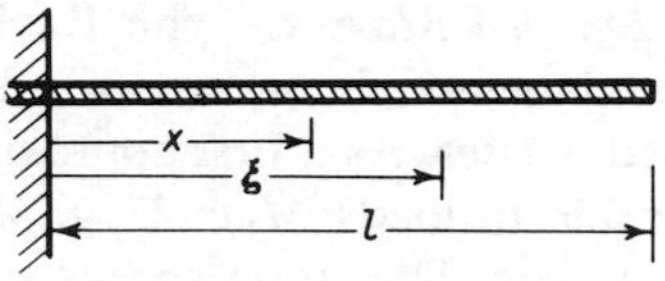

Figure 9.1-5.

Solution: If we use Eq. (9.1-12), we would find the result to be very much in error since the above curve does not satisfy the boundary conditions at the free end. By using Eq. (9.1-12) we obtain

$$\omega = 4.47\sqrt{EI/ml^4}$$

whereas the exact value is

$$\omega_1 = 3.52\sqrt{EI/ml^4}$$

Acceptable results using the given curve can be found by the procedure outlined in the previous section.

$$V(\xi) = \omega^2 \int_\xi^l mc\xi^2\, d\xi = \frac{\omega^2 mc}{3}(l^3 - \xi^3)$$

and the bending moment becomes

$$M(x) = \int_x^l V(\xi)\, d\xi = \frac{\omega^2 mc}{3} \int_x^l (l^3 - \xi^3)\, d\xi$$

$$= \frac{\omega^2 mc}{12} (3l^4 - 4l^3 x + x^4)$$

The maximum strain energy is found by substituting $M(x)$ into U_{max}.

$$U_{max} = \frac{1}{2EI} \left(\frac{\omega^2 mc}{12} \right)^2 \int_0^l (3l^4 - 4l^3 x + x^4)^2\, dx$$

$$= \frac{\omega^4}{2EI} \frac{m^2 c^2}{144} \frac{312}{135} l^9$$

The maximum kinetic energy is

$$T_{max} = \frac{1}{2} \int_0^l \dot{y}^2 m\, dx = \frac{1}{2} c^2 \omega^2 m \int_0^l x^4\, dx = \frac{1}{2} c^2 \omega^2 m \frac{l^5}{5}$$

By equating these results, we obtain

$$\omega_1 = \sqrt{12.47\, EI/ml^4} = 3.53\sqrt{EI/ml^4}$$

which is very close to the exact result.

Lumped Masses. The Rayleigh method can be used to determine the fundamental frequency of a beam or shaft represented by a series of lumped masses. As a first approximation we will assume a static deflection curve due to loads $M_1 g$, $M_2 g$, $M_3 g$, etc., with corresponding deflections y_1, y_2, y_3, etc. The strain energy stored in the beam is determined from the work done by these loads and the maximum potential and kinetic energies become

$$U_{max} = \frac{1}{2} g(M_1 y_1 + M_2 y_2 + M_3 y_3 + \cdots) \qquad (9.1\text{-}17)$$

$$T_{max} = \frac{1}{2} \omega^2 (M_1 y_1^2 + M_2 y_2^2 + M_3 y_3^2 + \cdots) \qquad (9.1\text{-}18)$$

Equating the two, the frequency equation is established as

$$\omega_1^2 = \frac{g \sum_i M_i y_i}{\sum_i M_i y_i^2} \qquad (9.1\text{-}19)$$

EXAMPLE 9.1-4

Calculate the first approximation to the fundamental frequency of lateral vibration for the system shown in Fig. 9.1-6.

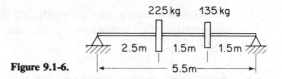

Figure 9.1-6.

Solution: Referring to the table at the end of Chapter 2, we see that the deflection of the beam at any point x from the left end due to a single load W at a distance b from the right end is

$$y(x) = \frac{Wbx}{6EIl}(l^2 - x^2 - b^2) \qquad x \le (l - b)$$

The deflections at the loads can be obtained from the superposition of the deflections due to each load acting separately.

Due to the 135 kg mass, we have

$$y_1' = \frac{(9.81 \times 135) \times 1.5 \times 2.5}{6 \times 5.5EI}(5.5^2 - 2.5^2 - 1.5^2) = 3.273 \times \frac{10^3}{EI} \ m$$

$$y_2' = \frac{(9.81 \times 135) \times 1.5 \times 4}{6 \times 5.5EI}(5.5^2 - 4.0^2 - 1.5^2) = 2.889 \times \frac{10^3}{EI} \ m$$

Due to the 225 kg mass, the deflections at the corresponding points are

$$y_1'' = \frac{(9.81 \times 225) \times 2.5 \times 3.0}{6 \times 5.5EI}(5.5^2 - 3.0^2 - 2.5^2) = 7.524 \times \frac{10^3}{EI} \ m$$

$$y_2'' = \frac{(9.81 \times 225) \times 2.5 \times 1.5}{6 \times 5.5EI}(5.5^2 - 1.5^2 - 2.5^2) = 5.455 \times \frac{10^3}{EI} \ m$$

Adding y' and y'' the deflections at 1 and 2 become

$$y_1 = 10.797 \times \frac{10^3}{EI} \ m, \qquad y_2 = 8.344 \times \frac{10^3}{EI} \ m$$

Substituting into Eq. (9.1-19), the first approximation to the fundamental frequency is

$$\omega_1 = \sqrt{\frac{9.81(225 \times 10.797 + 135 \times 8.344)EI}{(225 \times 10.797^2 + 135 \times 8.344^2)10^3}}$$

$$= 0.03129\sqrt{EI} \ \text{rad/sec}$$

If further accuracy is desired, a better approximation to the dynamic curve can be made by using the dynamic loads $m\omega^2 y$. Since the dynamic loads are proportional to the deflection y, we can recalculate the deflection with the modified loads gm_1 and $gm_2(y_2/y_1)$.

9.2 DUNKERLEY'S EQUATION

The Rayleigh method which gives the upper bound to the fundamental frequency can now be complemented by Dunkerley's* equation, which results in a lower bound to the fundamental frequency. For the basis of the Dunkerley equation, we examine the characteristic equation (6.4-2) formulated from the flexibility coefficients, which is

$$\begin{vmatrix} \left(a_{11}m_1 - \dfrac{1}{\omega^2}\right) & a_{12}m_2 & a_{13}m_3 \\ a_{21}m_1 & \left(a_{22}m_2 - \dfrac{1}{\omega^2}\right) & a_{23}m_3 \\ a_{31}m_1 & a_{32}m_2 & \left(a_{33}m_3 - \dfrac{1}{\omega^2}\right) \end{vmatrix} = 0 \qquad (6.4\text{-}2)$$

Expanding this determinant, we obtain the third degree equation in $1/\omega^2$.

$$\left(\frac{1}{\omega^2}\right)^3 - (a_{11}m_1 + a_{22}m_2 + a_{33}m_3)\left(\frac{1}{\omega^2}\right)^2 + \cdots = 0 \qquad (9.2\text{-}1)$$

If the roots of this equation are $1/\omega_1^2$, $1/\omega_2^2$, and $1/\omega_3^2$, the above equation can be factored into the following form

$$\left(\frac{1}{\omega^2} - \frac{1}{\omega_1^2}\right)\left(\frac{1}{\omega^2} - \frac{1}{\omega_2^2}\right)\left(\frac{1}{\omega^2} - \frac{1}{\omega_3^2}\right) = 0$$

or

$$\left(\frac{1}{\omega^2}\right)^3 - \left(\frac{1}{\omega_1^2} + \frac{1}{\omega_2^2} + \frac{1}{\omega_3^2}\right)\left(\frac{1}{\omega^2}\right)^2 \cdots = 0 \qquad (9.2\text{-}2)$$

As is well known in algebra, the coefficient of the second highest power is equal to the sum of the roots of the characteristic equation. It is also equal to the sum of the diagonal terms of the matrix A^{-1}, which is called the *trace* of the matrix. (See Appendix C.)

$$\text{trace } A^{-1} = \sum_{i=1}^{3}\left(\frac{1}{\omega_i^2}\right)$$

---

*S. Dunkerley, "On the Whirling and Vibration of Shafts," *Phil. Trans. Roy. Soc.*, 185 (1895), pp. 269–360.

These relationships are true for n greater than 3, and we can write for an n degree of freedom system the following equation

$$\frac{1}{\omega_1^2} + \frac{1}{\omega_2^2} + \cdots + \frac{1}{\omega_n^2} = a_{11}m_1 + a_{22}m_2 + \cdots + a_{nn}m_n \quad (9.2\text{-}3)$$

The estimate to the fundamental frequency is made by recognizing that ω_2, ω_3, etc., are natural frequencies of higher modes and hence $1/\omega_2^2$, $1/\omega_3^2$, etc., can be neglected in the left side of Eq. (9.2-3). The term $1/\omega_1^2$ is consequently larger than the true value and therefore ω_1 is smaller than the exact value of the fundamental frequency. Dunkerley's estimate of the fundamental frequency is then made from the equation

$$\frac{1}{\omega_1^2} < a_{11}m_1 + a_{22}m_2 + \cdots a_{nn}m_n \quad (9.2\text{-}4)$$

Since the left side of the equation has the dimension of the reciprocal of the frequency squared, each term on the right side must also be of the same dimension. Each term on the right side must then be considered to be the contribution to $1/\omega_1^2$ in the absence of all other masses, and for convenience we will let $a_{ii}m_i = 1/\omega_{ii}^2$, or

$$\frac{1}{\omega_1^2} < \frac{1}{\omega_{11}^2} + \frac{1}{\omega_{22}^2} + \cdots \frac{1}{\omega_{nn}^2} \quad (9.2\text{-}5)$$

Thus, the right side becomes the sum of the effect of each mass acting in the absence of all other masses.

EXAMPLE 9.2-1

Dunkerley's equation is useful for estimating the fundamental frequency of a structure undergoing vibration testing. Natural frequencies of structures are often determined by attaching to the structure an eccentric mass exciter, and noting the frequencies corresponding to the maximum amplitude. The frequencies so measured represent those of the structure plus exciter and may deviate considerably from the natural frequencies of the structure itself when the mass of the exciter is a substantial percentage of the total mass. In such cases the fundamental frequency of the structure by itself may be determined by the following equation

$$\frac{1}{\omega_1^2} = \frac{1}{\omega_{11}^2} + \frac{1}{\omega_{22}^2} \quad (a)$$

where ω_1 = fundamental frequency of structure plus exciter,

ω_{11} = fundamental frequency of the structure by itself,

ω_{22} = natural frequency of exciter mounted on the structure in the absence of other masses.

It is sometimes convenient to express the equation in another form, for instance

$$\frac{1}{\omega_1^2} = \frac{1}{\omega_{11}^2} + a_{22}m_2 \tag{b}$$

where m_2 is the mass of the concentrated weight or exciter and a_{22} the influence coefficient of the structure at the point of attachment of the exciter.

EXAMPLE 9.2-2

An airplane rudder tab showed a resonant frequency of 30 cps when vibrated by an eccentric mass shaker weighing 1.5 lb. By attaching an additional weight of 1.5 lb to the shaker, the resonant frequency was lowered to 24 cps. Determine the true natural frequency of the tab.

Solution: The measured resonant frequencies are those due to the total mass of the tab and shaker. Letting f_{11} be the true natural frequency of the tab, and substituting into Eq. (b) of Example 9.2-1, we obtain

$$\frac{1}{(2\pi \times 30)^2} = \frac{1}{(2\pi f_{11})^2} + \frac{1.5}{386}a_{22}$$

$$\frac{1}{(2\pi \times 24)^2} = \frac{1}{(2\pi f_{11})^2} + \frac{3.0}{386}a_{22}$$

Eliminating a_{22}, the true natural frequency is

$$f_{11} = 45.3 \text{ cps}$$

The rigidity of stiffness of the tab at the point of attachment of the shaker may be determined from $1/a_{22}$ which from the same equations is found to be

$$k_2 = \frac{1}{a_{22}} = \frac{1}{0.00407} = 246 \text{ lb/in.}$$

EXAMPLE 9.2-3

Determine the fundamental frequency of a uniformly loaded cantilever beam with a concentrated mass M at the end, equal to the mass of the uniform beam (see Fig. 9.2-1).

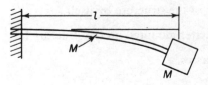

Figure 9.2-1.

Solution: The frequency equation for the uniformly loaded beam by itself is

$$\omega_{11}^2 = 3.515^2\left(\frac{EI}{Ml^3}\right)$$

For the concentrated mass by itself attached to a weightless canti-lever beam we have

$$\omega_{22}^2 = 3.00\left(\frac{EI}{Ml^3}\right)$$

Substituting into Dunkerley's formula rearranged in the following form, the natural frequency of the system is determined as

$$\omega_1^2 = \frac{\omega_{11}^2\omega_{22}^2}{\omega_{11}^2 + \omega_{22}^2} = \frac{3.515^2 \times 3.0}{3.515^2 + 3.0}\left(\frac{EI}{Ml^3}\right) = 2.41\left(\frac{EI}{Ml^3}\right)$$

This result may be compared to the frequency equation obtained by Rayleigh's method which is

$$\omega_1^2 = \frac{3\,EI}{\left(1 + \frac{33}{140}\right)Ml^3} = 2.43\left(\frac{EI}{Ml^3}\right)$$

EXAMPLE 9.2-4

The natural frequency of a given airplane wing in torsion is 1600 cpm. What will be the new torsional frequency if a 1000-lb fuel tank is hung at a position one-sixth of the semi-span from the center line of the airplane such that its moment of inertia about the torsional axis is 1800 lb in sec^2? The torsional stiffness of the wing at this point is 60×10^6 lb in/rad.

Solution: The frequency of the tank attached to the weightless wing is

$$f_{22} = \frac{1}{2\pi}\sqrt{\frac{60 \times 10^6}{1800}} = 29.1 \text{ cps} = 1745 \text{ cpm}$$

The new torsional frequency with the tank, from Eq. (a) of Example 9.2-1 then becomes

$$\frac{1}{f_1^2} = \frac{1}{1600^2} + \frac{1}{1745^2}, \qquad f_1 = 1180 \text{ cpm}$$

EXAMPLE 9.2-5

The fundamental frequency of a uniform beam of mass M, simply supported as in Fig. 9.2-2 is equal to $\pi^2\sqrt{EI/Ml^3}$. If a lumped mass m_0 is attached to the beam at $x = l/3$, determine the new fundamental frequency.

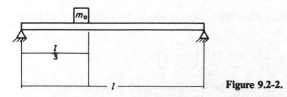

Figure 9.2-2.

Solution: Starting with Eq. (b) of Example 9.2-1, we let ω_{11} be the fundamental frequency of the uniform beam and ω_1 the new fundamental frequency with m_0 attached to the beam. Multiplying through Eq. (b) by ω_1^2, we have

$$1 = \left(\frac{\omega_1}{\omega_{11}}\right)^2 + a_{22}m_0\omega_{11}^2\left(\frac{\omega_1}{\omega_{11}}\right)^2$$

or

$$\left(\frac{\omega_1}{\omega_{11}}\right)^2 = \frac{1}{1 + a_{22}m_0\omega_{11}^2}$$

The quantity a_{22} is the influence coefficient at $x = l/3$ due to a unit load applied at the same point. It can be found from the beam formula in Example 9.1-4 to be

$$a_{22} = \frac{8}{6 \times 81}\frac{l^3}{EI}$$

Substituting $\omega_{11}^2 = \pi^4 EI/Ml^3$ together with a_{22}, we obtain the convenient equation

$$\left(\frac{\omega_1}{\omega_{11}}\right)^2 = \frac{1}{1 + \dfrac{8\pi^4}{6 \times 81}\dfrac{m_0}{M}} = \frac{1}{1 + 1.6\dfrac{m_0}{M}}$$

EXAMPLE 9.2-6

Determine the fundamental frequency of the three-story building shown in Fig. 9.2-3 where the foundation is capable of translation.

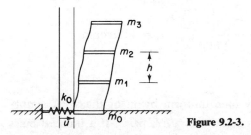

Figure 9.2-3.

Solution: If a unit force is placed at each floor, the influence coefficients are found to be

$$a_{00} = \frac{1}{k_0}$$

$$a_{11} = \frac{1}{k_0} + \frac{h^3}{24EI_1}$$

$$a_{22} = \frac{1}{k_0} + \frac{h^3}{24EI_1} + \frac{h^3}{24EI_2}$$

$$a_{33} = \frac{1}{k_0} + \frac{h^3}{24EI_1} + \frac{h^3}{24EI_2} + \frac{h^3}{24EI_3}$$

Dunkerley's equation then becomes

$$\frac{1}{\omega_1^2} = \frac{m_0}{k_0} + \left(\frac{1}{k_0} + \frac{h^3}{24EI_1}\right)m_1 + \left(\frac{1}{k_0} + \frac{h^3}{24EI_1} + \frac{h^3}{24EI_2}\right)m_2$$

$$+ \left(\frac{1}{k_0} + \frac{h^3}{24EI_1} + \frac{h^3}{24EI_2} + \frac{h^3}{24EI_3}\right)m_3$$

If the columns are of equal stiffness, the above equation reduces to

$$\frac{1}{\omega_1^2} = \frac{1}{k_0}(m_0 + m_1 + m_2 + m_3) + m_1\frac{h^3}{24EI} + m_2\frac{2h^3}{24EI} + m_3\frac{3h^3}{24EI}$$

9.3   RAYLEIGH-RITZ METHOD

W. Ritz developed a method which is an extension of Rayleigh's method. It not only provides a means of obtaining a more accurate value for the fundamental frequency, but it also gives approximations to the higher frequencies and mode shapes.

The Ritz method is essentially the Rayleigh method in which the single shape function is replaced by a series of shape functions multiplied by constant coefficients. The coefficients are adjusted by minimizing the frequency with respect to each of the coefficients, which results in n algebraic equations in ω^2. The solution of these equations then gives the natural frequencies and mode shapes of the system. As in Rayleigh's method, the success of the method depends on the choice of the shape functions which should satisfy the geometric boundary conditions of the problem. The method should also be differentiable, at least to the order of the derivatives appearing in the energy equations. The functions, however, can disregard discontinuities such as those of shear due to concentrated masses which involve third derivatives in beams.

We now outline in a general manner the procedure of the Rayleigh-Ritz method, starting with Rayleigh's equation

$$\omega^2 = \frac{U_{max}}{T^*_{max}} \qquad (9.3\text{-}1)$$

where the kinetic energy is expressed as $\omega^2 T^*_{max}$. In the Rayleigh method a single function is chosen for the deflection; Ritz, however, assumed the deflection to be a sum of several functions multiplied by constants as follows:

$$y(x) = C_1\phi_1(x) + C_2\phi_2(x) + \cdots C_n\phi_n(x) \qquad (9.3\text{-}2)$$

where $\phi_i(x)$ are any admissible functions satisfying the boundary conditions. U_{max} and T_{max} are expressible in the form of Eqs. (8.3-5) and (8.3-9).

$$U = \tfrac{1}{2}\sum_i \sum_j k_{ij} C_i C_j$$

$$T^* = \tfrac{1}{2}\sum_i \sum_j m_{ij} C_i C_j \qquad (9.3\text{-}3)$$

where k_{ij} and m_{ij} depend on the type of problem. For example, for the beam we have

$$k_{ij} = \int EI\phi_i'' \phi_j'' \, dx \quad \text{and} \quad m_{ij} = \int m\phi_i\phi_j \, dx$$

whereas for the longitudinal oscillation of slender rods

$$k_{ij} = \int EA\phi_i'\phi_j' \, dx \quad \text{and} \quad m_{ij} = \int m\phi_i\phi_j \, dx$$

We now minimize ω^2 by differentiating it with respect to each of the constants. For example, the derivative of ω^2 with respect to C_i is

$$\frac{\partial \omega^2}{\partial C_i} = \frac{\partial}{\partial C_i}\left(\frac{U_{max}}{T^*_{max}}\right) = \frac{T^*_{max}\dfrac{\partial U_{max}}{\partial C_i} - U_{max}\dfrac{\partial T^*_{max}}{\partial C_i}}{T^{*2}_{max}} = 0 \qquad (9.3\text{-}4)$$

which is satisfied by

$$\frac{\partial U_{max}}{\partial C_i} - \frac{U_{max}}{T^*_{max}}\frac{\partial T^*_{max}}{\partial C_i} = 0$$

or since $U_{max}/T^*_{max} = \omega^2$,

$$\frac{\partial U_{max}}{\partial C_i} - \omega^2\frac{\partial T^*_{max}}{\partial C_i} = 0 \qquad (9.3\text{-}5)$$

The two terms in this equation are then

$$\frac{\partial U_{max}}{\partial C_i} = \sum_j^n k_{ij} C_j \quad \text{and} \quad \frac{\partial T^*_{max}}{\partial C_i} = \sum_j^n m_{ij} C_j$$

and so Eq. (9.3-5) becomes

$$C_1(k_{i1} - \omega^2 m_{i1}) + C_2(k_{i2} - \omega^2 m_{i2}) + \cdots C_n(k_{in} - \omega^2 m_{in}) = 0$$

(9.3-6)

With i varying from 1 to n there will be n such equations which can be arranged in matrix form as

$$\begin{bmatrix} (k_{11} - \omega^2 m_{11}) & (k_{12} - \omega^2 m_{12}) \cdots & (k_{1n} - \omega^2 m_{1n}) \\ (k_{21} - \omega^2 m_{21}) & \cdots & \\ \vdots & & \\ (k_{n1} - \omega^2 m_{n1}) & \cdots & (k_{nn} - \omega^2 m_{nn}) \end{bmatrix} \begin{Bmatrix} C_1 \\ C_2 \\ \vdots \\ C_n \end{Bmatrix} = 0$$

(9.3-7)

The determinant of this equation is an n degree algebraic equation in ω^2 and its solution results in the n natural frequencies. The mode shape is also obtained by solving for the C_s for each natural frequency and substituting into Eq. (9.3-2) for the deflection.

EXAMPLE 9.3-1

Figure 9.3-1 shows a wedge-shaped plate of constant thickness fixed into a rigid wall. Determine the first two natural frequencies and mode shapes in longitudinal oscillation by using the Rayleigh-Ritz method.

Figure 9.3-1.

Solution: For the displacement function we will choose the first two longitudinal modes of a uniform rod clamped at one end.

$$u(x) = C_1 \sin \frac{\pi x}{2l} + C_2 \sin \frac{3\pi x}{2l}$$

(a)

$$= C_1 \phi_1(x) + C_2 \phi_2(x)$$

The mass per unit length and the stiffness at x are

$$m(x) = m_0\left(1 - \frac{x}{l}\right) \quad \text{and} \quad EA(x) = EA_0\left(1 - \frac{x}{l}\right)$$

The k_{ij} and the m_{ij} for the longitudinal modes are calculated from the equations

$$k_{ij} = \int_0^l EA(x)\phi_i'\phi_j' \, dx$$

$$m_{ij} = \int_0^l m(x)\phi_i\phi_j \, dx$$

(b)

$$k_{11} = \frac{\pi^2}{4l^2} EA_0 \int_0^l \left(1 - \frac{x}{l}\right) \cos^2\frac{\pi x}{2l} \, dx = \frac{EA_0}{2l}\left(\frac{\pi^2}{8} + \frac{1}{2}\right)$$

$$= 0.86685 \frac{EA_0}{l}$$

$$k_{12} = \frac{3\pi^2}{4l^2} EA_0 \int_0^l \left(1 - \frac{x}{l}\right) \cos\frac{\pi x}{2l} \cos\frac{3\pi x}{2l} \, dx = 0.750\frac{EA_0}{l}$$

$$k_{22} = \frac{9\pi^2}{4l^2} EA_0 \int_0^l \left(1 - \frac{x}{l}\right) \cos^2\frac{3\pi x}{2l} \, dx = \frac{EA_0}{2l}\left(\frac{9\pi^2}{8} + \frac{1}{2}\right)$$

$$= 5.80165 \frac{EA_0}{l}$$

$$m_{11} = m_0 \int_0^l \left(1 - \frac{x}{l}\right) \sin^2\frac{\pi x}{2l} \, dx = m_0 l\left(\frac{1}{4} - \frac{1}{\pi^2}\right) = 0.148679 m_0 l$$

$$m_{12} = m_0 \int_0^l \left(1 - \frac{x}{l}\right) \sin\frac{\pi x}{2l} \sin\frac{3\pi x}{2l} \, dx = m_0 l\left(\frac{1}{\pi^2}\right) = 0.101321 m_0 l$$

$$m_{22} = m_0 \int_0^l \left(1 - \frac{x}{l}\right) \sin^2\frac{3\pi x}{l} \, dx = m_0 l\left(\frac{1}{4} - \frac{1}{9\pi^2}\right) = 0.238742 m_0 l$$

Substituting into Eq. (9.3-7), we obtain

$$
\begin{bmatrix}
\left(0.86685\dfrac{EA_0}{l} - 0.14868 m_0 l\omega^2\right) & \left(0.750\dfrac{EA_0}{l} - 0.10132 m_0 l\omega^2\right) \\[2mm]
\left(0.750\dfrac{EA_0}{l} - 0.10132 m_0 l\omega^2\right) & \left(5.80165\dfrac{EA_0}{l} - 0.23874 m_0 l\omega^2\right)
\end{bmatrix}
$$

$$\times \begin{Bmatrix} C_1 \\ C_2 \end{Bmatrix} = 0$$

(c)

Setting the determinant of the above equation to zero, we obtain the frequency equation

$$\omega^4 - 36.3676\alpha\omega^2 + 177.0377\alpha^2 = 0$$

(d)

where

$$\alpha = \frac{EA_0}{m_0 l^2} \tag{e}$$

The two roots of this equation are

$$\omega_1^2 = 5.7898\alpha \quad \text{and} \quad \omega_2^2 = 30.5778\alpha$$

Using these results in Eq. (c), we obtain

$$C_2 = 0.03689 \, C_1 \quad \text{for mode 1}$$
$$C_1 = -0.63819 \, C_2 \quad \text{for mode 2}$$

The two natural frequencies and mode shapes are then

$$\omega_1 = 2.4062\sqrt{\frac{EA_0}{m_0 l^2}} \qquad u_1(x) = 1.0 \sin\frac{\pi x}{2l} + 0.03689 \sin\frac{3\pi x}{2l}$$

$$\omega_2 = 5.5297\sqrt{\frac{EA_0}{m_0 l^2}} \qquad u_2(x) = -0.63819 \sin\frac{\pi x}{2l} + 1.0 \sin\frac{3\pi x}{2l}$$

9.4 METHOD OF MATRIX ITERATION

The equations of motion, formulated either on the basis of the stiffness equation or the flexibility equation, are similar in form and appear as

$$
\begin{Bmatrix} x_1 \\ x_2 \\ \cdot \\ \cdot \\ x_n \end{Bmatrix} = \lambda
\begin{bmatrix} a_{11} & a_{12} & \cdots & a_{1n} \\ a_{21} & a_{22} & \cdots & \\ \cdot & & & \\ \cdot & & & \\ a_{n1} & a_{n2} & \cdots & a_{nn} \end{bmatrix}
\begin{Bmatrix} x_1 \\ x_2 \\ \cdot \\ \cdot \\ x_n \end{Bmatrix} \tag{9.4-1}
$$

where λ is equal to $1/\omega^2$ for the stiffness formulation, and ω^2 for the flexibility formulation.

The iteration procedure is started by assuming a set of deflections for the right column of Eq. (9.4-1) and performing the indicated operation, which results in a column of numbers. This is then normalized by making one of the amplitudes equal to unity and dividing each term of the column by the particular amplitude which was normalized. The procedure is then repeated with the normalized column until the amplitudes stabilize to a definite pattern.

As will be shown in Sec. 9.5, the iteration process converges to the lowest value of λ so that for the equation formulated on the flexibility influence coefficients, the fundamental or the lowest mode of vibration is found. Likewise, for the equation formulated on the basis of the stiffness influence coefficients, the convergence is to the highest mode which corresponds to the lowest value of $\lambda = 1/\omega^2$.

EXAMPLE 9.4-1

The uniform beam of Fig. 9.4-1, free to vibrate in the plane shown, has two concentrated masses $m_1 = 500$ kg and $m_2 = 100$ kg. Determine the fundamental frequency of the system.

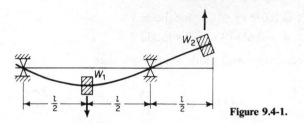

Figure 9.4-1.

Solution: The influence coefficients for this problem, determined from deflection equations of beams by placing a unit load at positions 1 and 2, are

$$a_{11} = \frac{l^3}{48EI} = \tfrac{1}{6}a_{22}, \quad a_{12} = a_{21} = \frac{l^3}{32EI} = \tfrac{1}{4}a_{22}, \quad a_{22} = \frac{l^3}{8EI}$$

Substituting into Eq. 9.4-1,

$$
\begin{bmatrix} x_1 \\ x_2 \end{bmatrix} = \frac{\omega^2 l^3}{8EI}
\begin{bmatrix} \dfrac{500}{6} & \dfrac{100}{4} \\ \dfrac{500}{4} & 100 \end{bmatrix}
\begin{bmatrix} x_1 \\ x_2 \end{bmatrix}
$$

Starting with $x_1 = x_2 = 1.0$ for the right column, we obtain

$$
\begin{bmatrix} x_1 \\ x_2 \end{bmatrix} = \frac{\omega^2 l^3}{8EI}
\begin{bmatrix} 108.3 \\ 225.0 \end{bmatrix} = \frac{108.3\omega^2 l^3}{8EI}
\begin{bmatrix} 1.00 \\ 2.08 \end{bmatrix}
$$

If the procedure is repeated with $x_1 = 1.0$ and $x_2 = 2.08$, the second result is

$$
\begin{bmatrix} x_1 \\ x_2 \end{bmatrix} = \frac{\omega^2 l^3}{8EI}
\begin{bmatrix} 135.3 \\ 333.0 \end{bmatrix} = \frac{135.3\omega^2 l^3}{8EI}
\begin{bmatrix} 1.00 \\ 2.46 \end{bmatrix}
$$

By repeating the procedure a few more times the deflections will converge to

$$
\begin{bmatrix} 1.00 \\ 2.60 \end{bmatrix} = \frac{148.3\omega^2 l^3}{8EI}
\begin{bmatrix} 1.00 \\ 2.60 \end{bmatrix}
$$

The fundamental frequency from the above equation is

$$\omega = \sqrt{8EI/148.3l^3} = 0.232\sqrt{EI/l^3}$$

and the amplitude ratio is found to be

$$\frac{x_1}{x_2} = \frac{1.0}{2.60}$$

If only the fundamental frequency is of interest, sufficient accuracy can be obtained from the results of the first and second iterations. From the first iteration the inertia forces are $500\omega^2$ and $208\omega^2$. These forces produce deflections obtained in the second iteration, which are $x_1 = 135.3\omega^2 l^3/8EI = 16.92\omega^2 l^3/EI$ and $x_2 = 2.46x_1$. The work done by these forces is then

$$U = \frac{1}{2}(500 + 208 \times 2.46)\omega^2 x_1 = \frac{1}{2} \times 1012 \times \omega^2 x_1$$

and the corresponding kinetic energy is

$$T = \frac{1}{2}(500 + 100 \times 2.46^2)\omega^2 x_1^2 = \frac{1}{2} \times 1105 \times \omega^2 x_1^2$$

Equating the two, the fundamental frequency is found as

$$\omega = \sqrt{\frac{1012}{1105 \times 16.92} \frac{EI}{l^3}} = 0.232\sqrt{\frac{EI}{l^3}}$$

9.5 CALCULATION OF HIGHER MODES

When the equations of motion are formulated in terms of the flexibility influence coefficients, the iteration procedure converges to the lowest mode present in the assumed deflection. It is evident that if the lowest mode is absent in the assumed deflection, the iteration technique will converge to the next lowest, or the second, mode.

Letting the assumed curve X be expressed by the sum of the normal modes X_i

$$X = C_1 X_1 + C_2 X_2 + C_3 X_3 + \cdots \qquad (9.5\text{-}1)$$

To distinguish between the assumed curve X and the normal modes X_i in the above equation, we will designate the normal modes as

$$X_i = \begin{Bmatrix} x_1 \\ x_2 \\ x_3 \end{Bmatrix}_i$$

and the assumed curve as

$$X = \begin{Bmatrix} \bar{x}_1 \\ \bar{x}_2 \\ \bar{x}_3 \end{Bmatrix}$$

We will now impose the condition $C_1 = 0$ to remove the first mode from the assumed deflection X. For this, we introduce the orthogonality relationship by premultiplying Eq. (9.5-1) by $X_1'M$, which eliminates all terms on the right side except the first term.

$$X_1'MX = C_1 X_1'MX_1 \qquad (9.5\text{-}2)$$

Equating the left side of the above equation to zero, C_1 becomes zero and the first mode is eliminated from Eq. (9.5-1)

$$X_1'MX = (x_1 \quad x_2 \quad x_3) \begin{bmatrix} m_1 & 0 & 0 \\ 0 & m_2 & 0 \\ 0 & 0 & m_3 \end{bmatrix} \begin{Bmatrix} \bar{x}_1 \\ \bar{x}_2 \\ \bar{x}_3 \end{Bmatrix} = 0 \qquad (9.5\text{-}3)$$

$$= m_1 x_1 \bar{x}_1 + m_2 x_2 \bar{x}_2 + m_3 x_3 \bar{x}_3 = 0$$

From the above equation, we obtain

$$\bar{x}_1 = -\frac{m_2}{m_1}\left(\frac{x_2}{x_1}\right)\bar{x}_2 - \frac{m_3}{m_1}\left(\frac{x_3}{x_1}\right)\bar{x}_3$$

$$\bar{x}_2 = \bar{x}_2 \qquad (9.5\text{-}4)$$

$$\bar{x}_3 = \bar{x}_3$$

where the last two equations in the above set are introduced merely as identities. Rewriting in matrix form, Eq. (9.5-4) becomes

$$\{X\} = \begin{bmatrix} 0 & -\dfrac{m_2}{m_1}\left(\dfrac{x_2}{x_1}\right) & -\dfrac{m_3}{m_1}\left(\dfrac{x_3}{x_1}\right) \\ 0 & 1 & 0 \\ 0 & 0 & 1 \end{bmatrix} \{X\} \qquad (9.5\text{-}5)$$

$$= SX$$

Since this equation is the result of $C_1 = 0$, the first mode has been swept out of the assumed deflection by the *sweeping matrix* S. When this equation is substituted into the original matrix equation

$$X = \omega^2 aMX \qquad (9.5\text{-}6)$$

the result is

$$X = \omega^2 aMSX \qquad (9.5\text{-}7)$$

The iteration procedure applied to Eq. (9.5-7) will converge to the second mode.

For the third and higher modes, the sweeping procedure is repeated, making $C_1 = C_2 = 0$, etc. This reduces the order of the matrix equation by one each time; however, the convergence for higher modes becomes more critical if impurities are introduced through the sweeping matrices. It is well to check the highest mode by the inversion of the original matrix equation, which should be equal to the equation formulated in terms of the stiffness influence coefficients.

EXAMPLE 9.5-1

Write the matrix equation, based on flexibility influence coefficients, for the system shown in Fig. 9.5-1 and determine all the natural modes.

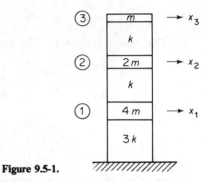

Figure 9.5-1.

Solution: The flexibility coefficients are found by applying a unit load, one at a time, to points 1, 2, and 3.

$$a_{11} = a_{21} = a_{12} = a_{31} = a_{13} = \frac{1}{3k}$$

$$a_{22} = a_{32} = a_{23} = \left(\frac{1}{3k} + \frac{1}{k} \right) = \frac{4}{3k}$$

$$a_{33} = \frac{1}{3k} + \frac{1}{k} + \frac{1}{k} = \frac{7}{3k}$$

The equations of motion in matrix form are then

$$\begin{Bmatrix} x_1 \\ x_2 \\ x_3 \end{Bmatrix} = \frac{\omega^2}{3k} \begin{bmatrix} 1 & 1 & 1 \\ 1 & 4 & 4 \\ 1 & 4 & 7 \end{bmatrix} \begin{bmatrix} 4m & 0 & 0 \\ 0 & 2m & 0 \\ 0 & 0 & m \end{bmatrix} \begin{Bmatrix} x_1 \\ x_2 \\ x_3 \end{Bmatrix}$$

$$\begin{Bmatrix} x_1 \\ x_2 \\ x_3 \end{Bmatrix} = \frac{\omega^2 m}{3k} \begin{bmatrix} 4 & 2 & 1 \\ 4 & 8 & 4 \\ 4 & 8 & 7 \end{bmatrix} \begin{Bmatrix} x_1 \\ x_2 \\ x_3 \end{Bmatrix}$$

Starting with arbitrary values of $x_1 x_2 x_3$, the above equation converges to the first mode, which is

$$\begin{Bmatrix} x_1 \\ x_2 \\ x_3 \end{Bmatrix} = \frac{\omega^2 m}{3k} 14.32 \begin{Bmatrix} 0.25 \\ 0.79 \\ 1.00 \end{Bmatrix}$$

The fundamental frequency is then found to be

$$\omega_1 = \sqrt{\frac{3k}{14.32m}} = 0.457 \sqrt{\frac{k}{m}}$$

To determine the second mode, we form the sweeping matrix given by Eq. (9.5-5)

$$S = \begin{bmatrix} 0 & -\frac{1}{2}\left(\frac{0.79}{0.25}\right) & -\frac{1}{4}\left(\frac{1.00}{0.25}\right) \\ 0 & 1 & 0 \\ 0 & 0 & 1 \end{bmatrix} = \begin{bmatrix} 0 & -1.58 & -1 \\ 0 & 1 & 0 \\ 0 & 0 & 1 \end{bmatrix}$$

The new equation for the second mode iteration is from Eq. (9.5-6).

$$\begin{Bmatrix} x_1 \\ x_2 \\ x_3 \end{Bmatrix} = \frac{\omega^2 m}{3k} \begin{bmatrix} 4 & 2 & 1 \\ 4 & 8 & 4 \\ 4 & 8 & 7 \end{bmatrix} \begin{bmatrix} 0 & -1.58 & -1 \\ 0 & 1 & 0 \\ 0 & 0 & 1 \end{bmatrix} \begin{Bmatrix} x_1 \\ x_2 \\ x_3 \end{Bmatrix}$$

$$= \frac{\omega^2 m}{3k} \begin{bmatrix} 0 & -4.32 & -3.0 \\ 0 & 1.67 & 0 \\ 0 & 1.67 & 3.0 \end{bmatrix} \begin{Bmatrix} x_1 \\ x_2 \\ x_3 \end{Bmatrix}$$

Starting the iteration process with arbitrary amplitudes, the above equation converges to the second mode, which is

$$\begin{Bmatrix} x_1 \\ x_2 \\ x_3 \end{Bmatrix} = \frac{\omega^2 m}{3k} 3 \begin{Bmatrix} -1.0 \\ 0 \\ 1.0 \end{Bmatrix}$$

The natural frequency of the second mode is therefore found to be

$$\omega_2 = \sqrt{\frac{k}{m}}$$

For the determination of third mode, we impose the conditions $C_1 = C_2 = 0$ from the orthogonality equation (9.5-3)

$$C_1 = \sum_{i=1}^{3} m_i(x_i)_1 \bar{x}_i = 4(0.25)\bar{x}_1 + 2(0.79)\bar{x}_2 + 1(1.0)\bar{x}_3 = 0$$

$$C_2 = \sum_{i=1}^{3} m_i(x_i)_2 \bar{x}_i = 4(-1.0)\bar{x}_1 + 2(0)\bar{x}_2 + 1(1.0)\bar{x}_3 = 0$$

From these two equations we obtain

$$\bar{x}_1 = 0.25\bar{x}_3 \quad \bar{x}_2 = -0.79\bar{x}_3$$

which can be expressed by the matrix equation

$$\begin{Bmatrix} x_1 \\ x_2 \\ x_3 \end{Bmatrix} = \begin{bmatrix} 0 & 0 & 0.25 \\ 0 & 0 & -0.79 \\ 0 & 0 & 1.00 \end{bmatrix} \begin{Bmatrix} \bar{x}_1 \\ \bar{x}_2 \\ \bar{x}_3 \end{Bmatrix}$$

This matrix is devoid of the first two modes and can be used as a sweeping matrix for the third mode. Applying this to the original equation, we obtain

$$\begin{Bmatrix} x_1 \\ x_2 \\ x_3 \end{Bmatrix} = \frac{\omega^2 m}{3k} \begin{bmatrix} 4 & 2 & 1 \\ 4 & 8 & 4 \\ 4 & 8 & 7 \end{bmatrix} \begin{bmatrix} 0 & 0 & 0.25 \\ 0 & 0 & -0.79 \\ 0 & 0 & 1.00 \end{bmatrix} \begin{Bmatrix} \bar{x}_1 \\ \bar{x}_2 \\ \bar{x}_3 \end{Bmatrix}$$

The above equation results immediately in the third mode, which is

$$\begin{Bmatrix} x_1 \\ x_2 \\ x_3 \end{Bmatrix} = \frac{\omega^2 m}{3k} 1.68 \begin{Bmatrix} 0.25 \\ -0.79 \\ 1.00 \end{Bmatrix}$$

The natural frequency of the third mode is then found to be

$$\omega_3 = \sqrt{\frac{3k}{1.68m}} = 1.34\sqrt{\frac{k}{m}}$$

These natural frequencies were checked by solving the stiffness equation, which is

$$m \begin{bmatrix} 4 & 0 & 0 \\ 0 & 2 & 0 \\ 0 & 0 & 1 \end{bmatrix} \begin{Bmatrix} \ddot{x}_1 \\ \ddot{x}_2 \\ \ddot{x}_3 \end{Bmatrix} + k \begin{bmatrix} 4 & -1 & 0 \\ -1 & 2 & -1 \\ 0 & -1 & 1 \end{bmatrix} \begin{Bmatrix} x_1 \\ x_2 \\ x_3 \end{Bmatrix} = 0$$

With $\lambda = m\omega^2/k$, the determinant of this equation, set equal to zero gives

$$8(1-\lambda)^3 - 5(1-\lambda) = 0$$

Its solution is found to be

$$\lambda_1 = 0.2094 \qquad \omega_1 = 0.4576\sqrt{\frac{k}{m}}$$

$$\lambda_2 = 1.0000 \qquad \omega_2 = 1.0000\sqrt{\frac{k}{m}}$$

$$\lambda_3 = 1.7906 \qquad \omega_3 = 1.3381\sqrt{\frac{k}{m}}$$

PROBLEMS

9-1 Write the kinetic and potential energy expressions for the system of Fig. P9-1 and determine the equation for ω^2 by equating the two energies. Letting $x_2/x_1 = n$, plot ω^2 versus n. Pick off the maximum and minimum values of ω^2 and the corresponding values of n, and show that they represent the two natural modes of the system.

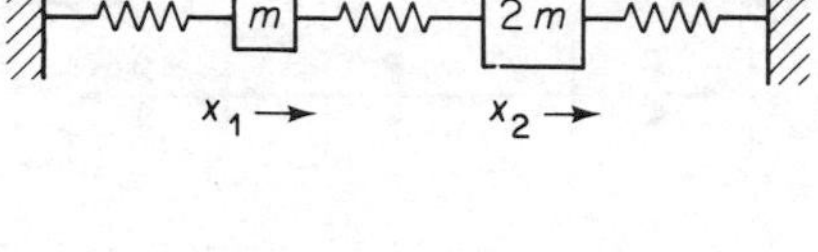

Figure P9-1.

9-2   Using Rayleigh's method, estimate the fundamental frequency of the lumped mass system shown in Fig. P9-2.

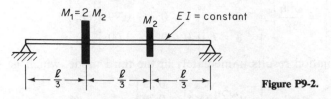

Figure P9-2.

9-3   Estimate the fundamental frequency of the lumped mass cantilever beam shown in Fig. P9-3.

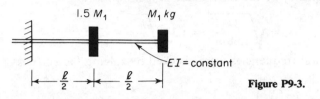

Figure P9-3.

9-4   Verify the results of Example 9.1-4 by using Eq. (9.1-3).

9-5   Another form of Rayleigh's quotient for the fundamental frequency can be obtained by starting from the equation of motion based on the flexibility influence coefficient

$$X = aM\ddot{X}$$
$$= \omega^2 aMX$$

Premultiplying by $X'M$ we obtain

$$X'MX = \omega^2 X'MaMX$$

and the Rayleigh quotient becomes

$$\omega^2 = \frac{X'MX}{X'MaMX}$$

Solve for ω_1 in Example 9.1-4 by using the above equation and compare results with Prob. 9-4.

9-6   Using the curve

$$y(x) = \frac{l^3}{3EI}\left(\frac{x}{l}\right)^2$$

solve Prob. 9-3 by using the method of integration. *Hint*: Draw shear and moment diagrams based on inertia loads.

9-7   Using the deflection $y(x) = y_{max}\sin(\pi x/l)$, determine the fundamental frequency of the beam shown in Fig. P9-7 (a) if $EI_2 = EI_1$ and (b) if $EI_2 = 4EI_1$.

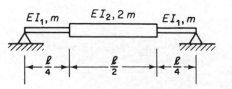

Figure P9-7.

9-8   Repeat Prob. 9-7 but use the curve

$$y(x) = y_{max}\frac{4x}{l}\left(1 - \frac{x}{l}\right)$$

9-9   A uniform cantilever beam of mass m per unit length has its free end pinned to two springs of stiffness k and mass m_0 each as shown in Fig. P9-9. Using Rayleigh's method, find its natural frequency ω_1.

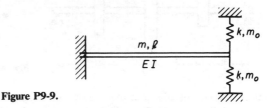

Figure P9-9.

9-10   A uniform beam of mass M and stiffness $K = EI/l^3$, shown in Fig. P9-10, is supported on equal springs with total vertical stiffness of k lb/in. Using Rayleigh's method with the deflection $y_{max} = \sin(\pi x/l) + b$, show that the frequency equation becomes

$$\omega^2 = \frac{2k}{M}\left[\frac{\dfrac{K}{k}\dfrac{\pi^4}{4} + \dfrac{b^2}{2}}{\dfrac{1}{2} + \dfrac{4b}{\pi} + b^2}\right]$$

By $\partial\omega^2/\partial b = 0$, show that the lowest frequency results when

$$b = -\frac{\pi}{4}\left(\frac{1}{2} - \frac{K\pi^4}{2k}\right) \pm \sqrt{\left[\frac{\pi}{2}\left(\frac{1}{2} - \frac{K\pi^4}{2k}\right)\right]^2 + \frac{\pi^4 K}{2k}}$$

Figure P9-10.

l, M

$k/2$ $k/2$

9-11   Assuming a static deflection curve

$$y(x) = y_{max}\left[3\left(\frac{x}{l}\right) - 4\left(\frac{x}{l}\right)^3\right], \qquad 0 \le x \le \frac{l}{2}$$

determine the lowest natural frequency of a simply supported beam of constant EI and a mass distribution of

$$m(x) = m_0\frac{x}{l}\left(1 - \frac{x}{l}\right)$$

by the Rayleigh method.

9-12   Using Dunkerley's equation, determine the fundamental frequency of the three-mass cantilever beam shown in Fig. P9-12.

$l \quad\quad m \quad\quad l \quad\quad m \quad\quad l \quad\quad m$

Figure P9-12.

9-13 Using Dunkerley's equation, determine the fundamental frequency of the beam shown in Fig. P9-13.

$$W_1 = W, \qquad W_2 = 4W, \qquad W_3 = 2W$$

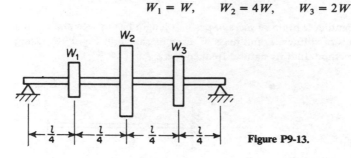

Figure P9-13.

9-14 A load of 100 lb at the wing tip of a fighter plane produced a corresponding deflection of 0.78 in. If the fundamental bending frequency of the same wing is 622 cpm, approximate the new bending frequency when a 320-lb fuel tank (including fuel) is attached to the wing tip.

9-15 A given beam was vibrated by an eccentric mass shaker of mass 5.44 kg at the mid-span, and resonance was found at 435 cps. With an additional mass of 4.52 kg, the resonant frequency was lowered to 398 cps. Determine the natural frequency of the beam.

9-16 Using the Rayleigh-Ritz method and assuming modes x/l and $\sin(\pi x/l)$; determine the two natural frequencies and modes of a uniform beam pinned at the right end and attached to a spring of stiffness k at the left end.

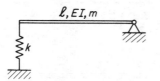

Figure P9-16.

9-17 For the wedge-shaped plate of Example 9.3-1, determine the first two natural frequencies and mode shapes for bending vibration by using the Ritz deflection function $y = C_1 x^2 + C_2 x^3$.

9-18 Using the Rayleigh-Ritz method, determine the first two natural frequencies and mode shapes for the longitudinal vibration of a uniform rod with a spring of stiffness k_0 attached to the free end as shown in Fig. P9-18. Use the first two normal modes of the fixed-free rod in longitudinal motion.

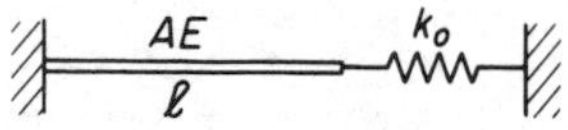

Figure P9-18.

9-19 Repeat Prob. 9-18 but this time the spring is replaced by a mass m_0 as shown in Fig. P9-19.

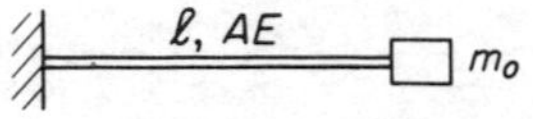

Figure P9-19.

9-20   For the simply supported variable mass beam of Prob. 9-11, assume the deflection to be made up of the first two modes of the uniform beam and solve for the two natural frequencies and mode shapes by the Rayleigh-Ritz method.

9-21   A uniform rod hangs freely from a hinge at the top. Using the three modes $\phi_1 = x/l$, $\phi_2 = \sin(\pi x/l)$, and $\phi_3 = \sin(2\pi x/l)$ determine the characteristic equation by using the Rayleigh-Ritz method.

9-22   Using matrix iteration, determine the three natural frequencies and modes for the cantilever beam of Prob. 9-12.

9-23   Determine the influence coefficients for the three-mass system of Fig. P9-23, and calculate the principal modes by matrix iteration.

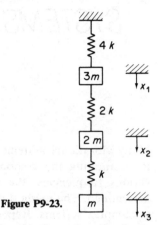

Figure P9-23.

9-24   Using matrix iteration, determine the natural frequencies and mode shapes of the torsional system of Fig. P9-24.

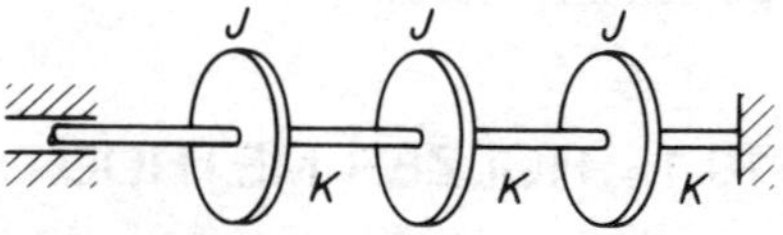

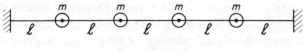

Figure P9-24.

9-25   In Fig. P9-25 four masses are strung along strings of equal lengths. Assuming the tension to be constant, determine the natural frequencies and mode shapes by matrix iteration.

Figure P9-25.

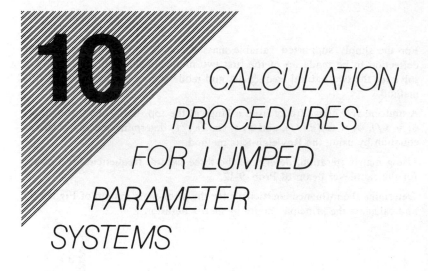

CALCULATION PROCEDURES FOR LUMPED PARAMETER SYSTEMS

Many vibrational systems can be modeled as lumped parameter systems. By examining the response of such systems to harmonic excitation of various frequencies, the natural frequencies and mode shapes can be determined. The method of transfer matrices expand the scope of analysis to complex systems. Repetitive systems lend themselves to a more general matrix analysis. Finally, the difference equation enables one to simply calculate by equations the natural frequencies and mode shapes of repeated structures.

10.1 HOLZER METHOD

When an undamped system is vibrating freely at any one of its natural frequencies, no external force, torque, or moment is necessary to maintain the vibration. Also, the amplitude of the mode shape is immaterial to the vibration. Recognizing these facts, Holzer* proposed a method of calculation for the natural frequencies and mode shapes of torsional systems by assuming a frequency and starting with a unit amplitude at one end of the system and progressively calculating the torque and angular displacement to the other end. The frequencies that result in zero external torque or compatible boundary conditions at the other end are the natural frequencies of the system. The method can be applied to any lumped mass system,

*H. Holzer, *Die Berechnung der Drehschwingungen* (Berlin: Springer-Verlag, 1921).

linear spring-mass systems, beams modeled by discrete masses and beam springs, etc.

Holzer's Procedure for Torsional Systems. Figure 10.1-1 shows a torsional system represented by a series of disks connected by shafts. Assuming a frequency ω and amplitude $\theta_1 = 1$, the inertia torque of the first disk is

$$- J_1 \ddot{\theta}_1 = J_1 \omega^2 \theta_1 = J_1 \omega^2 1$$

where harmonic motion is implied. This torque acts through shaft 1 and twists it by

$$\frac{J_1 \omega^2}{K_1} = \theta_1 - \theta_2 = 1 - \theta_2$$

or

$$\theta_2 = 1 - \frac{J_1 \omega^2}{K_1}$$

With θ_2 known, the inertia torque of the second disk is calculated as $J_2 \omega^2 \theta_2$. The sum of the first two inertia torques acts through the shaft K_2, causing it to twist by

$$\frac{J_1 \omega^2 + J_2 \omega^2 \theta_2}{K_2} = \theta_2 - \theta_3$$

In this manner, the amplitude and torque at every disk can be calculated. The resulting torque at the far end

$$T_{\text{ext}} = \sum_{i=1}^{4} J_i \omega^2 \theta_i$$

can then be plotted for the chosen ω. By repeating the calculation with other values of ω, the natural frequencies are found when $T_{\text{ext}} = 0$. The angular displacements θ_i corresponding to the natural frequencies are the mode shapes.

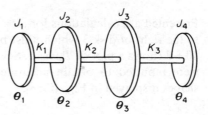

Figure 10.1-1.

EXAMPLE 10.1-1

Determine the natural frequencies and mode shapes of the system shown in Fig. 10.1-2.

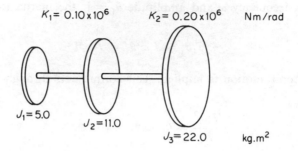

$K_1 = 0.10 \times 10^6$ $K_2 = 0.20 \times 10^6$ Nm/rad

$J_1 = 5.0$

$J_2 = 11.0$

$J_3 = 22.0$ kg.m^2

Figure 10.1-2.

Solution: The following table defines the parameters of the system and the sequence of calculations, which can be easily carried out on any programmable calculator.

	Parameters of the System		
	Station 1	Station 2	Station3
	$J_1 = 5$ $K_1 = 0.10 \times 10^6$	$J_2 = 11$ $K_2 = 0.20 \times 10^6$	$J_3 = 22$ $K_3 = 0$
	Calculation Program		
ω ω^2	$\theta_1 = 1.0$ $T_1 = \omega^2 \theta_1 J_1$	$\theta_2 = 1 - T_1/k_1$ $T_2 = T_1 + \omega^2 \theta_2 J_2$	$\theta_3 = \theta_2 - T_2/k_2$ $T_3 = T_2 + \omega^2 \theta_3 J_3$
20 400	1.0 2.0×10^3	0.980 6.312×10^3	0.9484 14.66×10^3
40 1600	1.0 8.0×10^3	0.920 24.19×10^3	0.799 52.32×10^3

Presented are calculations for $\omega = 20$ and 40. The quantity T_3 is the torque to the right of disk 3 which must be zero at the natural frequencies. Figure 10.1-3 shows a plot of T_3 vs. ω. Several frequencies in the vicinity of $T_3 = 0$ were inputted to obtain accurate values of the first and second mode shapes displayed in Fig. 10.1-4.

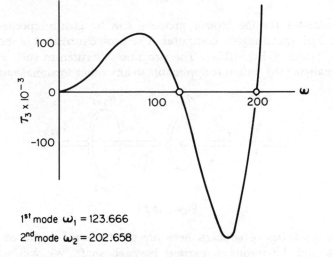

1st mode $\omega_1 = 123.666$

2nd mode $\omega_2 = 202.658$

Figure 10.1-3.

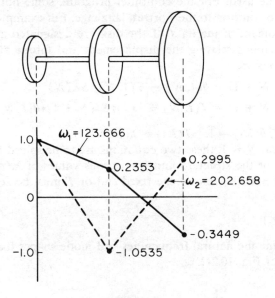

$\omega_1 = 123.666$

0.2353

$\leftarrow \omega_2 = 202.658$

0.2995

−0.3449

−1.0535

Figure 10.1-4.

10.2 DIGITAL COMPUTER PROGRAM FOR THE TORSIONAL SYSTEM

The calculations for the Holzer problem can be greatly speeded up by using the high-speed digital computer. The problem treated is the general torsional system of Fig. 10.2-1. The program is written in such a manner that by changing the data it is applicable to any other torsional system.

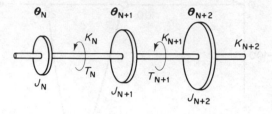

Figure 10.2-1.

The quantities of concern here are the torsional displacement θ of each disk and the torque T carried by each shaft. We will adopt two indexes: N to define the position along the structure and I for the frequency to be used. For the computer program, some notation changes are required to conform to the Fortran language. For example, the stiffness K and the moment of inertia J of the disk are designated as SK and SJ.

The equations relating the displacement and torque at the Nth and $N + 1$st stations are

$$\theta(I, N + 1) = \theta(I, N) - T(I, N)/SK(N) \qquad (10.2\text{-}1)$$
$$T(I, N + 1) = T(I, N) + \lambda(I)*SJ(N + 1)*\theta(I, N + 1) \quad (10.2\text{-}2)$$

where $\lambda = \omega^2$, $\theta(I,1) = 1$, $T(I,1) = \lambda(I)*SJ(1)$.

Starting at $N = 1$, these two equations are to be solved for θ and T at each point N of the structure and for various values of λ. At the natural frequencies, θ must be zero at the fixed end or T must be zero at the free end.

EXAMPLE 10.2-1

Determine the natural frequencies and mode shapes for the torsional system of Fig. 10.2-2.

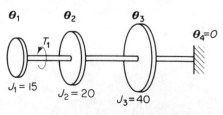

$K_1 = 2 \times 10^6 \qquad K_2 = 2 \times 10^6 \qquad K_3 = 3 \times 10^6$ **Figure 10.2-2.**

Solution: The frequency range can be scanned by choosing an initial ω and an increment $\Delta\omega$. We choose for this problem the frequencies

$$\omega = 40, 60, 80, \cdots 620$$

which can be programmed as

$$\omega(I) = 40 + (I - 1)*20, \qquad I = 1 \text{ to } 30$$

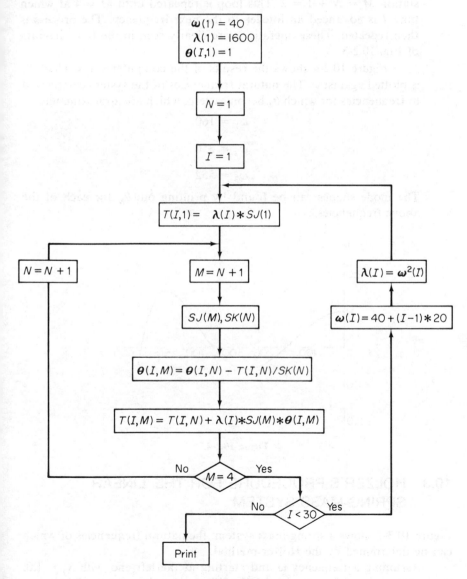

Figure 10.2-3.

The corresponding $\lambda(I)$ is computed as

$$\lambda(I) = \omega(I)**2$$

The computation is started with the boundary conditions, $N = 1$

$$\theta(I, 1) = 1$$

$$T(I, 1) = \lambda(I)*SJ(1)$$

Eqs. (10.2-1) and (10.2-2) then give the values of θ and T at the next station $M = N + 1 = 2$. This loop is repeated until $M = 4$ at which time I is advanced an integer to the next frequency. The process is then repeated. These operations are clearly seen in the flow diagram of Fig. 10.2-3.

Figure 10.2-4 shows the results of the computer study where θ_4 is plotted against ω. The natural frequencies of the system correspond to frequencies for which θ_4 becomes zero, which are approximately

$$\omega_1 = 160$$

$$\omega_2 = 356$$

$$\omega_3 = 552$$

The mode shapes can be found by printing out θ_N for each of the above frequencies.

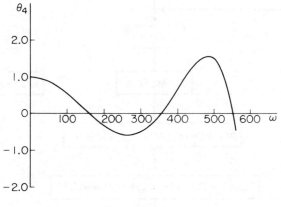

Figure 10.2-4.

10.3   HOLZER'S PROCEDURE FOR THE LINEAR SPRING-MASS SYSTEM

Figure 10.3-1 shows a spring-mass system, the natural frequencies of which can be determined by the Holzer method.

Assuming a frequency ω and starting at the left end with $x_1 = 1.0$, the inertia force of mass m_1 is $m_1\omega^2 1$. This force acts on spring k_1 causing

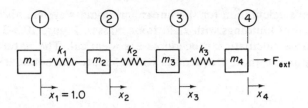

Figure 10.3-1.

it to deform by

$$\frac{m_1\omega^2}{k_1} = 1 - x_2$$

or

$$x_2 = 1 - \frac{m_1\omega^2}{k_1}$$

The inertia force of m_2 can now be found as $m_2\omega^2 x_2$ and the sum of the inertia forces $m_1\omega^2 + m_2\omega^2 x_2$ acts on the spring k_2 causing it to deform by

$$\frac{m_1\omega^2 + m_2\omega^2 x_2}{k_2} = x_2 - x_3$$

Thus, x_3 is found and the procedure can be repeated until all the displacements are found. The external force

$$F_{ext} = \sum_{i=1}^{4} m_i\omega^2 x_i$$

necessary to maintain the vibration for the assumed frequency can now be plotted against ω.

Repeating the calculations with other frequencies, the natural frequencies of the system are found when $F_{ext} = 0$. It is evident then that the procedure for the linear spring-mass system is identical to that of the torsional system.

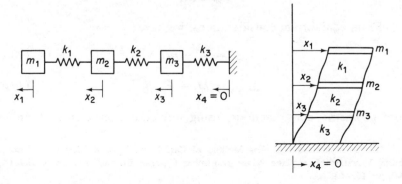

Figure 10.3-2.   Two similar systems for Holzer's method.

Holzer's calculation for the linear spring-mass system also applies to the vibration of buildings with rigid floor masses. Figure 10.3-2 shows two systems whose calculation procedures are identical. The natural frequencies correspond to those frequencies which lead to the boundary condition $x_4 = 0$.

10.4   MYKLESTAD'S METHOD FOR BEAMS

When a beam is replaced by lumped masses connected by massless beam sections, a method developed by N. O. Myklestad* can be used to progressively compute the deflection, slope, moment, and shear from one section to the next, in a manner similar to the Holzer method.

Uncoupled Flexural Vibration.   Figure 10.4-1 shows a typical section of an idealized beam with lumped masses. By taking the free-body section in the manner indicated, it will be possible to write equations for the shear and moment at $i + 1$ entirely in terms of quantities at i. These can then be substituted into the geometric equations for θ and y.

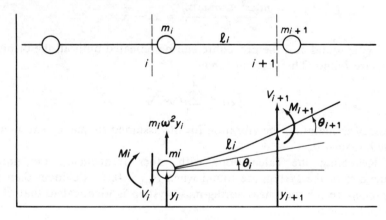

Figure 10.4-1.

From equilibrium considerations, we have

$$V_{i+1} = V_i - m_i\omega^2 y_i \qquad (10.4\text{-}1)$$

$$M_{i+1} = M_i - V_{i+1}l_i \qquad (10.4\text{-}2)$$

From geometric considerations, using influence coefficients of uniform

*N. O. Myklestad, "A New Method of Calculating Natural Modes of Uncoupled Bending Vibration of Airplane Wings and Other Types of Beams," *Jour. Aero. Sci.* (April 1944), pp. 153–62.

beam sections, we have

$$\theta_{i+1} = \theta_i + M_{i+1}\left(\frac{l}{EI}\right)_i + V_{i+1}\left(\frac{l^2}{2EI}\right)_i \qquad (10.4\text{-}3)$$

$$y_{i+1} = y_i + \theta_i l_i + M_{i+1}\left(\frac{l^2}{2EI}\right)_i + V_{i+1}\left(\frac{l^3}{3EI}\right)_i \qquad (10.4\text{-}4)$$

where $(l/EI)_i$ = slope at $i + 1$ measured from a tangent at i due to a unit moment at $i + 1$;

$(l^2/2EI)_i$ = slope at $i + 1$ measured from a tangent at i due to a unit shear at $i + 1$ = deflection at $i + 1$ measured from a tangent at i due to a unit moment at $i + 1$;

$(l^3/3EI)_i$ = deflection at $i + 1$ measured from a tangent at i due to a unit shear at $i + 1$.

Thus Eqs. (10.4-1) through (10.4-4) in the sequence given enables the calculations to proceed from i to $i + 1$.

Boundary Conditions.   Of the four boundary conditions at each end, two are generally known. For example, a cantilever beam with $i = 1$ at the free end would have $V_1 = M_1 = 0$. Since the amplitude is arbitrary, we can choose $y_1 = 1.0$. Having done so, the slope θ_1 is fixed to a value which is yet to be determined. Because of the linear character of the problem, the four quantities at the far end will be in the form

$$V_n = a_1 + b_1\theta_1$$

$$M_n = a_2 + b_2\theta_1$$

$$\theta_n = a_3 + b_3\theta_1$$

$$y_n = a_4 + b_4\theta_1$$

where a_i, b_i are constants and θ_1 is unknown. Thus the frequencies which satisfy the boundary condition $\theta_n = y_n = 0$ for the cantilever beam will establish θ_1 and the natural frequencies of the beam, i.e., $\theta_1 = -a_3/b_3$ and $y_n = a_4 - (a_3/b_3)b_4 = 0$. Hence, by plotting y_n vs. ω, the natural frequencies of the beam can be found.

EXAMPLE 10.4-1

To illustrate the computational procedure we will determine the natural frequencies of the cantilever beam shown in Fig. 10.4-1. The massless beam sections are assumed to be identical so that the influence coefficients for each section are equal. The numerical

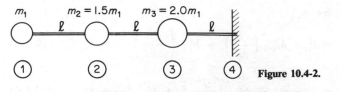

Figure 10.4-2.

constants for the problem are given as

$$m_1 = 100 \text{ kg}$$

$$l = 0.5\text{m}$$

$$EI = 0.10 \times 10^{-6} \text{ Nm}^2$$

$$\frac{l}{EI} = 5 \times 10^{-6} \frac{1}{Nm}$$

$$\frac{l^2}{2EI} = 1.25 \times 10^{-6} \frac{1}{N}$$

$$\frac{l^3}{3EI} = 0.41666 \times 10^{-6} \frac{m}{N}$$

The computation is started at 1. Since each of the quantities V, M, θ, and y will be in the form $a + b$, they are arranged into two columns, each of which can be computed separately. The calculation for the left column is started with $V_1 = 0$, $M_1 = 0$, $\theta_1 = 0$, and $y_1 = 1.0$. The right columns, which are proportional to θ, are started with the initial values of $V_1 = 0$, $M_1 = 0$, $\theta_1 = 1\theta$, and $y_1 = 0$.

Table 10.4-1 shows how the computation for Eqs. (10.4-1) through (10.4-4) can be carried out with any programmable calculator. The frequency chosen for this table is $\omega = 10$.

To start the computation we note that the moment and shear at station 1 are zero. We can choose the deflection at station 1 to be 1.0, in which case the slope at this point becomes an unknown θ. We therefore carry out two columns of calculations for each quantity starting with $y_1 = 1.0$, $\theta_1 = 0$, and $y_1 = 0$, $\theta_1 = \theta$. The unknown slope $\theta_1 = \theta$ is found by forcing θ_4 at the fixed end to be zero, after which the deflection y_4 can be calculated and plotted against ω. The natural frequencies of the system are those for which $y_4 = 0$.

TABLE 10.4-1

$$\Omega = 10. \qquad \Omega^2 = 100.$$

i	V Newtons		M Newton meters		θ Radians		y Meters	
1	0	0	0	0	0	θ	1.0	0
2	$-10,000.$	0	5000.	0	0.0125	1.0θ	1.002084	0.5θ
3	$-25031.$	-7500θ	17515.	3750θ	0.06879	1.00937θ	1.0198	1.001563θ
4	$-45427.$	-275320	40228.	175160	0.21315	1.06250	1.08555	1.51670

$0_4 = 0.21315 + 1.0625\theta = 0 \qquad \theta_1 = -0.2006117$

$y_4 = 1.08555 + 1.5167(-0.2006117) = 0.78128$ plot vs. $\omega = 10$

Digital Computer Program.   The calculation with the programmable calculator, although feasible, is still laborious. The following section presents a more practical approach using the digital computer.

Figure 10.4-3 shows a flow diagram for the natural frequency search. The initial frequency survey is made using the Computer Program 10.4 with $\Delta\omega = 10$ and ω between 10 and 400. The numerical results given in Table 10.4-2 show three natural frequencies located between $20 \leq \omega_1 \leq 30$, $130 \leq \omega_2 \leq 140$ and $340 \leq \omega_3 \leq 350$. Further computations were carried out in each of these regions with a much smaller $\Delta\omega$. Since the lumped mass model of only three masses could hardly give reliable results for the third mode, only the first two modes were recomputed and found to be $\omega_1 = 25.03$ and $\omega_2 = 138.98$. The mode shape at ω_2 is plotted in Fig. 10.4-4.

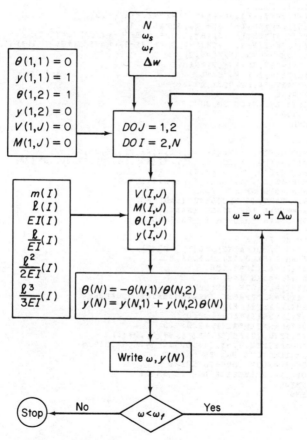

Flow diagram for natural frequency search of beams.

Figure 10.4-3.

```
1            DIMENSION AM(20),AL(20),EI(20),V(20,2),BM(20,2),TH(20,2),
             1 Y(20,2),THI(2),YI(2), SM(20),SV(20),DV(20)
2            DATA THI/0.0,1.0/,YI/1.0,0.0/
      C      V(I,J)=SHEAR AT SECTION I DUE TO INITIAL CONDITION J
      C      BM(I,J) = BENDING MOMENT
      C      TH(I,J)=SLOPE
      C      Y(I,J)=DEFLECTION
3            CALL INFL(AM,AL,EI,N,WS,WF,DW,SM,SV,DV)
4            W=WS
5     30 CONTINUE
      C LOOP ON INITIAL CONDITIONS
6            DO 10 J= 1,2
7            TH(1,J)=THI(J)
8            Y(1,J)=YI(J)
9            V(1,J)=0.0
10           BM(1,J)=0.0
      C LOOP ON NUMBER OF SECTIONS
11           DO 10 I=2,N
12           V(I,J)=V(I-1,J)-AM(I-1)*W**2.*Y(I-1,J)
13           BM(I,J)=BM(I-1,J)-V(I,J)*AL(I-1)
14           TH(I,J)=TH(I-1,J)+BM(I,J)*SM(I-1)+V(I,J)*SV(I-1)
15      10 Y(I,J)=Y(I-1,J)+TH(I-1,J)*AL(I-1)+BM(I,J)*SV(I-1)+V(I,J)*DV(I-1)
16           SLOPE=-TH(N,1)/TH(N,2)
17           DEFL=Y(N,1)+Y(N,2)*SLOPE
18           WRITE(6,20) W, DEFL
19      20 FORMAT(1X,F6.0,3X,E14.6)
20           W=W+DW
21           IF(W.LE.WF)GO TO 30
22            STOP
23           END


24           SUBROUTINE INFL(AM,AL,EI,N,WS,WF,DW,SM,SV,DV)
25           DIMENSION AM(20),AL(20),EI(20),SM(20),SV(20),DV(20)
26           READ(5,10) N,WS,WF,DW
27      10 FORMAT(I2,1X,F6.2,1X,F6.2,1X,F4.1)
      C      N=NUMBER OF SECTIONS
      C      WS=STARTING FREQUENCY(RADIANS)
      C      WF=FINAL FREQUENCY (RADIANS)
      C      DW=DELTA FREQUENCY (RADIANS)
28           M=N-1
29           DO 20 I=1,M
30           READ(5,30) AM(I),AL(I), EI(I)
31      30 FORMAT(F4.0,1X,F3.1,1X,F5.2)
32           EI(I)=EI(I)*1.0E6
      C      AM(I)=MASS (KG)
      C      AL(I)=LENGTH (METERS)
      C      EI(I)=STIFFNESS (NEWTON METER SQUARED)
33           SM(I)=AL(I)/EI(I)
34           SV(I)=SM(I)*AL(I)/2.0
35           DV(I)=SV(I)*AL(I)*2.0/3.0
36      20 WRITE(6,40)AM(I),SM(I),SV(I),DV(I)
37      40 FORMAT(F10.4,5X,E14.6,5X,E14.6,5X,E14.6)
      C      SM(I)=SLOPE DUE TO MOMENT
      C      SV(I)=SLOPE DUE TO SHEAR
      C      DV(I)=DEFLECTION DUE TO SHEAR
38           RETURN
39           END
```

TABLE 10.4-2 PART I AND II

m	ℓ/EI	$\ell^2/2EI$	$\ell^3/3EI$
100.0000	0.500000E-05	0.125000E-05	0.416666E-06
150.0000	0.500000E-05	0.125000E-05	0.416666E-06
200.0000	0.500000E-05	0.125000E-05	0.416666E-06

ω	y_4	
10.	0.781295E 00	
20.	0.265886E 00	$\leftarrow \omega_1$
30.	-0.292211E 00	
40.	-0.737204E 00	
50.	-0.102850E 01	
60.	-0.118001E 01	
70.	-0.121920E 01	
80.	-0.117191E 01	
90.	-0.105892E 01	
100.	-0.896440E 00	
110.	-0.697179E 00	
120.	-0.471451E 00	
130.	-0.227783E 00	
140.	0.263367E-01	$\leftarrow \omega_2$
150.	0.284378E 00	
160.	0.540451E 00	
170.	0.789139E 00	
180.	0.102559E 01	
190.	0.124516E 01	
200.	0.144368E 01	
210.	0.161714E 01	
220.	0.176195E 01	
230.	0.187518E 01	
240.	0.195322E 01	
250.	0.199371E 01	
260.	0.199416E 01	
270.	0.195226E 01	
280.	0.186609E 01	
290.	0.173399E 01	
300.	0.155418E 01	
310.	0.132565E 01	
320.	0.104712E 01	
330.	0.718018E 00	
340.	0.336182E 00	
350.	-0.974121E-01	$\leftarrow \omega_3$
360.	-0.584229E 00	
370.	-0.112500E 01	
380.	-0.171924E 01	
390.	-0.236646E 01	
400.	-0.306812E 01	

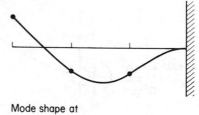

Figure 10.4-4. Mode shape at $\omega_2 = 138.98$

309

10.5 MYKLESTAD'S METHOD FOR ROTATING BEAMS

The Myklestad method can be extended to the rotating beam problem, i.e., propellers and turbine blades vibrating in a plane perpendicular to the axis of rotation. The centrifugal force will, in this case, introduce terms in addition to those in the beam analysis of the previous section.

Figure 10.5-1 shows the same beam section of the previous section. With the axis of rotation at the right end, the centrifugal force at i and $i + 1$ are

$$F_i = \Omega^2 \sum_{j=1}^{i-1} m_j x_j \qquad (10.5\text{-}1)$$

$$F_{i+1} = F_i + \Omega^2 m_i x_i \qquad (10.5\text{-}2)$$

where Ω is the angular velocity of rotation of the beam.

Since the shear is the force in the plane of the beam perpendicular to the beam center line and the centrifugal force is perpendicular to the rotation axis, the shear equation of the previous section is altered to

$$V_{i+1} = V_i - m_i \omega^2 y_i - F_{i+1} \theta_i \qquad (10.5\text{-}3)$$

The moment equation is also altered by the additional moment of the centrifugal force and the new shear

$$M_{i+1} = M_i - V_{i+1} l_i + F_{i+1}(y_{i+1} - y_i)$$

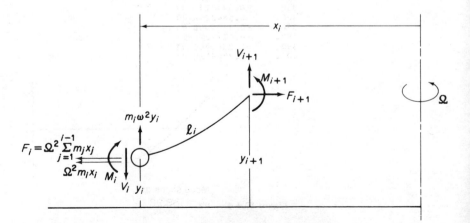

Figure 10.5-1.

Substituting for $(y_{i+1} - y_i)$ from Eq. (10.5-3), the above equation becomes

$$M_{i+1} = M_i - V_{i+1}l_i + F_{i+1}\left(\theta_i l_i + M_{i+1}\frac{l^2}{2EI} + V_{i+1}\frac{l^3}{3EI}\right)$$

$$= \frac{M_i}{\left(1 - F_{i+1}\dfrac{l^2}{2EI}\right)} - V_{i+1}\left[\frac{l_i - F_{i+1}\dfrac{l^3}{3EI}}{1 - F_{i+1}\dfrac{l^2}{2EI}}\right] + \theta_i l_i \frac{F_{i+1}}{\left(1 - F_{i+1}\dfrac{l^2}{2EI}\right)}$$

$$\text{(10.5-4)}$$

Since F_{i+1} can be calculated in advance for all i, V_{i+1} and M_{i+1}, including the effect of the centrifugal force, can be determined by Eqs. (10.5-3) and (10.5-4) in the order given.

The effect of the centrifugal force on the slope and deflection are due to the new V_{i+1} and M_{i+1}, and Eqs. (10.4-3) and (10.4-4) for θ_{i+1} and y_{i+1} need not be changed.

10.6 COUPLED FLEXURE-TORSION VIBRATION

Natural modes of vibration of airplane wings and other beam structures are often coupled flexure-torsion vibration which for higher modes differ considerably from those of uncoupled modes. To treat such problems we must model the beam as shown in Fig. 10.6-1. The elastic axis of the beam about which the torsional rotation takes place is assumed to be initially straight. It is able to twist, but its bending displacement is restricted to the vertical plane. The principal axes of bending for all cross sections are parallel in the undeformed state. Masses are lumped at each station with its center of gravity at distance c_i from the elastic axis and J_i is the mass moment of inertia of the section about the elastic axis, i.e., $J_i = J_{cg} + m_i c_i^2$.

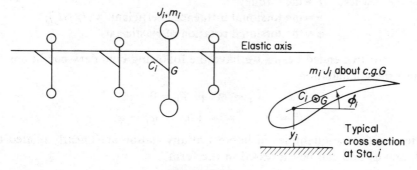

Figure 10.6-1.

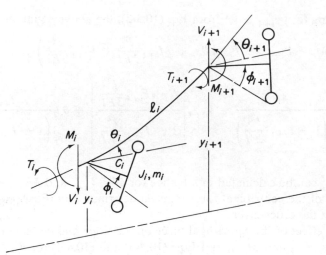

Figure 10.6-2.

Figure 10.6-2 shows the ith section from which the following equations can be written:

$$V_{i+1} = V_i - m_i\omega^2(y_i + c_i\varphi_i) \tag{10.6-1}$$

$$M_{i+1} = M_i - V_{i+1}l_i \tag{10.6-2}$$

$$T_{i+1} = T_i + J_i\omega^2\varphi_i + m_ic_i\omega^2y_i \tag{10.6-3}$$

$$\theta_{i+1} = \theta_i + V_{i+1}\left(\frac{l^2}{2EI}\right)_i + M_{i+1}\left(\frac{l}{EI}\right)_i \tag{10.6-4}$$

$$y_{i+1} = y_i + \theta_il_i + V_{i+1}\left(\frac{l^3}{3EI}\right)_i + M_{i+1}\left(\frac{l^2}{2EI}\right)_i \tag{10.6-5}$$

$$\varphi_{i+1} = \varphi_i + T_{i+1}h_i \tag{10.6-6}$$

where $T =$ the torque;
$h =$ the torsional influence coefficient $= (l/GI_p)$;
$\varphi =$ the torsional rotation of elastic axis.

For free-ended beams we have the following boundary conditions to start the computation.

$$V_1 = M_1 = T_1 = 0$$

$$\theta_1 = \theta, \qquad y_1 = 1.0, \qquad \varphi_1 = \varphi_1$$

Here again, the quantities of interest at any station are linearly related to θ_1 and φ_1 and can be expressed in the form

$$a + b\theta_1 + c\varphi_1 \tag{10.6-7}$$

Natural frequencies are formed by the satisfaction of the boundary conditions at the other end. Often for symmetric beams, such as the airplane wing, only one-half the beam need be considered. The satisfaction of the boundary conditions for the symmetric and antisymmetric modes enables sufficient equations for the solution.

10.7 TRANSFER MATRICES

The transfer matrix method* offers another approach to the analysis of lumped parameter systems. It is exceptionally suitable for large systems made up of several subsystems. The subsystem is made up of simple elastic and dynamic elements assembled as the field matrix and the point matrix. The formulation is in terms of the state vector which is a column matrix of the displacements and internal forces. In fact, the method of transfer matrix is simply the matrix systemization of the Holzer or the Myklestad's procedure. Its advantages are in the assembly of complex systems, branched systems, and in the simplification for the identification of the boundary equations.

The Spring-Mass System. Figure 10.7-1 shows a part of a linear spring-mass system with one of the subsections isolated. The nth section consists of the mass m_n with displacement x_n and the spring of stiffness k_n, whose ends have displacements x_n and x_{n-1}. When necessary to do so, we designate quantities to the left and right of the element by superscripts L and R.

For the mass m_n, Newton's second law is

$$m_n \ddot{x}_n = F_n^R - F_n^L$$

which for harmonic motion becomes

$$F_n^R = -\omega^2 m_n x_n + F_n^L \qquad (10.7\text{-}1)$$

Since the displacement on either side of m_n is the same, we have the identity

$$x_n = x_n^R = x_n^L \qquad (10.7\text{-}2)$$

Equations (10.7-1) and (10.7-2) can now be assembled into a single matrix equation

$$\left\{ \begin{matrix} x \\ F \end{matrix} \right\}_n^R = \begin{bmatrix} 1 & 0 \\ -\omega^2 m & 1 \end{bmatrix} \left\{ \begin{matrix} x \\ F \end{matrix} \right\}_n^L \qquad (10.7\text{-}3)$$

where $\left\{ \begin{matrix} x \\ F \end{matrix} \right\}$ is the *state vector* and the square matrix is the point matrix.

*E. C. Pestel and F. A. Leckie, *Matrix Methods in Elastomechanics* (New York: McGraw-Hill Book Co., 1963).

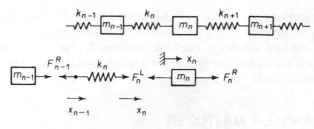

Figure 10.7-1.

Next we examine the spring k_n whose end forces are equal

$$F_{n-1}^R = F_n^L \qquad (10.7\text{-}4)$$

The spring force is related to the spring modulus k_n by the equation

$$x_n^L - x_{n-1}^R = \frac{F_{n-1}^R}{k_n} \qquad (10.7\text{-}5)$$

Equations (10.7-4) and (10.7-5) are now assembled in matrix form

$$\begin{Bmatrix} x \\ F \end{Bmatrix}_n^L = \begin{bmatrix} 1 & \dfrac{1}{k} \\ 0 & 1 \end{bmatrix} \begin{Bmatrix} x \\ F \end{Bmatrix}_{n-1}^R \qquad (10.7\text{-}6)$$

where the square matrix above is the field matrix.

We now relate the quantities at station n in terms of quantities at station $n - 1$ by substituting Eq. (10.7-6) into (10.7-3)

$$\begin{Bmatrix} x \\ F \end{Bmatrix}_n^R = \begin{bmatrix} 1 & 0 \\ -\omega^2 m & 1 \end{bmatrix} \begin{bmatrix} 1 & \dfrac{1}{k} \\ 0 & 1 \end{bmatrix} \begin{Bmatrix} x \\ F \end{Bmatrix}_{n-1}^R$$

$$\rightarrow \;\; = \begin{bmatrix} 1 & \dfrac{1}{k} \\ -\omega^2 m & \left(1 - \dfrac{\omega^2 m}{k}\right) \end{bmatrix}_n \begin{Bmatrix} x \\ F \end{Bmatrix}_{n-1}^R$$

$$(10.7\text{-}7)$$

Since the state vector at $n - 1$ is transferred to the state vector at n through the square matrix above, it is called the *transfer matrix* for section n. With known values of the state vector at station 1 and a chosen value of ω^2, it is possible to progressively compute the state vectors to the last station n. Depending on the boundary conditions, either x_n or F_n can be plotted as a function of ω^2; the natural frequencies of the system are established when the boundary conditions are satisfied.

Torsional System.   Signs are often a source of confusion in rotating systems, and it is necessary to clearly define the sense of positive quantities. The coordinate along the rotational axis is considered positive towards the right. If a cut is made across the shaft, the face with the outward normal towards the positive coordinate direction is called the positive face. Positive torques and positive angular displacements are indicated on the positive face by arrows pointing positively according to the right hand screw rule as shown in Fig. 10.7-2.

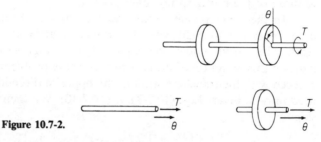

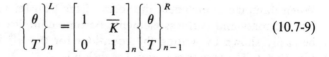

Figure 10.7-2.

With this definition, the development of the transfer matrix of the torsional system is identical to that of the linear spring-mass system with $\left\{ \begin{matrix} \theta \\ T \end{matrix} \right\}$ as the state vector. We isolate the nth section as in Fig. 10.7-3 and write the dynamical equation for the point matrix and the elastic equation for the field matrix. They are

$$\left\{ \begin{matrix} \theta \\ T \end{matrix} \right\}_n^R = \begin{bmatrix} 1 & 0 \\ -\omega^2 J & 1 \end{bmatrix}_n \left\{ \begin{matrix} \theta \\ T \end{matrix} \right\}_n^L \tag{10.7-8}$$

and

$$\left\{ \begin{matrix} \theta \\ T \end{matrix} \right\}_n^L = \begin{bmatrix} 1 & \dfrac{1}{K} \\ 0 & 1 \end{bmatrix}_n \left\{ \begin{matrix} \theta \\ T \end{matrix} \right\}_{n-1}^R \tag{10.7-9}$$

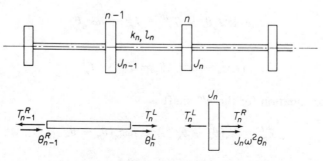

Figure 10.7-3.

which combine to

$$\left\{ \begin{array}{c} \theta \\ T \end{array} \right\}_n^R \underset{\rightarrow}{=} \left[\begin{array}{cc} 1 & \dfrac{1}{K} \\ -\omega^2 J & \left(1 - \dfrac{\omega^2 J}{K}\right) \end{array} \right]_n \left\{ \begin{array}{c} \theta \\ T \end{array} \right\}_{n-1}^R \qquad (10.7\text{-}10)$$

We thus note that each of Eqs. (10.7-8), (10.7-9), and (10.7-10) is identical to those of the linear spring-mass system.

In the development so far, the stations were numbered in increasing order from left to right with the transfer matrix also progressing to the right. The arrow under the equal sign in Eqs. (10.7-7) and (10.7-10) indicate this direction of progression. In some problems it is convenient to proceed with the transfer matrix in the opposite direction, in which case we need only to invert Eqs. (10.7-7) or (10.7-10). We then obtain the relationship

$$\left\{ \begin{array}{c} \theta \\ T \end{array} \right\}_{n-1}^R \underset{\leftarrow}{=} \left[\begin{array}{cc} \left(1 - \dfrac{\omega^2 J}{K}\right) & -\dfrac{1}{K} \\ \omega^2 J & 1 \end{array} \right] \left\{ \begin{array}{c} \theta \\ T \end{array} \right\}_n^R \qquad (10.7\text{-}11)$$

The arrow now indicates that the transfer matrix progresses from right to left with the order of the station numbering unchanged. The student should verify this equation, starting with the free-body development.

10.8 SYSTEMS WITH DAMPING

When damping is included, the form of the transfer matrix is not altered, but the mass and stiffness elements become complex quantities. This can be easily shown by writing the equations for the nth subsystem shown in Fig. 10.8-1. The torque equation for disk n is

$$- \omega^2 J_n \theta_n = T_n^R - T_n^L - i\omega c_n \theta_n$$

or

$$\left(i\omega c_n - \omega^2 J_n\right)\theta_n = T_n^R - T_n^L \qquad (10.8\text{-}1)$$

The elastic equation for the nth shaft is

$$T_n^L = K_n(\theta_n - \theta_{n-1}) + i\omega g_n(\theta_n - \theta_{n-1})$$

$$= (K_n + i\omega g_n)(\theta_n - \theta_{n-1}) \qquad (10.8\text{-}2)$$

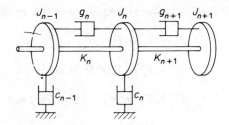

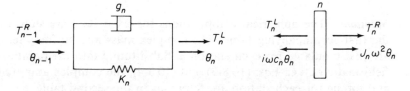

Figure 10.8-1. Torsional system with damping.

Thus the point matrix and the field matrix for the damped system become

$$\begin{Bmatrix} \theta \\ T \end{Bmatrix}_n^R = \begin{bmatrix} 1 & 0 \\ (i\omega c - \omega^2 J) & 1 \end{bmatrix}_n \begin{Bmatrix} \theta \\ T \end{Bmatrix}_n^L \tag{10.8-3}$$

$$\begin{Bmatrix} \theta \\ T \end{Bmatrix}_n^L = \begin{bmatrix} 1 & \dfrac{1}{(K + i\omega g)} \\ 0 & 1 \end{bmatrix}_n \begin{Bmatrix} \theta \\ T \end{Bmatrix}_{n-1}^R \tag{10.8-4}$$

which are identical to the undamped case except for the mass and stiffness elements; these elements are now complex.

EXAMPLE 10.8-1

The torsional system of Fig. 10.8-2 is excited by a harmonic torque at a point to the right of disk 4. Determine the torque-frequency curve

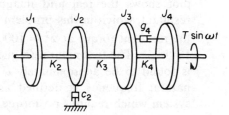

Figure 10.8-2.

and establish the first natural frequency of the system.

$$J_1 = J_1 = 500 \text{ lb in.sec}^2$$

$$J_3 = J_4 = 1000 \text{ lb in.sec}^2$$

$$K_2 = K_3 = K_4 = 10^6 \text{ lb in./rad}$$

$$c_2 = 10^4 \text{ lb in.sec/rad}$$

$$g_4 = 2 \times 10^4 \text{ lb in.sec/rad}$$

Solution: The numerical computations for $\omega^2 = 1000$ are shown in the first accompanying table. The complex mass and stiffness terms are first tabulated for each station n. Substituting into the point and field matrices, i.e., Eqs. (10.8-3) and (10.8-4), the complex amplitude and torque for each station are found, as in the second table.

n	$(\omega^2 J_n - i\omega c_n)10^{-6}$	$(K_n + i\omega g_n)10^{-6}$
1	$0.50 + 0.0i$	
2	$0.50 - 0.316i$	$1.0 + 0.0i$
3	$1.0 + 0.0i$	$1.0 + 0.0i$
4	$1.0 + 0.0i$	$1.0 + 0.635i$

n	θ_n	T_n^R(for $\omega^2 = 1000$)
1	$1.0 + 0.0i$	$(-0.50 + 0.0i) \times 10^6$
2	$0.50 + 0.0i$	$(-0.750 + 0.158i) \times 10^6$
3	$-0.250 + 0.158i$	$(-0.50 + 0.0i) \times 10^6$
4	$-0.607 + 0.384i$	$(0.107 - 0.384i) \times 10^4$

The above computations are repeated for a sufficient number of frequencies to plot the torque-frequency curve of Fig. 10.8-3. The plot shows the real and imaginary parts of T_4^R as well as their resultant, which in this problem is the exciting torque. For example, the resultant torque at $\omega^2 = 1000$ is $10^6 \sqrt{0.107^2 + 0.384^2} = 0.394 \times 10^6$ in. lb. The first natural frequency of the system from this diagram is found to be approximately $\omega = \sqrt{930} = 30.5$ rad/sec, where the natural frequency is defined as that frequency of the undamped system which requires no torque to sustain the motion.

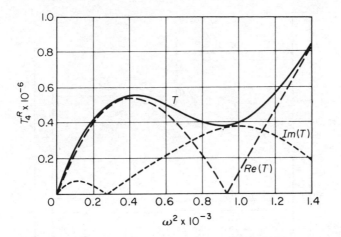

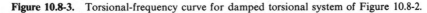

Figure 10.8-3. Torsional-frequency curve for damped torsional system of Figure 10.8-2.

EXAMPLE 10.8-2

In Fig. 10.8-2 if $T = 2000$ in. lb and $\omega = 31.6$ rad/sec, determine the amplitude of the second disk.

Solution: The table above indicates that a torque of 394,000 in. lb will produce an amplitude of $\theta_2 = 0.50$ radian. Since amplitude is proportional to torque, the amplitude of the second disk for the specified torque is $0.50 \times \frac{2}{394} = 0.00254$ radian.

10.9 GEARED SYSTEM

Consider the geared torsional system of Fig. 10.9-1, where the speed ratio of shaft 2 to shaft 1 is n. The system can be reduced to an equivalent single shaft system as follows.

With the speed of shaft 2 equal to $\dot{\theta}_2 = n\dot{\theta}_1$, the kinetic energy of the system is

$$T = \tfrac{1}{2}J_1\dot{\theta}_1^2 + \tfrac{1}{2}J_2n^2\dot{\theta}_1^2 \qquad (10.9\text{-}1)$$

Thus the equivalent inertia of disk 2 referred to shaft 1 is n^2J_2.

To determine the equivalent stiffness of shaft 2 referred to shaft 1, clamp disks 1 and 2 and apply a torque to gear 1, rotating it through an angle θ_1. Gear 2 will then rotate through the angle $\theta_2 = n\theta_1$, which will also be the twist in shaft 2. The potential energy of the system is then

$$U = \tfrac{1}{2}K_1\theta_1^2 + \tfrac{1}{2}K_2n^2\theta_1^2 \qquad (10.9\text{-}2)$$

and the equivalent stiffness of shaft 2 referred to shaft 1 is n^2K_2.

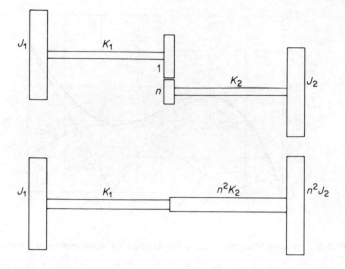

Figure 10.9-1.   Geared system and its equivalent single shaft system.

The rule for geared systems is thus quite simple: *multiply all stiffness and inertias of the geared shaft by n^2*, where n is the speed ratio of the geared shaft to the reference shaft.

10.10   BRANCHED SYSTEMS

Branched systems are frequently encountered; some common examples are the dual propeller system of a marine installation and the drive shaft and differential of an automobile, which are shown in Fig. 10.10-1.

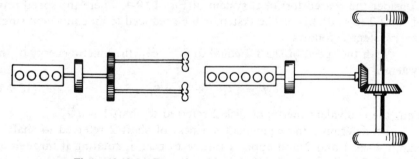

Figure 10.10-1.   Examples of branched torsional systems.

Such systems can be reduced to the form with one-to-one gears shown in Fig. 10.10-2 by multiplying all the inertias and stiffnesses of the branches by the squares of their speed ratios.

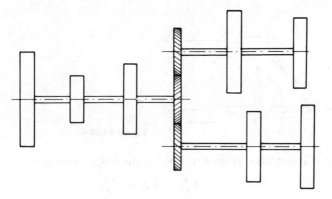

Figure 10.10-2. Branched system reduced to common speeds by 1 to 1 gears.

EXAMPLE 10.10-1

Outline the matrix procedure for solving the torsional branched system of Fig. 10.10-3.

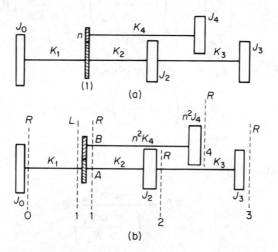

Figure 10.10-3. Branched system and reduced system.

Solution: We first convert to a system having one-to-one gears by multiplying the stiffness and inertia of branch B by n^2, as shown in Fig. 10.10-3b. We can then proceed from station 0 through to station 3, taking note of the fact that gear B introduces a torque T_{B1}^R on gear A.

Figure 10.10-4 shows the free-body diagram of the two gears. With T_{B1}^R shown as positive torque, the torque exerted on gear A by

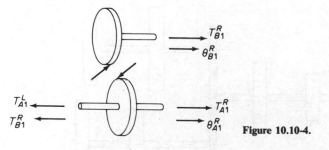

Figure 10.10-4.

gear B is negative as shown. The torque balance on gear A is then

$$T_{A1}^R = T_{A1}^L + T_{B1}^R \tag{a}$$

and we need now to express T_{B1}^R in terms of the angular displacement θ_1 of shaft A.

Using Eq. (10.7-11) and noting that $T_{B4}^R = 0$, we have for shaft B

$$\left\{ \begin{matrix} \theta_B \\ T_B \end{matrix} \right\}_1^R = \left[\begin{matrix} \left(1 - \dfrac{\omega^2 n^2 J_4}{n^2 K_4} \right) & -\dfrac{1}{n^2 K_4} \\ \omega^2 n^3 J_4 & 1 \end{matrix} \right] \left\{ \begin{matrix} \theta_B \\ 0 \end{matrix} \right\}_4^R \tag{b}$$

Since $\theta_{B1}^R = - \theta_{A1}^L = - \theta_{A1}^R$, we obtain

$$\theta_{B1}^R = \left(1 - \dfrac{\omega^2 J_4}{K_4} \right) \theta_{B4}^R = - \theta_{A1}^L \tag{c}$$

$$T_{B1}^R = \omega^2 n^2 J_4 \theta_{B4}^R \tag{d}$$

Eliminating θ_{B4}^R

$$T_{B1}^R = \dfrac{-\omega^2 n^2 J_4}{\left(1 - \dfrac{\omega^2 J_4}{K_4} \right)} \theta_{A1}^L \tag{e}$$

Substituting Eq. (e) into Eq. (a), the transfer function of shaft A across the gears becomes

$$\left\{ \begin{matrix} \theta_A \\ T_A \end{matrix} \right\}_1^R \rightarrow \left[\begin{matrix} 1 & 0 \\ \dfrac{-\omega^2 J_4}{\left(1 - \dfrac{\omega^2 J_4}{K_4} \right)} & 1 \end{matrix} \right] \left\{ \begin{matrix} \theta_A \\ T_A \end{matrix} \right\}_1^L \tag{f}$$

It is now possible to proceed along shaft A from $1R$ to $3R$ in the usual manner.

10.11 TRANSFER MATRICES FOR BEAMS

The algebraic equations of Sec. 10.4 can be rearranged so that the four quantities at station $i + 1$ are expressed in terms of the same four quantities at station i. When such equations are presented in matrix form they are known as *transfer matrices*. In this section we present a procedure for the formulation and assembly of the matrix equation in terms of its boundary conditions.

Figure 10.11-1 shows the same i^{th} section of the beam of Fig. 10.4-1 broken down further into a point mass and a massless beam by cutting the beam just right of the mass. We designate the quantities to the left and right of the mass by superscripts L and R, respectively.

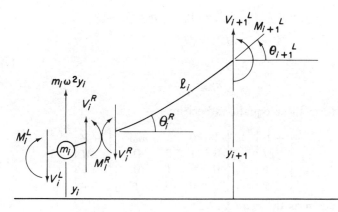

Figure 10.11-1. Beam sections for transfer matrices.

Considering first the massless beam section, the following equations can be written:

$$V_{i+1}^L = V_i^R$$

$$M_{i+1}^L = M_i^R - V_i^R l_i$$

$$\theta_{i+1}^L = \theta_i^R + M_{i+1}^L \left(\frac{l}{EI}\right)_i + V_{i+1}^L \left(\frac{l^2}{2EI}\right)_i$$

$$y_{i+1}^L = y_i^R + \theta_i^R l_i + M_{i+1}^L \left(\frac{l^2}{2EI}\right)_i + V_{i+1}^L \left(\frac{l^3}{3EI}\right)_i$$

(10.11-1)

Substituting for V_{i+1}^L and M_{i+1}^L from the first two equations into the last two and arranging the results in matrix form, we obtain what is referred to as the *field matrix*

$$
\left\{ \begin{array}{c} -V \\ M \\ \theta \\ y \end{array} \right\}_{i+1}^{L}
=
\begin{bmatrix}
1 & 0 & 0 & 0 \\
l & 1 & 0 & 0 \\
\dfrac{l^2}{2EI} & \dfrac{l}{EI} & 1 & 0 \\
\dfrac{l^3}{6EI} & \dfrac{l^2}{2EI} & l & 1
\end{bmatrix}
\left\{ \begin{array}{c} -V \\ M \\ \theta \\ y \end{array} \right\}_{i}^{R}
\qquad (10.11\text{-}2)
$$

In the above equation a minus sign has been inserted for V in order to make the elements of the field matrix all positive.

Next consider the point mass for which the following equations can be written:

$$
\begin{aligned}
V_i^R &= V_i^L - m_i \omega^2 y_i \\
M_i^R &= M_i^L \\
\theta_i^R &= \theta_i^L \\
y_i^R &= y_i^L
\end{aligned}
\qquad (10.22\text{-}3)
$$

In matrix form these equations become

$$
\left\{ \begin{array}{c} -V \\ M \\ \theta \\ y \end{array} \right\}_{i}^{R}
=
\begin{bmatrix}
1 & 0 & 0 & m\omega^2 \\
0 & 1 & 0 & 0 \\
0 & 0 & 1 & 0 \\
0 & 0 & 0 & 1
\end{bmatrix}
\left\{ \begin{array}{c} -V \\ M \\ \theta \\ y \end{array} \right\}_{i}^{L}
\qquad (10.11\text{-}4)
$$

which is known as the *point matrix*.

Substituting Eq. (10.11-4) into Eq. (10.11-2) and multiplying, we obtain the assembled equation for the ith section

$$
\left\{ \begin{array}{c} -V \\ M \\ \theta \\ y \end{array} \right\}_{i+1}^{R}
=
\begin{bmatrix}
1 & 0 & 0 & m\omega^2 \\
l & 1 & 0 & m\omega^2 l \\
\dfrac{l^2}{2EI} & \dfrac{l}{EI} & 1 & m\omega^2 \dfrac{l^2}{2EI} \\
\dfrac{l^3}{6EI} & \dfrac{l^2}{2EI} & l & \left(1 + \dfrac{m\omega^2 l^3}{6EI}\right)
\end{bmatrix}
\left\{ \begin{array}{c} -V \\ M \\ \theta \\ y \end{array} \right\}_{i}^{L}
\qquad (10.11\text{-}5)
$$

The square matrix in this equation is called the *transfer matrix* since the state vector at i is transferred to the state vector at $i + 1$ through this matrix. It is evident then that it is possible to progress through the structure so that the state vector at the far end is related to the state vector at the starting end by an equation of the form

$$
\left\{ \begin{array}{c} -V \\ M \\ \theta \\ y \end{array} \right\}_{n}
=
\begin{bmatrix}
u_{11} & u_{12} & \cdots & u_{14} \\
 & \vdots & & \\
u_{41} & & \cdots & u_{44}
\end{bmatrix}
\left\{ \begin{array}{c} -V \\ M \\ \theta \\ y \end{array} \right\}_{1}
\qquad (10.11\text{-}6)
$$

where the matrix $[u]$ is the product of all the transfer matrices of the structure.

The advantage of the transfer matrix lies in the fact that the unknown quantity at 1, i.e., θ_1, for the cantilever beam, need not be carried through each station as in the algebraic set of equations. The multiplication of the 4×4 matrices by the digital computer is a routine problem. Also, the boundary equations are clearly evident in the matrix equation. For example, the assembled equation for the cantilever beam is

$$
\begin{Bmatrix} -V \\ M \\ 0 \\ 0 \end{Bmatrix}_n = \begin{bmatrix} - & - & - & - \\ - & - & - & - \\ - & - & u_{33} & u_{34} \\ - & - & u_{43} & u_{44} \end{bmatrix} \begin{Bmatrix} 0 \\ 0 \\ \theta \\ 1 \end{Bmatrix}_1 \tag{10.11-7}
$$

and the natural frequencies must satisfy the equations

$$
0 = u_{33}\theta + u_{34}
$$
$$
0 = u_{43}\theta + u_{44}
$$

or

$$
y_n = -\frac{u_{34}}{u_{33}} \cdot u_{43} + u_{44} = 0 \tag{10.11-8}
$$

By plotting y_n vs. ω the natural frequencies correspond to the zeros of the curve.

10.12 TRANSFER MATRIX FOR REPEATED STRUCTURES

The transfer matrix of the previous chapter, when applied to repeated identical sections, leads to some interesting results. It should be noted that the determinant of the transfer matrix is unity regardless of whether or not the system is damped. The following three cases are presented to verify the above statement. Case 1, with Fig. 10.12-1.

$$
\begin{Bmatrix} F \\ x \end{Bmatrix}_n = \begin{bmatrix} \left(1 - \dfrac{\omega^2 m}{k}\right) & -\omega^2 m \\ \dfrac{1}{k} & 1 \end{bmatrix} \begin{Bmatrix} F \\ x \end{Bmatrix}_{n-1} \tag{10.12-1}
$$

This is the same equation as Eq. (10.7-7) with the state vector inverted.

Case 1

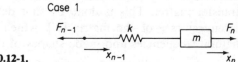

Figure 10.12-1.

Case 2, with Fig. 10.12-2.

$$\left\{ \begin{array}{c} F \\ x \end{array} \right\}_n = \left[\begin{array}{cc} \left(1 - \dfrac{\omega^2 m}{k + i\omega c}\right) & -\omega^2 m \\ \dfrac{1}{k + i\omega c} & 1 \end{array} \right] \left\{ \begin{array}{c} F \\ x \end{array} \right\}_{n-1} \qquad (10.12\text{-}2)$$

Case 2

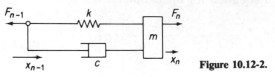

Figure 10.12-2.

Case 3, with Fig. 10.12-3.

$$\left\{ \begin{array}{c} F \\ x \end{array} \right\}_n = \left[\begin{array}{cc} \left(1 - \dfrac{\omega^2 m [k_1 + \omega c]}{kk_1 + i\omega c(k_1 + k)}\right) & -\omega^2 m \\ \dfrac{k_1 + i\omega c}{kk_1 + i\omega c(k_1 + k)} & 1 \end{array} \right] \left\{ \begin{array}{c} F \\ x \end{array} \right\}_{n-1} \qquad (10.12\text{-}3)$$

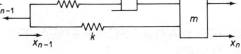

Figure 10.12-3.   Viscoelastic system.

The intermediate coordinate y has been eliminated in the above equation.
In each of the above cases, the transfer matrix is in the form

$$[T] = \left[\begin{array}{cc} A & B \\ C & D \end{array} \right] \qquad (10.12\text{-}4)$$

and the determinant $AD - BC = 1.0$. Even for the transfer matrix of the beam section (i.e., Eq. 10.11-5), the determinant of the 4×4 matrix is unity, as one can easily show.

When the system has n identical sections, the transfer matrix procedure leads to the equation

$$\left\{ \begin{array}{c} F \\ x \end{array} \right\}_n = [T]^n \left\{ \begin{array}{c} F \\ x \end{array} \right\}_0 \qquad (10.12\text{-}5)$$

and hence it is of interest to be able to calculate the nth power of the transfer matrix. This is done by first determining the eigenvalues μ and eigenvectors ξ of the matrix $[T]$, which must not be confused with the natural frequencies and mode shapes of the system previously discussed.

The eigenvalues and eigenvectors of the matrix $[T]$ satisfy the equation

$$[T]\{\xi\} - \mu\{\xi\} = 0 \tag{10.12-6}$$

For $[T] = \begin{bmatrix} A & B \\ C & D \end{bmatrix}$, the eigenvalues are found from the characteristic equation

$$\begin{vmatrix} (A - \mu) & B \\ C & (D - \mu) \end{vmatrix} = 0 \tag{10.12-7}$$

which as a result of $AD - BC = 1$ gives

$$\mu_{1,2} = \tfrac{1}{2}(A + D) \pm \sqrt{\tfrac{1}{4}(A + D)^2 - 1} \tag{10.12-8}$$

The eigenvectors can only be determined in terms of its ratio

$$\left(\frac{\xi_1}{\xi_2}\right)_1 = \frac{B}{\mu_1 - A} = r_1$$

$$\left(\frac{\xi_1}{\xi_2}\right)_2 = \frac{B}{\mu_2 - A} = r_2 \tag{10.12-9}$$

We next form the modal matrix P of the eigenvector columns

$$[P] = \begin{bmatrix} r_1 & r_2 \\ 1 & 1 \end{bmatrix} \tag{10.12-10}$$

The two equations

$$[T]\{\xi\} = \mu\{\xi\}$$

for μ_1 and μ_2 may now be assembled as a single matrix equation.

$$[T][P] = [P][\Lambda] \tag{10.12-11}$$

where $[\Lambda] = \begin{bmatrix} \mu_1 & 0 \\ 0 & \mu_2 \end{bmatrix}$ = diagonal matrix of the eigenvalues.

By post-multiplying by $[P]^{-1}$ we obtain

$$[T] = [P][\Lambda][P]^{-1}$$

The square of the above equation is

$$[T]^2 = [T][T] = [P][\Lambda][P]^{-1}[P][\Lambda][P]^{-1} = [P][\Lambda]^2[P]^{-1}$$

where $[\Lambda]^2 = \begin{bmatrix} \mu_1^2 & 0 \\ 0 & \mu_2^2 \end{bmatrix}$

Repeated multiplication leads to the nth power

$$[T]^n = [P][\Lambda]^n[P]^{-1} \tag{10.12-12}$$

The boundary conditions can now be applied to the equation

$$\left\{ \begin{matrix} F \\ x \end{matrix} \right\}_n = [T]^n \left\{ \begin{matrix} F \\ x \end{matrix} \right\}_0 = \begin{bmatrix} t_{11} & t_{12} \\ t_{21} & t_{22} \end{bmatrix} \left\{ \begin{matrix} F \\ x \end{matrix} \right\}_0 \qquad (10.12\text{-}13)$$

For example, if the end 0 is fixed and the end n is free, $x_0 = 0$ and $F_n = 0$, and we obtain

$$0 = t_{11} F_0$$

Since $F \neq 0$, the natural frequencies are found from $t_{11} = 0$. In case damping is present, the elements of the transfer matrix are complex quantities. In this case, the end displacement x_n may be chosen as unity, and the force F_0 is found from

$$1 = t_{21} F_0$$

10.13   DIFFERENCE EQUATION

The difference equation offers another approach to the problem of repeated identical sections. As an example of repeating sections, consider the N-story building shown in Fig. 10.13-1 where the mass of each floor is m and the lateral or shear stiffness of each section between floors is k lb/in. The equation of motion for the nth mass is then

$$m\ddot{x}_n = k(x_{n+1} - x_n) - k(x_n - x_{n-1}) \qquad (10.13\text{-}1)$$

which for harmonic motion can be represented in terms of the amplitudes as

$$X_{n+1} - 2\left(1 - \frac{\omega^2 m}{2k}\right) X_n + X_{n-1} = 0 \qquad (10.13\text{-}2)$$

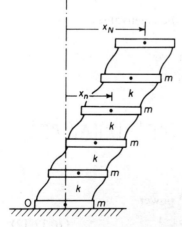

Figure 10.13-1.   Repeated structure for difference equation analysis.

The solution to this equation is found by substituting

$$X_n = e^{i\beta n} \tag{10.13-3}$$

which leads to the relationship

$$\left(1 - \frac{\omega^2 m}{2k}\right) = \frac{e^{i\beta} + e^{-i\beta}}{2} = \cos \beta \tag{10.13-4}$$

$$\frac{\omega^2 m}{k} = 2(1 - \cos \beta) = 4 \sin^2 \frac{\beta}{2}$$

The general solution for X_n is

$$X_n = A \cos \beta n + B \sin \beta n \tag{10.13-5}$$

where A and B are evaluated from the boundary conditions.

The difference equation (10.13-2) is restricted to $1 \lesssim n \lesssim (N - 1)$ and must be extended to $n = 0$ and $n = N$ by the boundary conditions.

At the ground the amplitude of motion is zero. With $n = 1$, Eq. (10.13-2) with $X_0 = 0$ becomes

$$X_2 - 2\left(1 - \frac{m\omega^2}{2k}\right)X_1 = 0$$

Substituting Eq. (10.13-4) and Eq. (10.13-5) into the above equation, we obtain

$$(A \cos 2\beta + B \sin 2\beta) - 2 \cos \beta(A \cos \beta + B \sin \beta) = 0$$

$$A(\cos 2\beta - 2 \cos^2 \beta) - B(\sin 2\beta - 2 \sin \beta \cos \beta) = 0$$

$$A(1) - B(0) = 0$$

$$\therefore A = 0$$

Thus the general solution is reduced to $X_n = B \sin \beta n$.

At the top the boundary equation is

$$m\ddot{x}_N = -k(x_N - x_{N-1})$$

which, in terms of the amplitude, becomes

$$X_{N-1} = \left(1 - \frac{\omega^2 m}{k}\right)X_N \tag{10.13-6}$$

Substituting from the general solution, we obtain the following relationship for the evaluation of β

$$\sin \beta(N - 1) = [1 - 2(1 - \cos \beta)] \sin \beta N$$

This result can be reduced to the product form

$$2 \cos \beta\left(N + \tfrac{1}{2}\right) \sin \frac{\beta}{2} = 0 \tag{10.13-7}$$

which is satisfied by

$$\cos \beta\left(N + \tfrac{1}{2}\right) = 0, \quad \frac{\beta}{2} = \frac{\pi}{2(2N+1)}, \ \frac{3\pi}{2(2N+1)}, \ \frac{5\pi}{2(2N+1)}, \cdots$$

$$(10.13\text{-}8)$$

The natural frequencies are then available from Eq. 10.13-4 as

$$\omega = 2\sqrt{k/m}\ \sin\frac{\beta}{2} \qquad (10.13\text{-}9)$$

which lead to

$$\omega_1 = 2\sqrt{\frac{k}{m}}\ \sin\frac{\pi}{2(2N+1)}$$

$$\omega_2 = 2\sqrt{\frac{k}{m}}\ \sin\frac{3\pi}{2(2N+1)}$$

$$\vdots$$

$$\omega_N = 2\sqrt{\frac{k}{m}}\ \sin\frac{(2N-1)\pi}{2(2N+1)}$$

Figure 10.13-2 shows a graphical representation of these natural frequencies when $N = 4$.

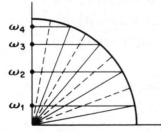

Figure 10.13-2. Natural frequencies of a repeated structure with $N = 4$.

The method of difference equation presented here is applicable to many other dynamical systems where repeating sections are present. The natural frequencies are always given by Eq. (10.13-9); however, the quantity β must be established for each problem from its boundary conditions.

PROBLEMS

10-1   Write a computer program for your programmable calculator for the torsional system given in Sec. 10.1. Fill in the actual algebraic operations performed in the program steps.

10-2   Using Holzer's method, determine the natural frequencies and mode shapes of the torsional system of Fig. P10-2 when $J = 1.0$ kg m^2 and $K = 0.20 \times 10^6$ Nm/rad.

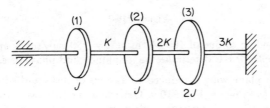

Figure P10-2.

10-3   Using Holzer's method, determine the first two natural frequencies and mode shapes of the torsional system shown in Fig. P10-3 with the following values of J and K.

$$J_1 = J_2 = J_3 = 1.13 \text{ kg m}^2$$

$$J_4 = 2.26 \text{ kg m}^2$$

$$K_1 = K_2 = 0.169 \text{ Nm/rad} \times 10^6$$

$$K_3 = 0.226 \text{ Nm/rad} \times 10^6$$

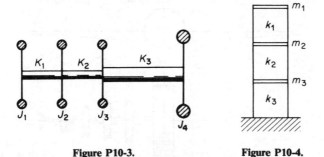

Figure P10-3.   **Figure P10-4.**

10-4   Determine the natural frequencies and mode shapes of the three-story building of Fig. P10-4 by using Holzer's method for all $m_s = m$ and all $k_s = k$.

10-5   Repeat Prob. 10-4 when $m_1 = m$, $m_2 = 2m$, $m_3 = 3m$, $k_1 = k$, $k_2 = k$, and $k_3 = 2k$.

10-6   Compare the equations of motion for the linear spring-mass system vs. torsional system with same mass and stiffness distribution. Show that they are similar.

10-7   Determine the natural frequencies and mode shapes of the spring-mass system of Fig. P10-7 by the Holzer method when all masses are equal and all stiffnesses are equal.

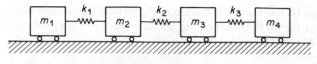

Figure P10-7.

10-8 If a harmonic torque of 1000 N m at $\omega = 150$ rad/sec is applied to disk 3 of Example 10.1-1, determine the amplitude and phase of each disk.

10-9 A fighter-plane wing is reduced to a series of disks and shafts for Holzer's analysis as shown in Fig. P10-9. Determine the first two natural frequencies for symmetric and antisymmetric torsional oscillations of the wings, and plot the torsional mode corresponding to each.

n	J lb. in. sec.^{2}	K lb. in./rad.
1	50	15×10^6
2	138	30
3	145	22
4	181	36
5	260	120
6	$\frac{1}{2} \times 140,000$	

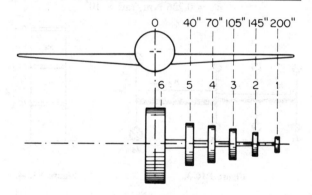

Figure P10-9.

10-10 Determine the natural modes of the simplified model of an airplane shown in Fig. P10-10 where $M/m = n$ and the beam of length l is uniform.

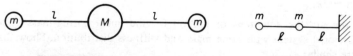

Figure P10-10.   **Figure P10-11.**

10-11 Using Myklestad's method, determine the natural frequencies and mode shapes of the two-lumped-mass cantilever beam of Fig. P10-11. Compare with previous results by using influence coefficients.

10-12 Determine the first two natural frequencies and mode shapes of the three mass cantilever of Fig. P10-12.

Figure P10-12.   Figure P10-13.

10-13 Using Myklestad's method, determine the boundary equations for the simply supported beam of Fig. P10-13.

10-14 The beam of Fig. P10-14 has been previously solved by the method of matrix iteration. Check that the boundary condition of zero deflection at the left end is satisfied for these natural frequencies when Myklestad's method is used. That is, check the deflection for change in sign when frequencies above and below the natural frequency are used.

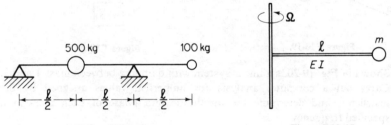

Figure P10-14.   Figure P10-15.

10-15 Using Myklestad's method, determine the equation for the natural frequency of the single mass beam hinged at the rotation axis as shown in Fig. P10-15.

10-16 A rotating beam, such as a helicopter blade, is sometimes considered as pinned at the hub. Establish the boundary equations for such a case.

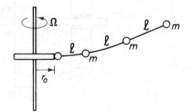

Figure P10-16 and P10-17.

10-17 Assume a helicopter blade to be represented by three lumped masses equally spaced as shown in Fig. P10-17. On the basis of constant bending stiffness, determine the natural frequencies for rotational speed Ω.

10-18 Repeat Prob. 10-17 if the blade at the hub is clamped.

10-19 Determine the flexure-torsion vibration for the system shown in Fig. P10-19.

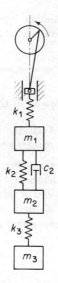

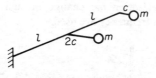

Figure P10-19.   **Figure P10-20.**

10-20   Shown in Fig. 10-20 is a linear system with damping between mass 1 and 2. Carry out a computer analysis for numerical values assigned by the instructor, and determine the amplitude and phase of each mass at a specified frequency.

10-21   A torsional system with a torsional damper is shown in Fig. 10-21. Determine the torque-frequency curve for the system.

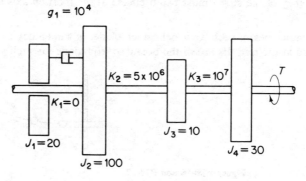

Figure P10-21.

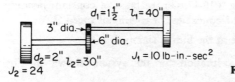

Figure P10-22.

10-22 Determine the equivalent torsional system for the geared system shown in Fig. P10-22 and find its natural frequency.

10-23 If the small and large gears of Prob. 10-22 have the following inertias, $J' = 2$, $J'' = 6$, determine the equivalent single shaft system and establish the natural frequencies.

10-24 Determine the two lowest natural frequencies of the torsional system shown in Fig. P10-24 for the following values of J, K, and n

$$J_1 = 15 \text{ lb in. sec}^2 \quad K_1 = 2 \times 10^6 \text{ lb in./rad}$$
$$J_2 = 10 \text{ lb in. sec}^2 \quad K_2 = 1.6 \times 10^6 \text{ lb in./rad}$$
$$J_3 = 18 \text{ lb in. sec}^2 \quad K_3 = 1 \times 10^6 \text{ lb in./rad}$$
$$J_4 = 6 \text{ lb in. sec}^2 \quad K_4 = 4 \times 10^6 \text{ lb in./rad}$$
$$\text{Speed ratio of drive shaft to axle} = 4 \text{ to } 1.$$

What are the amplitude ratios of J_2 to J_1 at the natural frequencies?

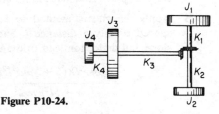

Figure P10-24.

10-25 Reduce the torsional system of the automobile shown in Fig. P10-25a to the equivalent torsional system shown in Fig. P10-25b. The necessary information is given as follows

J of each rear wheel = 9.2 lb in. sec^2
J of flywheel = 12.3 lb in. sec^2
transmission speed ratio (drive shaft to engine speed) = 1.0 to 3.0
differential speed ratio (axle to drive shaft) = 1.0 to 3.5
axle dimensions = $1\frac{1}{4}$ in. diameter, 25 in. long (each)
drive shaft dimensions = $1\frac{1}{2}$ in. diameter, 74 in. long
stiffness of crankshaft between cylinders, measured experimentally
   = 6.1×10^6 lb in./rad
stiffness of crankshaft between cylinder 4 and flywheel = 4.5×10^6 lb in./rad

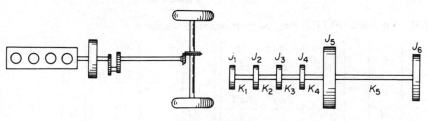

Figure P10-25.

10-26 Assume that the J of each cylinder of Prob. 10-25 = 0.20 lb in. sec^2 and determine the natural frequencies of the system.

10-27 Determine the equations of motion for the torsional system shown in Fig. P10-27, and arrange them into the matrix iteration form. Solve for the principal modes of oscillation.

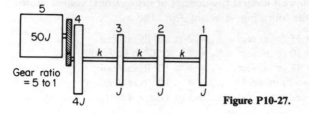

Figure P10-27.

10-28 Apply the matrix method to a cantilever beam of length l and mass m at the end, and show that the natural frequency equation is directly obtained.

10-29 Apply the matrix method to a cantilever beam with two equal masses spaced equally a distance l. Show that the boundary conditions of zero slope and deflection lead to the equation

$$\theta_1 = \frac{\frac{1}{2}m\omega^2 lK\left(5 + \frac{1}{6}m\omega^2 l^2 K\right)}{1 + \frac{1}{2}l^2 K m\omega^2} = \frac{1 + \frac{3}{2}m\omega^2 l^2 K + \left(\frac{1}{6}m\omega^2 l^2 K\right)^2}{2l + \frac{1}{6}m\omega^2 l^3 K}$$

where $K = l/EI$.

Obtain the frequency equation from the above relationship and determine the two natural frequencies.

10-30 Using the matrix formulation, establish the boundary conditions for the symmetric and antisymmetric bending modes for the system shown in Fig. P10-30. Plot the boundary determinant against the frequency ω to establish the natural frequencies, and draw the first two mode shapes.

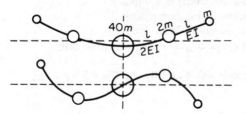

Figure P10-30.

10-31 Equation (10.11-2) may be rearranged to the form

$$\left\{\begin{array}{c} \theta \\ y \\ M \\ -V \end{array}\right\}_n = \left[\begin{array}{cc|cc} 1 & 0 & \dfrac{l}{EI} & \dfrac{l^2}{2EI} \\ l & 1 & \dfrac{l^2}{2EI} & \dfrac{l^3}{6EI} \\ \hline 0 & 0 & 1 & l \\ 0 & 0 & 0 & 1 \end{array}\right] \left\{\begin{array}{c} \theta \\ y \\ M \\ -V \end{array}\right\}_{n-1} = \left[\begin{array}{c|c} A' & B \\ \hline 0 & A \end{array}\right] \left\{\begin{array}{c} \theta \\ y \\ M \\ V \end{array}\right\}_{n-1}$$

where B is symmetric about its diagonal. Letting $\delta = (\theta, y)'$ and $L = (M, V)'$, show that the stiffness matrix is

$$\left\{ \begin{array}{c} L_{n-1} \\ L_n \end{array} \right\} = \left[\begin{array}{c|c} B^{-1} & -B^{-1}A' \\ \hline AB^{-1} & -AB^{-1}A' \end{array} \right] \left\{ \begin{array}{c} \delta_n \\ \delta_{n-1} \end{array} \right\}$$

10-32   Evaluate the partitioned matrices of Prob. 10-31 and show that they are in the form (primes indicate transpose)

$$\left\{ \begin{array}{c} L_{n-1} \\ L_n \end{array} \right\} = \left[\begin{array}{c|c} \mathcal{C} & \mathcal{B} \\ \hline -\mathcal{B}' & \mathcal{C} \end{array} \right] \left\{ \begin{array}{c} \delta_n \\ \delta_{n-1} \end{array} \right\}$$

which is expected due to the reciprocity theorem.

10-33   Using the notation of Prob. 10-32, rewrite Eq. (10.11-5) in the form

$$\left\{ \begin{array}{c} \delta \\ L \end{array} \right\}_n^R = \left[\begin{array}{c|c} A' & B \\ \hline Q & S \end{array} \right] \left\{ \begin{array}{c} \delta \\ L \end{array} \right\}_{n-1}^R$$

and show that the determinant of the transfer matrix is equal to unity.

10-34   From the boundary equation, Eq. (10.11-6), establish the boundary determinant $D(\omega)$ for a simply supported beam.

10-35   Determine the boundary determinant $D(\omega)$ for a clamped-clamped beam.

10-36   Determine the boundary determinant $D(\omega)$ for a clamped-hinged beam.

10-37   Determine the boundary determinant $D(\omega)$ for a hinged-free beam.

10-38   Prove that the elements of the modal matrix $[P]$ for a two degree of freedom system are

$$r_1 = \frac{\omega^2 m}{\mu_1 - 1} \qquad \text{and} \qquad r_2 = \frac{\omega^2 m}{\mu_2 - 1}$$

where the system section is a spring and a mass as in Fig. 10.12-1.

10-39   Show that for the system shown in Fig. P10-39, the natural frequency equation using the procedure of Sec. 10.12 reduces to

$$-\mu_1^n r_2 + \mu_2^n r_1 = 0$$

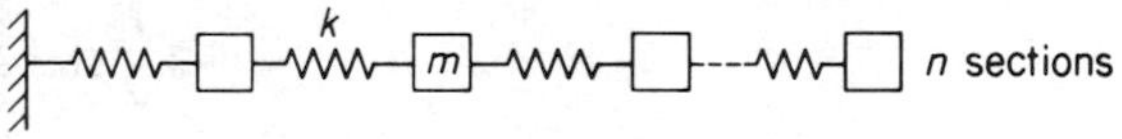

Figure P10-39.

10-40   Letting $\mu_1 = e^\alpha$ and $\mu_2 = e^{-\alpha}$, in Eq. (10.12-8), $(A + D)/2 = \cosh \alpha$. Develop the frequency equation in terms of this substitution.

10-41   Reduce the system of Fig. 10.12-3 to an equivalent of the system shown in Fig. 10.12-2.

10-42 Set up the difference equations for the torsional system shown in Fig. P10-42. Determine the boundary equations and solve for the natural frequencies.

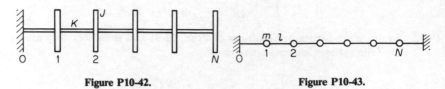

Figure P10-42.   Figure P10-43.

10-43 Set up the difference equations for N equal masses on a string with tension T, as shown in Fig. P10-43. Determine the boundary equations and the natural frequencies.

10-44 Write the difference equations for the spring-mass system shown in Fig. P10-44 and find the natural frequencies of the system.

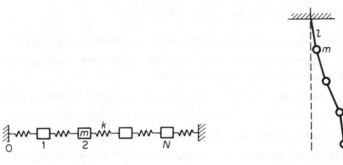

Figure P10-44.   Figure P10-45.

10-45 An N-mass pendulum is shown in Fig. P10-45. Determine the difference equations, boundary conditions, and the natural frequencies.

10-46 If the left end of the system of Prob. 10-42 is connected to a heavy flywheel, as shown in Fig. P10-46, show that the boundary conditions lead to the equation

$$(-\sin N\beta \cos \beta + \sin N\beta)\left(1 + 4\frac{K}{K_a}\frac{J_a}{J}\sin^2\frac{\beta}{2}\right)$$

$$= -2\frac{J_a}{J}\sin^2\frac{\beta}{2}\sin \beta \cos N\beta$$

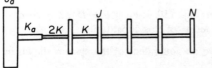

Figure P10-46.

10-47   If the top story of a building is restrained by a spring of stiffness K_N, as shown in Fig. P10-47, determine the natural frequencies of the N-story building.

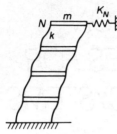

Figure P10-47.

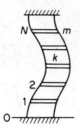

Figure P10-48.

10-48   A ladder-type structure is fixed at both ends, as shown in Fig. P10-48. Determine the natural frequencies.

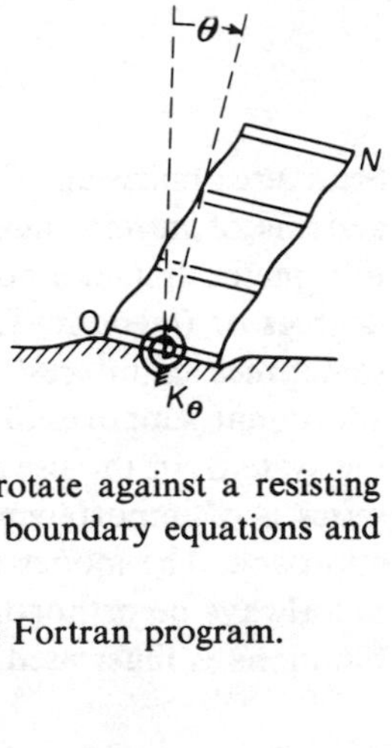

Figure P10-49.

10-49   If the base of an N-story building is allowed to rotate against a resisting spring K_θ, as shown in Fig. P10-49, determine the boundary equations and the natural frequencies.

10-50   Draw a flow diagram for Prob. 10-3 and write the Fortran program.

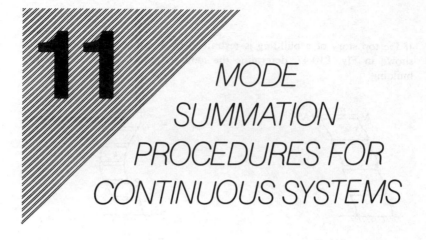

MODE SUMMATION PROCEDURES FOR CONTINUOUS SYSTEMS

Structures made up of beams are common in engineering. They constitute systems of infinite number of degrees of freedom, and the mode summation methods make possible their analysis as systems of finite number of degrees of freedom. The effect of rotary inertia and shear deformation is sometimes of interest in beam problems. Constraints are often found as additional supports of the structure, and they alter the normal modes of the system. In the use of the mode summation method, convergence of the series is of importance, and the mode acceleration method offers a varied approach. The modes used in representing the deflection of a system need not always be orthogonal. The synthesis of a system using non-orthogonal functions is illustrated.

11.1 MODE SUMMATION METHOD

In Section 6.8 the equations of motion were decoupled by the modal matrix to obtain the solution of forced vibration in terms of the normal coordinates of the system. In this section, we apply a similar technique to continuous systems by expanding the deflection in terms of the normal modes of the system.

Consider, for example, the general motion of a beam loaded by a distributed force $p(x, t)$, whose equation of motion is

$$[EIy''(x, t)]'' + m(x)\ddot{y}(x, t) = p(x, t) \qquad (11.1\text{-}1)$$

The normal modes $\phi_i(x)$ of such a beam must satisfy the equation

$$(EI\phi_i'')'' - \omega_i^2 m(x)\phi_i = 0 \qquad (11.1\text{-}2)$$

and its boundary conditions. The normal modes $\phi_i(x)$ are also orthogonal functions satisfying the relation

$$\int_0^l m(x)\phi_i\phi_j \, dx = \begin{cases} 0 & \text{for } j \neq i \\ M_i & \text{for } j = i \end{cases} \qquad (11.1\text{-}3)$$

By representing the solution to the general problem in terms of $\phi_i(x)$

$$y(x, t) = \sum_i \phi_i(x)q_i(t) \qquad (11.1\text{-}4)$$

the generalized coordinate $q_i(t)$ can be determined from Lagrange's equation by first establishing the kinetic and potential energies.

Recognizing the orthogonality relation, Eq. (11.1-3), the kinetic energy is

$$T = \tfrac{1}{2}\int_0^l \dot{y}^2(x, t)m(x) \, dx = \tfrac{1}{2}\sum_i \sum_j \dot{q}_i \dot{q}_j \int_0^l \phi_i\phi_j m(x) \, dx$$

$$\qquad (11.1\text{-}5)$$

$$= \tfrac{1}{2}\sum_i M_i \dot{q}_i^2$$

where the generalized mass M_i is defined as

$$M_i = \int_0^l \phi_i^2(x)m(x) \, dx \qquad (11.1\text{-}6)$$

Similarly, the potential energy is

$$U = \tfrac{1}{2}\int_0^l EI y''^2(x, t) \, dx = \tfrac{1}{2}\sum_i \sum_j q_i q_j \int_0^l EI\phi_i''\phi_j'' \, dx$$

$$\qquad (11.1\text{-}7)$$

$$= \tfrac{1}{2}\sum_i K_i q_i^2 = -\tfrac{1}{2}\sum \omega_i^2 M_i q_i^2$$

where the generalized stiffness is

$$K_i = \int_0^l EI[\phi_i''(x)]^2 \, dx \qquad (11.1\text{-}8)$$

In addition to T and U, we need the generalized force Q_i, which is determined from the work done by the applied force $p(x, t) \, dx$ in the virtual displacement δq_i.

$$\delta W = \int_0^l p(x, t)\left(\sum_i \phi_i \delta q_i\right) dx$$

$$\qquad (11.1\text{-}9)$$

$$= \sum_i \delta q_i \int_0^l p(x, t)\phi_i(x) \, dx$$

or

$$Q_i = \int_0^l p(x, t)\phi_i(x)\, dx \qquad (11.1\text{-}10)$$

Substituting into Lagrange's equation

$$\frac{d}{dt}\left(\frac{\partial T}{\partial \dot{q}_i}\right) - \frac{\partial T}{\partial q_i} + \frac{\partial U}{\partial q_i} = Q_i \qquad (11.1\text{-}11)$$

the differential equation for $q_i(t)$ is found as

$$\ddot{q}_i + \omega_i^2 q_i = \frac{1}{M_i}\int_0^l p(x, t)\phi_i(x)\, dx \qquad (11.1\text{-}12)$$

It is convenient at this point to consider the case where the loading per unit length $p(x, t)$ is separable in the form

$$p(x, t) = \frac{P_0}{l}p(x)f(t) \qquad (11.1\text{-}13)$$

Eq. (11.1-12) then reduces to

$$\ddot{q}_i + \omega_i^2 q_i = \frac{P_0}{M_i}\Gamma_i f(t) \qquad (11.1\text{-}14)$$

where

$$\Gamma_i = \frac{1}{l}\int_0^l p(x)\phi_i(x)\, dx \qquad (11.1\text{-}15)$$

is defined as the *mode participation factor* for mode i. The solution of Eq. (11.1-14) is then

$$q_i(t) = q_i(0)\cos \omega_i t + \frac{1}{\omega_i}\dot{q}_i(0)\sin \omega_i t$$

$$+ \left(\frac{P_0\Gamma_i}{M_i\omega_i^2}\right)\omega_i\int_0^t f(\xi)\sin \omega_i(t - \xi)\, d\xi \qquad (11.1\text{-}16)$$

Since the ith mode statical deflection (with $\ddot{q}_i(t) = 0$) expanded in terms of $\phi_i(x)$ is $P_0\Gamma_i/M_i\omega_i^2$, the quantity

$$D_i(t) = \omega_i\int_0^t f(\xi)\sin \omega_i(t - \xi)\, d\xi \qquad (11.1\text{-}17)$$

can be called the *dynamic load factor* for the ith mode.

EXAMPLE 11.1-1

A simply supported uniform beam of mass M_0 is suddenly loaded by the force shown in Fig. 11.1-1. Determine the equation of motion.

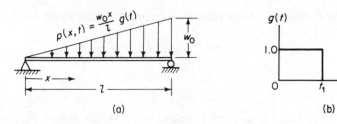

Figure 11.1-1.

Solution: The normal modes of the beam are

$$\phi_n(x) = \sqrt{2}\,\sin\frac{n\pi x}{l}$$

$$\omega_n = (n\pi)^2\sqrt{EI/M_0 l^3}$$

and the generalized mass is

$$M_n = \frac{M_0}{l}\int_0^l 2\sin^2\frac{n\pi x}{l}\,dx = M_0$$

The generalized force is

$$\int_0^l p(x,t)\phi_n\,dx = g(t)\int_0^l \frac{w_0 x}{l}\sqrt{2}\,\sin\frac{n\pi x}{l}\,dx$$

$$= g(t)\frac{w_0\sqrt{2}}{l}\left[\frac{\sin(n\pi x/l)}{(n\pi/l)^2} - \frac{x\cos(n\pi x/l)}{(n\pi/l)}\right]_0^l$$

$$= -g(t)\frac{w_0\sqrt{2}}{n\pi}\frac{l}{}\cos n\pi$$

$$= -\frac{\sqrt{2}\,lw_0}{n\pi}g(t)(-1)^n$$

where $g(t)$ is the time history of the load. The equation for q_n is then

$$\ddot{q}_n + \omega_n^2 q_n = -\frac{\sqrt{2}\,lw_0}{n\pi M_0}(-1)^n g(t)$$

which has the solution

$$q_n(t) = \frac{-\sqrt{2}\,lw_0}{n\pi M_0}\frac{(-1)^n}{\omega_n^2}(1 - \cos\omega_n t)\qquad 0 \le t \le t_1$$

$$= \frac{-\sqrt{2}\,lw_0}{n\pi M_0}\frac{(-1)^n}{\omega_n^2}(1 - \cos\omega_n t)$$

$$+ \frac{2\sqrt{2}\,lw_0(-1)^n}{n\pi M_0\omega_n^2}\big[1 - \cos\omega_n(t - t_1)\big]\qquad t_1 \le t \le \infty$$

Thus the deflection of the beam is expressed by the summation

$$y(x, t) = \sum_{n=1}^{\infty} q_n(t)\sqrt{2} \sin\frac{\pi n x}{l}$$

EXAMPLE 11.1-2

A missile in flight is excited longitudinally by the thrust $F(t)$ of its rocket engine at the end $x = 0$. Determine the equation for the displacement $u(x, t)$ and the acceleration $\ddot{u}(x, t)$.

Solution: We assume the solution for the displacement to be

$$u(x, t) = \sum q_i(t)\varphi_i(x)$$

where $\varphi_i(x)$ are normal modes of the missile in longitudinal oscillation. The generalized coordinate q_i satisfies the differential equation

$$\ddot{q}_i + \omega_i^2 q_i = \frac{F(t)\varphi_i(0)}{M_i}$$

If, instead of $F(t)$, a unit impulse acted at $x = 0$, the above equation would have the solution $(\varphi_i(0)/M_i\omega_i) \sin \omega_i t$ for initial conditions $q_i(0) = \dot{q}(0) = 0$. Thus the response to the arbitrary force $F(t)$ is

$$q_i(t) = \frac{\varphi_i(0)}{M_i\omega_i} \int_0^t F(\xi) \sin \omega_i(t - \xi)\, d\xi$$

and the displacement at any point x is

$$u(x, t) = \sum_i \frac{\varphi_i(x)\varphi_i(0)}{M_i\omega_i} \int_0^t F(\xi) \sin \omega_i(t - \xi)\, d\xi$$

The acceleration $\ddot{q}_i(t)$ of mode i can be determined by rewriting the differential equation and substituting the former solution for $q_i(t)$

$$\ddot{q}_i(t) = \frac{F(t)\varphi_i(0)}{M_i} - \omega_i^2 q_i$$

$$= \frac{F(t)\varphi_i(0)}{M_i} - \frac{\varphi_i(0)\omega_i}{M_i} \int_0^t F(\xi) \sin \omega_i(t - \xi)\, d\xi$$

Thus the equation for the acceleration of any point x is found as

$$\ddot{u}(x, t) = \sum_i \ddot{q}_i(t)\varphi_i(x)$$

$$= \sum_i \left\{ \frac{F(t)\varphi_i(0)\varphi_i(x)}{M_i} - \frac{\varphi_i(0)\varphi_i(x)\omega_i}{M_i} \int_0^t F(\xi) \sin \omega_i(t - \xi)\, d\xi \right\}$$

EXAMPLE 11.1-3

Determine the response of a cantilever beam when its base is given a motion $y_b(t)$ normal to the beam axis as shown in Fig. 11.1-2.

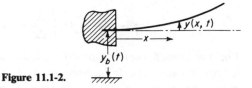

Figure 11.1-2.

Solution: The differential equation for the beam with base motion is

$$[EIy''(x, t)]'' + m(x)[\ddot{y}_b(t) + \ddot{y}(x, t)] = 0$$

which can be rearranged to

$$[EIy''(x, t)]'' + m(x)\ddot{y}(x, t) = -m(x)\ddot{y}_b(t)$$

Thus, instead of the force per unit length $F(x, t)$ we have the inertial force per unit length $-m(x)\ddot{y}_b(t)$. Assuming the solution in the form

$$y(x, t) = \sum_i q_i(t)\varphi_i(x)$$

the equation for the generalized coordinate q_i becomes

$$\ddot{q}_i + \omega_i^2 q_i = -\ddot{y}_b(t)\frac{1}{M_i}\int_0^l \varphi_i(x)\, dx$$

The solution for q_i then differs from that of a simple oscillator only by the factor $-1/M_i\int_0^l \varphi_i(x)\, dx$ so that for the initial conditions $y(0) = \dot{y}(0) = 0$

$$q_i(t) = \left\{-\frac{1}{M_i}\int_0^l \varphi_i(x)dx\right\}\frac{1}{\omega_i}\int_0^t \ddot{y}_b(\xi)\sin\omega_i(t - \xi)d\xi$$

11.2   BEAM ORTHOGONALITY INCLUDING ROTARY INERTIA AND SHEAR DEFORMATION

The equations for the beam, including rotary inertia and shear deformation, were derived in Sec. 7.5. For such beams the orthogonality is no longer expressed by Eq. (11.1-3), but by the equation

$$\int [m(x)\varphi_j\varphi_i + J(x)\psi_j\psi_i]\, dx = \begin{cases} 0 & \text{if } j \neq i \\ M_i & \text{if } j = i \end{cases} \qquad (11.2\text{-}1)$$

which can be proved in the following manner.

For convenience we will rewrite Eqs. (7.5-5) and (7.5-6), including a distributed moment per unit length $\mathfrak{M}(x, t)$

$$\frac{d}{dx}\left(EI\frac{d\psi}{dx}\right) + kAG\left(\frac{dy}{dx} - \psi\right) - J\ddot{\psi} - \mathfrak{M}(x, t) = 0 \qquad (7.5\text{-}5)$$

$$m\ddot{y} - \frac{d}{dx}\left[kAG\left(\frac{dy}{dx} - \psi\right)\right] - p(x, t) = 0 \qquad (7.5\text{-}6)$$

For the forced oscillation with excitation $p(x, t)$ and $\mathfrak{M}(x, t)$ per unit length of beam, the deflection $y(x, t)$ and the bending slope $\psi(x, t)$ can be expressed in terms of the generalized coordinates

$$y = \sum_j q_j(t)\varphi_j(x)$$
$$\psi = \sum_j q_j(t)\psi_j(x) \qquad (11.2\text{-}2)$$

With these summations substituted into the two beam equations, we obtain

$$J\sum_j \ddot{q}_j\psi_j = \sum_j q_j\left\{\frac{d}{dx}(EI\psi_j') + kAG(\varphi_j' - \psi_j)\right\} + \mathfrak{M}(x, t)$$

$$m\sum_j \ddot{q}_j\varphi_j = \sum_j q_j\frac{d}{dx}\{kAG(\varphi_j' - \psi_j)\} + p(x, t) \qquad (11.2\text{-}3)$$

However, normal-mode vibrations are of the form

$$y = \varphi_j(x)e^{i\omega_j t}$$
$$\psi = \psi_j(x)e^{i\omega_j t} \qquad (11.2\text{-}4)$$

which, when substituted into the beam equations with zero excitation, lead to

$$-\omega_j^2 J\psi_j = \frac{d}{dx}(EI\psi_j') + kAG(\varphi_j' - \psi_j)$$

$$-\omega_j^2 m\varphi_j = \frac{d}{dx}\{kAG(\varphi_j' - \psi_j)\} \qquad (11.2\text{-}5)$$

The right sides of this set of equations are the coefficients of the generalized coordinates q_j in the forced vibration equations, so that we can write Eqs. (11.2-3) as

$$J\sum_j \ddot{q}_j\psi_j = -\sum_j q_j\omega_j^2 J\psi_j + \mathfrak{M}(x, t)$$

$$m\sum_j \ddot{q}_j\varphi_j = -\sum_j q_j\omega_j^2 m\varphi_j + p(x, t) \qquad (11.2\text{-}6)$$

Multiplying these two equations by $\varphi_i \, dx$ and $\psi_i \, dx$, adding, and integrating, we obtain

$$\sum_j \ddot{q}_j \int_0^l (m\varphi_j\varphi_i + J\psi_j\psi_i) \, dx + \sum_j q_j\omega_j^2 \int_0^l (m\varphi_j\varphi_i + J\psi_j\psi_i) \, dx$$
$$= \int_0^l p(x, t)\varphi_i \, dx + \int_0^l \mathfrak{M}(x, t)\psi_i \, dx$$

$$(11.2\text{-}7)$$

If the q's in these equations are generalized coordinates, they must be independent coordinates which satisfy the equation

$$\ddot{q}_i + \omega_i^2 q_i = \frac{1}{M_i} \left\{ \int_0^l p(x, t)\varphi_i \, dx + \int_0^l \mathfrak{M}(x, t)\psi_i \, dx \right\} \quad (11.2\text{-}8)$$

We see then that this requirement is satisfied only if

$$\int_0^l (m\varphi_j\varphi_i + J\psi_j\psi_i) \, dx = \begin{cases} 0 & \text{if} \quad j \neq i \\ M_i & \text{if} \quad j = i \end{cases} \quad (11.2\text{-}9)$$

which defines the orthogonality for the beam, including rotary inertia and shear deformation.

11.3 NORMAL MODES OF CONSTRAINED STRUCTURES

When a structure is altered by the addition of a mass or a spring, we refer to it as a *constrained structure*. For example, a spring will tend to act as a constraint on the motion of the structure at the point of its application, and possibly increase the natural frequencies of the system. An added mass, on the other hand, may decrease the natural frequencies of the system. Such problems can be formulated in terms of generalized coordinates and the mode-summation technique.

Consider the forced vibration of any one dimensional structure (i.e., the points on the structure defined by one coordinate x) excited by a force per unit length $f(x, t)$ and moment per unit length $M(x, t)$. If we know the normal modes of the structure, ω_i and $\varphi_i(x)$, its deflection at any point x can be represented by

$$y(x, t) = \sum_i q_i(t)\varphi_i(x) \quad (11.3\text{-}1)$$

where the generalized coordinate q_i must satisfy the equation

$$\ddot{q}_i(t) + \omega_i^2 q_i(t) = \frac{1}{M_i} \left[\int f(x, t)\varphi_i(x) \, dx + \int M(x, t)\varphi_i'(x) \, dx \right] \quad (11.3\text{-}2)$$

The right side of this equation is $1/M_i$ times the generalized force Q_i, which can be determined from the virtual work of the applied loads as $Q_i = \delta W / \delta q_i$.

If, instead of distributed loads, we have a concentrated force $F(a, t)$ and a concentrated moment $M(a, t)$ at some point $x = a$, the generalized force for such loads is found from

$$\delta W = F(a, t)\, \delta y(a, t) + M(a, t)\, \delta y'(a, t)$$

$$= F(a, t) \sum_i \varphi_i(a)\, \delta q_i + M(a, t) \sum_i \varphi_i'(a)\, \delta q_i \qquad (11.3\text{-}3)$$

$$Q_i = \frac{\delta W}{\delta q_i} = F(a, t)\varphi_i(a) + M(a, t)\varphi_i'(a)$$

Then, instead of Eq. (11.3-14), we obtain the equation

$$\ddot{q}_i(t) + \omega_i^2 q_i(t) = \frac{1}{M_i}\left[F(a, t)\varphi_i(a) + M(a, t)\varphi_i'(a)\right] \qquad (11.3\text{-}4)$$

These equations form the starting point for the analysis of constrained structures, provided the constraints are expressible as external loads on the structure.

As an example, let us consider attaching a linear and torsional spring to the simply supported beam of Fig. 11.3-1. The linear spring exerts a force on the beam equal to

$$F(a, t) = -ky(a, t) = -k \sum_j q_j(t)\varphi_j(a) \qquad (11.3\text{-}5)$$

whereas the torsional spring exerts a moment

$$M(a, t) = -Ky'(a, t) = -K \sum_j q_j(t)\varphi_j'(a) \qquad (11.3\text{-}6)$$

Substituting these equations into Eq. (11.3-4), we obtain

$$\ddot{q}_i + \omega_i^2 q_i = \frac{1}{M_i}\left[-k\varphi_i(a) \sum_j q_j\varphi_j(a) - K\varphi_i'(a) \sum_j q_j\varphi_j'(a)\right] \qquad (11.3\text{-}7)$$

The normal modes of the constrained modes are also harmonic and so we can write

$$q_i = \bar{q}_i e^{i\omega t}$$

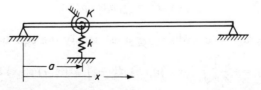

Figure 11.3-1.

The solution to the ith equation is then

$$\bar{q}_i = \frac{1}{M_i(\omega_i^2 - \omega^2)}\left[-k\varphi_i(a)\sum_j \bar{q}_j\varphi_j(a) - K\varphi_i'(a)\sum_j \bar{q}_j\varphi_j'(a)\right]$$

(11.3-8)

If we use n modes, there will be n values of $\bar{q}_j$ and n equations such as the one above. The determinant formed by the coefficients of the $\bar{q}_j$ will then lead to the natural frequencies of the constrained modes, and the mode shapes of the constrained structure are found by substituting the $\bar{q}_j$ into Eq. (11.3-1).

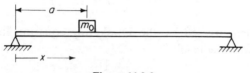

Figure 11.3-2.

If, instead of springs, a mass m_0 is placed at a point $x = a$, as shown in Fig. 11.3-2, the force exerted by m_0 on the beam is

$$F(a, t) = -m_0\ddot{y}(a, t) = -m_0\sum_j \ddot{\bar{q}}_j\varphi_j(a)$$

(11.3-9)

Thus, in place of Eq. (11.3-8), we would obtain the equation

$$\bar{q}_i = \frac{1}{M_i(\omega_i^2 - \omega^2)}\left[\omega^2 m_0\varphi_i(a)\sum_j \bar{q}_j\varphi_j(a)\right]$$

(11.3-10)

EXAMPLE 11.3-1

Give a single mode approximation for the natural frequency of a simply supported beam when a mass m_0 is attached to it at $x = l/3$.

Solution: When only a single mode is used, Eq. (11.3-10) reduces to

$$M_1(\omega_1^2 - \omega^2) = \omega^2 m_0\varphi_1^2(a)$$

Solving for ω^2, we obtain

$$\left(\frac{\omega}{\omega_1}\right)^2 = \frac{1}{1 + \dfrac{m_0}{M_1}\varphi_1^2(a)}$$

For the first mode of the unconstrained beam, we have

$$\omega_1 = \pi^2 \sqrt{\frac{EI}{Ml^3}} \;, \qquad \varphi_1(x) = \sqrt{2}\, \sin\frac{\pi x}{l}$$

$$\varphi_1\!\left(\frac{l}{3}\right) = \sqrt{2}\, \sin\frac{\pi}{3} = \sqrt{2} \times 0.866$$

$$M_1 = M = \text{mass of the beam}$$

Thus its substitution into the above equation gives the one-mode approximation for the constrained beam the value

$$\left(\frac{\omega}{\omega_1}\right)^2 = \frac{1}{1 + 1.5\dfrac{m_0}{M}}$$

The same problem treated by the Dunkerley equation in Example 9.2-5 gave, for this ratio, the result

$$\frac{1}{1 + 1.6\dfrac{m_0}{M}}$$

EXAMPLE 11.3-2

A missile is constrained in a test stand by linear and torsional springs, as shown in Fig. 11.3-3. Formulate the inverse problem of determining its free-free modes from the normal modes of the constrained missile, which are designated as Φ_i and Ω_i.

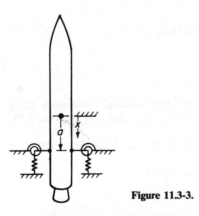

Figure 11.3-3.

Solution: The problem is approached in a manner similar to that of the direct problem where, in place of φ_i and ω_i, we use Φ_i and Ω_i. We now relieve the constraints at the supports by introducing opposing forces $-F(a)$ and $-M(a)$ equal to $ky(a)$ and $Ky'(a)$.

To carry out this problem in greater detail, we start with the equation

$$\bar{q}_i = \frac{-F(a)\Phi_i(a) - M(a)\Phi_i'(a)}{M_i\Omega_i^2\left[1 - (\Omega/\Omega_i)^2\right]}$$

which replaces Eq. (11.3-8). Letting $D_i(\omega) = M_i\Omega_i^2[1 - (\omega/\Omega_i)^2]$, the displacement at $x = a$ is

$$y(a) = \sum_i \Phi_i(a)\bar{q}_i = \sum_i \frac{-F(a)\Phi_i^2(a) - M(a)\Phi_i'(a)\Phi_i(a)}{D_i(\omega)}$$

We now replace $-F(a)$ and $-M(a)$ with $ky(a)$ and $Ky'(a)$ and write

$$y(a) = \sum_i \frac{ky(a)\Phi_i^2(a) + Ky'(a)\Phi_i'(a)\Phi_i(a)}{D_i(\omega)}$$

$$y'(a) = \sum_i \frac{ky(a)\Phi_i'(a)\Phi_i(a) + Ky'(a)\Phi_i'^2(a)}{D_i(\omega)}$$

These equations may now be rearranged as

$$y(a)\left[1 - k\sum_i \frac{\Phi_i^2(a)}{D_i(\omega)}\right] = y'(a)K\sum_i \frac{\Phi_i'(a)\Phi_i(a)}{D_i(\omega)}$$

$$y(a)k\sum_i \frac{\Phi_i'(a)\Phi_i(a)}{D_i(\omega)} = y'(a)\left[1 - K\sum_i \frac{\Phi_i'^2(a)}{D_i(\omega)}\right]$$

The frequency equation then becomes

$$\left[1 - k\sum_i \frac{\Phi_i^2(a)}{D_i(\omega)}\right]\left[1 - K\sum_i \frac{\Phi_i'^2(a)}{D_i(\omega)}\right] - kK\left[\sum_i \frac{\Phi_i'(a)\Phi_i(a)}{D_i(\omega)}\right]^2 = 0$$

The slope to deflection ratio at $x = a$ is

$$\frac{y'(a)}{y(a)} = \frac{1 - k\sum_i \dfrac{\Phi_i^2(a)}{D_i(\omega)}}{K\sum_i \dfrac{\Phi_i'(a)\Phi_i(a)}{D_i(\omega)}}$$

The free-free mode shape is then given by

$$\frac{y(x)}{y(a)} = \sum_i \frac{k\Phi_i(a)\Phi_i(x) + K\dfrac{y'(a)}{y(a)}\Phi_i'(a)\Phi_i(x)}{D_i(\omega)}$$

EXAMPLE 11.3-3

Determine the constrained modes of the missile of Fig. 11.3-3, using only the first free-free mode $\varphi_1(x)$, ω_1, together with translation $\varphi_T = 1$, $\Omega_T = 0$ and rotation $\varphi_R = x$, $\Omega_R = 0$, where x is measured positively toward the tail of the missile.

Solution: The generalized mass for each of the three modes is

$$M_T = \int dm = M$$

$$M_R = \int x^2\, dm = I = M\rho^2$$

$$M_1 = \int \varphi_1^2(x)\, dm = M$$

where the $\varphi_1(x)$ mode was normalized such that $M_1 = M =$ actual mass.

The frequency dependent factors D_i are

$$D_T = -M_T\omega^2 = -M\omega^2 = -M\omega_1^2\lambda$$

$$D_R = -M\rho^2\omega^2 = -M\rho^2\omega_1^2\lambda$$

$$D_1 = M\omega_1^2\left[1 - \left(\frac{\omega}{\omega_1}\right)^2\right] = M\omega_1^2(1 - \lambda)$$

$$\left(\frac{\omega}{\omega_1}\right)^2 = \lambda$$

The frequency equation for this problem is the same as that of Example 11.3-2, except that the minus k's are replaced by positive k's and $\varphi(x)$ and ω replace $\Phi(x)$ and Ω. Substituting the above quantities into the frequency equation, we have

$$\left\{1 - \frac{k}{M\omega_1^2}\left[\frac{1}{\lambda} + \frac{a^2}{\rho^2\lambda} - \frac{\varphi_1^2(a)}{(1-\lambda)}\right]\right\}\left\{1 - \frac{K}{M\omega_1^2}\left[\frac{1}{\rho^2\lambda} - \frac{\varphi_1^2(a)}{(1-\lambda)}\right]\right\}$$

$$- \frac{kK}{M^2\omega_1^4}\left\{\frac{-a}{\rho^2\lambda} + \frac{\varphi_1'(a)\varphi_1(a)}{(1-\lambda)}\right\}^2 = 0$$

which can be simplified to

$$\lambda^2(1 - \lambda) + \left(\frac{k}{M\omega_1^2}\right)\left[\varphi_1^2(a) + \frac{K}{k}\varphi_1'^2(a)\right]\lambda^2$$

$$- \left(\frac{k}{M\omega_1^2}\right)\left[1 + \frac{a^2}{\rho^2} + \frac{K}{k\rho^2}\right]\lambda(1 - \lambda) + \left(\frac{k}{M\omega_1^2}\right)^2\frac{K}{k\rho^2}(1 - \lambda)$$

$$- \left(\frac{k}{M\omega_1^2}\right)^2\frac{K}{k}\lambda\left\{\varphi_1'^2(a) + \frac{1}{\rho^2}[\varphi_1(a) - a\varphi_1'(a)]^2\right\} = 0$$

A number of special cases of the above equation are of interest, and we mention one of these. If $K = 0$, the frequency equation simplifies to

$$\lambda^2 - \left\{1 + \left(\frac{k}{M\omega_1^2}\right)\left[1 + \frac{a^2}{\rho^2} + \varphi_1^2(a)\right]\right\}\lambda + \left(\frac{k}{M\omega_1^2}\right)\left(1 + \frac{a^2}{\rho^2}\right) = 0$$

Here $x = a$ might be taken negatively so that the missile is hanging by a spring.

11.4   MODE ACCELERATION METHOD

One of the difficulties encountered in any mode summation method has to do with the convergence of the procedure. If this convergence is poor, a large number of modes must be used, thereby increasing the order of the frequency determinant. The mode acceleration method tends to overcome this difficulty by improving the convergence so that a fewer number of normal modes are needed.

The mode acceleration method starts with the same differential equation for the generalized coordinate q_i, but rearranged in order. For example, we can start with Eq. (11.3-4) and write it in the order

$$q_i(t) = \frac{F(a, t)\varphi_i(a)}{M_i\omega_i^2} + \frac{M(a, t)\varphi_i'(a)}{M_i\omega_i^2} - \frac{\ddot{q}_i(t)}{\omega_i^2} \qquad (11.4\text{-}1)$$

Substituting this into Eq. (11.3-1), we obtain

$$y(x, t) = \sum_i q_i(t)\varphi_i(x)$$

$$= F(a, t)\sum_i \frac{\varphi_i(a)\varphi_i(x)}{M_i\omega_i^2} + M(a, t)\sum_i \frac{\varphi_i'(a)\varphi_i(x)}{M_i\omega_i^2} - \sum_i \frac{\ddot{q}_i(t)\varphi_i(x)}{\omega_i^2}$$

$$(11.4\text{-}2)$$

We note here that, if $F(a, t)$ and $M(a, t)$ were static loads, the last term containing the acceleration would be zero. Thus the terms

$$\sum_i \frac{\varphi_i(a)\varphi_i(x)}{M_i \omega_i^2} = \alpha(a, x)$$

$$\sum_i \frac{\varphi_i'(a)\varphi_i(x)}{M_i \omega_i^2} = \beta(a, x)$$

$$(11.4\text{-}3)$$

must represent influence functions, where $\alpha(a, x)$ and $\beta(a, x)$ are the deflections at x due to a unit load and unit moment at a, respectively. We can therefore rewrite Eq. (11.4-2) as

$$y(x, t) = F(\alpha, t)\alpha(a, x) + M(a, t)\beta(a, x) - \sum \frac{\ddot{q}_i(t)\varphi_i(x)}{\omega_i^2}$$

$$(11.4\text{-}4)$$

Because of ω_i^2 in the denominator of the terms summed, the convergence is improved over the mode summation method.

In the forced vibration problem where $F(a, t)$ and $M(a, t)$ are excitations, Eq. (11.3-4) is first solved for $q_i(t)$ in the conventional manner, and then substituted into Eq. (11.4-4) for the deflection. For the normal modes of constrained structures, $F(a, t)$ and $M(a, t)$ are again the forces and moments exerted by the constraints, and the problem is treated in a manner similar to those of Sec. 11.3. However, because of the improved convergence, fewer number of modes will be found to be necessary.

EXAMPLE 11.4-1

Using the mode acceleration method, solve the problem of Fig. 11.3-2 of a concentrated mass m_0 attached to the structure.

Solution: Assuming harmonic oscillations

$$F(a, t) = \bar{F}(a)e^{i\omega t}$$

$$q_i(t) = \bar{q}_i e^{i\omega t}$$

$$y(x, t) = y(x)e^{i\omega t}$$

Substituting these equations into Eq. (11.4-4) and letting $x = a$,

$$\bar{y}(a) = \bar{F}(a)\alpha(a, a) + \omega^2 \sum_j \frac{\bar{q}_j \varphi_j(a)}{\omega_j^2}$$

Since the force exerted by m_0 on the structure is

$$\bar{F}(a) = m_0 \omega^2 \bar{y}(a)$$

we can eliminate $\bar{y}(a)$ between the above two equations, obtaining

$$\frac{\bar{F}(a)}{m_0\omega^2} = \bar{F}(a)x(a, a) + \omega^2\sum_j \frac{\bar{q}_j\varphi_j(a)}{\omega_j^2}$$

or

$$\bar{F}(a) = \frac{\omega^2\sum_j \dfrac{\bar{q}_j\varphi_j(a)}{\omega_j^2}}{\dfrac{1}{m_0\omega^2} - \alpha(a, a)}$$

If we now substitute this equation into Eq. (11.3-4) and assume harmonic motion, we obtain the equation

$$(\omega_i^2 - \omega^2)\bar{q}_i = \frac{\bar{F}(a)\varphi_i(a)}{M_i} = \frac{\omega^2\varphi_i(a)\sum_j \bar{q}_j\dfrac{\varphi_j(a)}{\omega_j^2}}{M_i\left[\dfrac{1}{m_0\omega^2} - \alpha(a, a)\right]}$$

Rearranging, we have

$$\left[1 - m_0\omega^2\alpha(a, a)\right](\omega_i^2 - \omega^2)\bar{q}_i = \frac{\omega^4 m_0\varphi_i(a)}{M_i}\sum_j \frac{\bar{q}_j\varphi_j(a)}{\omega_j^2}$$

which represents a set of linear equations in $\bar{q}_k$. The series represented by the summation will, however, converge rapidly because of ω_j^2 in the denominator. Offsetting this advantage of smaller number of modes is the disadvantage that these equations are now quartic rather than quadratic in ω.

11.5   COMPONENT MODE SYNTHESIS

The treatment of large structural systems may be simplified by breaking up the system into smaller subsystems which are related through the displacement and force conditions at their junction points. Each subsystem is represented by mode functions, the sum of which allows the satisfaction of the displacement and force conditions at the junctions. These functions need not be orthogonal or normal modes of the subsystem, and each mode used need not satisfy the junction conditions as long as their combined sum allows these conditions to be satisfied. Lagrange's equations, and in particular the method of superfluous coordinates, form the basis for the synthesis process.

To present the basic ideas of the method of modal synthesis, we will consider a simple beam with a 90° bend, an example which was used by W. Hurty.* The beam, shown in Fig. 11.5-1, is considered to vibrate only in the plane of the paper.

We separate the beam into two sections, ① and ②, whose coordinates are shown as w_1, x; w_2, x; and u_2, x. For part ① we assume the deflection to be

$$w_1(x, t) = \phi_1(x)p_1(t) + \phi_2(x)p_2(t) + \cdots$$

$$= \left(\frac{x}{l}\right)^2 p_1 + \left(\frac{x}{l}\right)^3 p_2 \tag{11.5-1}$$

Note that the two mode functions satisfy the geometric and force conditions at the boundaries of section ① as follows

$$w_1(0) = 0 \qquad\qquad w_1(l) = p_1 + p_2$$

$$w_1'(0) = 0 \qquad\qquad w_1'(l) = \frac{2}{l}p_1 + \frac{3}{l}p_3$$

$$w_1''(0) = \frac{M(0)}{EI} = \frac{2}{l^2}p_1 \qquad w_1''(l) = \frac{M(l)}{EI} = \frac{2}{l^2}p_1 + \frac{6}{l^2}p_2 \tag{11.5-2}$$

$$w_1'''(0) = \frac{V(0)}{EI} = \frac{6}{l^3}p_2 \qquad w_1'''(l) = \frac{V(l)}{EI} = \frac{6}{l^3}p_2$$

Next consider part ② with the origin of the coordinates w_2, x at the free end. The following functions will satisfy the boundary conditions of beam section ②

$$w_2(x, t) = \phi_3(x)p_3(t) + \phi_4(x)p_4(t) + \phi_5(x)p_5(t) + \cdots$$

$$= 1p_3 + \left(\frac{x}{l}\right)p_4 + \left(\frac{x}{l}\right)^4 p_5 \tag{11.5-3}$$

$$u_2(x, t) = \phi_6(x)p_6(t) + \cdots$$

$$= 1p_6 \tag{11.5-4}$$

where $u_2(x, t)$ is the displacement in the x direction.

The next step is to calculate the generalized mass from the equation

$$m_{ij} = \int_0^l m(x)\phi_i(x)\phi_j(x)\,dx$$

*Walter C. Hurty, "Vibrations of Structural Systems by Component Synthesis," *Jour. Engr. Mech. Div., Proc. of ASCE* (August 1960), pp. 51–69.

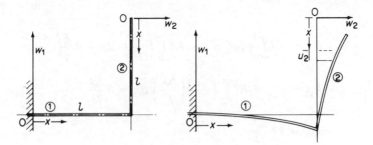

Figure 11.5-1. Beam sections 1 and 2 with their coordinates.

For subsection ① we have

$$m_{11} = \int_0^l m\phi_1\phi_1 \, dx = \int_0^l m\left(\frac{x}{l}\right)^4 dx = 0.20ml$$

$$m_{12} = \int_0^l m\phi_1\phi_2 \, dx = \int_0^l m\left(\frac{x}{l}\right)^5 dx = 0.166ml = m_{21}$$

$$m_{22} = \int_0^l m\phi_2\phi_2 \, dx = \int_0^l m\left(\frac{x}{l}\right)^6 dx = 0.1428ml$$

The generalized mass for subsection ② is computed in a similar manner using ϕ_3 to ϕ_6

$$m_{33} = 1.0ml$$

$$m_{34} = 0.50ml = m_{43}$$

$$m_{35} = 0.20ml = m_{53}$$

$$m_{44} = 0.333ml$$

$$m_{45} = 0.166ml = m_{54}$$

$$m_{55} = 0.111ml$$

$$m_{66} = 1.0ml$$

Since there is no coupling between the longitudinal displacement u_2 and the lateral displacement w_2, $m_{63} = m_{64} = m_{65} = 0$.

The generalized stiffness is found from the equation

$$k_{ij} = \int_0^l EI\phi_i''\phi_j'' \, dx$$

Thus

$$k_{11} = EI \int_0^l \phi_1'' \phi_1'' \, dx = EI \int_0^l \left(\frac{2}{l^2}\right)^2 dx = 4\frac{EI}{l^3}$$

$$k_{12} = k_{21} = EI \int_0^l \left(\frac{2}{l^2}\right)\left(\frac{6x}{l^3}\right) dx = 6\frac{EI}{l^3}$$

$$k_{22} = 12\frac{EI}{l^3}$$

$$k_{55} = 28.8\frac{EI}{l^3}$$

All other k_{ij} are zero.

The results computed for m_{ij} and k_{ij} can now be arranged in the mass and stiffness matrices partitioned as follows

$$[m] = ml \begin{bmatrix} 0.2000 & 0.1666 & 0 & 0 & 0 & 0 \\ 0.1666 & 0.1428 & 0 & 0 & 0 & 0 \\ 0 & 0 & 1.0000 & 0.5000 & 0.2000 & 0 \\ 0 & 0 & 0.5000 & 0.3333 & 0.1666 & 0 \\ 0 & 0 & 0.2000 & 0.1666 & 0.1111 & 0 \\ 0 & 0 & 0 & 0 & 0 & 1.0000 \end{bmatrix}$$

$$(11.5\text{-}5)$$

$$[k] = \frac{EI}{l^3} \begin{bmatrix} 4 & 6 & 0 & 0 & 0 & 0 \\ 6 & 12 & 0 & 0 & 0 & 0 \\ 0 & 0 & 0 & 0 & 0 & 0 \\ 0 & 0 & 0 & 0 & 0 & 0 \\ 0 & 0 & 0 & 0 & 28.8 & 0 \\ 0 & 0 & 0 & 0 & 0 & 0 \end{bmatrix}$$

$$(11.5\text{-}6)$$

where the upper left matrix refers to section ① and the remainder to section ②.

At the junction between sections ① and ② we have the following constraint equations

$$w_1(l) + u_2(l) = 0 \qquad \text{or} \qquad p_1 + p_2 + p_6 = 0$$

$$w_2(l) = 0 \qquad\qquad\qquad p_3 + p_4 + p_5 = 0$$

$$w_1'(l) - w_2'(l) = 0 \qquad\qquad 2p_1 + 3p_2 - p_4 - 4p_5 = 0$$

$$EI[w_1''(l) + w_2''(l)] = 0 \qquad 2p_1 + 6p_2 + 12p_5 = 0$$

Arranged in matrix form, these are

$$
\begin{bmatrix}
1 & 1 & 0 & 0 & 0 & 1 \\
0 & 0 & 1 & 1 & 1 & 0 \\
2 & 3 & 0 & -1 & -4 & 0 \\
2 & 6 & 0 & 0 & 12 & 0
\end{bmatrix}
\begin{Bmatrix}
p_1 \\ p_2 \\ p_3 \\ p_4 \\ p_5 \\ p_6
\end{Bmatrix} = 0
\tag{11.5-7}
$$

Since the total number of coordinates used are six and there are four constraint equations, the number of generalized coordinates for the system is two (i.e., there are four superfluous coordinates corresponding to the four constraint equations (see Sec. 8.1). We can thus choose any two of the coordinates to be the generalized coordinates q. Let $p_1 = q_1$ and $p_6 = q_6$ be the generalized coordinates and express $p_1 \ldots p_6$ in terms of q_1 and q_6. This is accomplished in the following steps.

Rearrange Eq. (11.5-7) by shifting columns 1 and 6 to the right side

$$
\begin{bmatrix}
1 & 0 & 0 & 0 \\
0 & 1 & 1 & 1 \\
3 & 0 & -1 & -4 \\
6 & 0 & 0 & 12
\end{bmatrix}
\begin{Bmatrix}
p_2 \\ p_3 \\ p_4 \\ p_5
\end{Bmatrix} =
\begin{bmatrix}
-1 & -1 \\
0 & 0 \\
-2 & 0 \\
-2 & 0
\end{bmatrix}
\begin{Bmatrix}
q_1 \\ q_6
\end{Bmatrix}
\tag{11.5-8}
$$

In abbreviated notation the above equation is

$$
[s]\{p_{2-5}\} = [Q]\{q_{1,6}\}
$$

Premultiply by $[s]^{-1}$ to obtain

$$
\{p_{2-5}\} = [s]^{-1}[Q]\{q_{1,6}\}
$$

Supply the identity $p_1 = q_1$ and $p_6 = q_6$ and write

$$
\{p_{1-6}\} = [C]\{q_{1,6}\}
$$

The above constraint equation is now in terms of the generalized coordinates q_1 and q_6 as follows

$$
\begin{Bmatrix}
p_1 \\ p_2 \\ p_3 \\ p_4 \\ p_5 \\ p_6
\end{Bmatrix} =
\begin{bmatrix}
1 & 0 \\
-1 & -1 \\
2 & 4.50 \\
-2.333 & -5.0 \\
0.333 & 0.50 \\
0 & 1
\end{bmatrix}
\begin{Bmatrix}
q_1 \\ q_6
\end{Bmatrix} = [C]\begin{Bmatrix}
q_1 \\ q_6
\end{Bmatrix}
\tag{11.5-9}
$$

Returning to the Lagrange equation for the system, which is

$$
ml[m]\{\ddot{p}\} + \frac{EI}{l^3}[k]\{p\} = 0
\tag{11.5-10}
$$

substitute for $\{p\}$ in terms of $\{q\}$ from the constraint equation (9.10-9)

$$ml[m][C]\{\ddot{q}\} + \frac{EI}{l^3}[k][C]\{q\} = 0$$

Premultiply by the transpose $[C]'$

$$ml[C]'[m][C]\{\ddot{q}\} + \frac{EI}{l^3}[C]'[k][C]\{q\} = 0 \qquad (11.5\text{-}11)$$

Comparing Eqs. (11.5-10) and (11.5-11), we note that in (11.5-10) the mass and stiffness matrices are 6×6 (see Eqs. 11.5-5 and 11.5-6), whereas the matrices $[C]'[m][C]$ and $[C]'[k][C]$ in Eq. (11.5-11) are 2×2. Thus we have reduced the size of the system from a 6×6 to a 2×2 problem.

Letting $\{\ddot{q}\} = -\omega^2\{q\}$, Eq. (11.5-11) is in the form

$$\left[-\omega^2 ml \begin{bmatrix} a_{11} & a_{12} \\ a_{21} & a_{22} \end{bmatrix} + \frac{EI}{l^3} \begin{bmatrix} b_{11} & b_{12} \\ b_{21} & b_{22} \end{bmatrix} \right] \begin{Bmatrix} q_1 \\ q_6 \end{Bmatrix} = 0 \qquad (11.5\text{-}12)$$

The numerical values of the matrix $[a_{ij}]$ and $[b_{ij}]$ from Eqs. (11.5-5), (11.5-6), and (11.5-9) are

$$[a_{ij}] = [C]'[m][C] = \begin{bmatrix} 1.1774 & 2.6614 \\ 2.6614 & 7.3206 \end{bmatrix}$$

$$[b_{ij}] = [C]'[k][C] = \begin{bmatrix} 7.200 & 10.800 \\ 10.800 & 19.200 \end{bmatrix}$$

Using these numerical results, we find the two natural frequencies of the system from the characteristic equation of Eq. (11.5-12)

$$\omega_1 = 1.172\sqrt{\frac{EI}{ml^4}}$$

$$\omega_2 = 3.198\sqrt{\frac{EI}{Ml^4}}$$

Figure 11.5-2 shows the mode shapes corresponding to the above frequencies. Since Eq. (11.5-12) enables the solution of the eigenvectors only in terms of an arbitrary reference, q_6 can be solved with $q_1 = 1.0$. The coordinates p are then found from Eq. (11.5-9) and the mode shapes are obtained from Eqs. (11.5-1), 11.5-3), and (11.5-4).

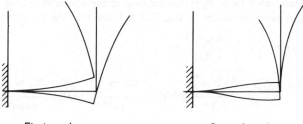

First mode   Second mode

Figure 11.5-2.   First and second mode shapes.

PROBLEMS

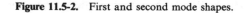

11-1   Show that the dynamic load factor for a suddenly applied constant force reaches a maximum value of 2.0.

11-2   If a suddenly applied constant force is applied to a system for which the damping factor of the ith mode is $\zeta = c/c_{cr}$, show that the dynamic load factor is given approximately by the equation

$$D_i = 1 - e^{-\zeta \omega_i t} \cos \omega_i t$$

11-3   Determine the mode participation factor for a uniformly distributed force.

11-4   If a concentrated force acts at $x = a$, the loading per unit length corresponding to it can be represented by a delta function $l \, \delta(x - a)$. Show that the mode-participation factor then becomes $K_i = \varphi_i(a)$ and the deflection is expressible as

$$y(x, t) = \frac{P_0 l_3}{EI} \sum_i \frac{\varphi_i(a)\varphi_i(x)}{(\beta_i l)^4} D_i(t)$$

where $\omega_i^2 = (\beta_i l)^4 (EI/Ml^3)$ and $(\beta_i l)$ is the eigenvalue of the normal-mode equation.

11-5   For a couple of moment M_0 acting at $x = a$, show that the loading $p(x)$ is the limiting case of two delta functions shown in Fig. P11-5 as $\epsilon \to 0$. Show also that the mode-participation factor for this case is

$$K_i = l \frac{d\varphi_i(x)}{dx}\bigg|_{x-a} = (\beta_i l)\varphi_i'(x)_{x-a}$$

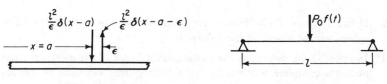

Figure P11-5.   **Figure P11-6.**

11-6   A concentrated force $P_0 f(t)$ is applied to the center of a simply supported uniform beam, as shown in Fig. P11-6. Show that the deflection is given by

$$y(x, t) = \frac{P_0 l^3}{EI} \sum_i \frac{K_i \varphi_i(x)}{(\beta_i l)^4} D_i$$

$$= \frac{2P_0 l^3}{EI} \left\{ \frac{\sin \pi \dfrac{x}{l}}{\pi^4} D_1(t) - \frac{\sin 3\pi \dfrac{x}{l}}{(3\pi)^4} D_3(t) + \frac{\sin 5\pi \dfrac{x}{l}}{(5\pi)^4} D_5(t) \cdots \right\}$$

11-7   A couple of moment M_0 is applied at the center of the beam of Prob. 11-6, as shown in Fig. P11-7. Show that the deflection at any point is given by the equation

$$y(x, t) = \frac{M_0 l^2}{EI} \sum_i \frac{\varphi_i'(a) \varphi_i(x)}{(\beta_i l)^3} D_i(t)$$

$$= \frac{2M_0 l^2}{EI} \left\{ -\frac{\sin 2\pi \dfrac{x}{l}}{(2\pi)^3} D_2(t) + \frac{\sin 4\pi \dfrac{x}{l}}{(4\pi)^3} D_4(t) - \frac{\sin 6\pi \dfrac{x}{l}}{(6\pi)^3} D_6(t) \cdots \right\}$$

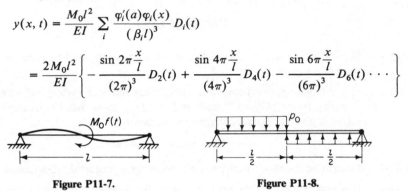

Figure P11-7.   **Figure P11-8.**

11-8   A simply supported uniform beam has suddenly applied to it the load distribution shown in Fig. P11-8, where the time variation is a step function. Determine the response $y(x, t)$ in terms of the normal modes of the beam. Indicate what modes are absent and write down the first two existing modes.

11-9   A slender rod of length l, free at $x = 0$ and fixed at $x = l$, is struck longitudinally by a time-varying force concentrated at the end $x = 0$. Show that all modes are equally excited (i.e., that the mode-participation factor is independent of the mode number), the complete solution being

$$u(x, t) = \frac{2F_0 l}{AE} \left\{ \frac{\cos \dfrac{\pi}{2} \dfrac{x}{l}}{\left(\dfrac{\pi}{2}\right)^2} D_1(t) + \frac{\cos \dfrac{3\pi}{2} \dfrac{x}{l}}{\left(\dfrac{3\pi}{2}\right)^2} D_3(t) + \cdots \right\}$$

11-10   If the force of Prob. 11-9 is concentrated at $x = l/3$, determine which modes will be absent in the solution.

11-11   In Prob. 11-10, determine the participation factor of the modes present and obtain a complete solution for an arbitrary time variation of the applied force.

11-12   Consider a uniform beam of mass M and length l supported on equal springs of total stiffness k, as shown in Fig. P11-12a. Assume the deflection to be

$$y(x, t) = \varphi_1(x)q_1(t) + \varphi_2(x)q_2(t)$$

and choose $\varphi_1 = \sin\dfrac{\pi x}{l}$ and $\varphi_2 = 1.0$.

Using Lagrange's equation, show that

$$\ddot{q}_1 + \frac{4}{\pi}\ddot{q}_2 + \omega_{11}^2 q_1 = 0$$

$$\frac{2}{\pi}\ddot{q}_1 + \ddot{q}_2 + \omega_{22}^2 q_2 = 0$$

where $\omega_{11}^2 = \pi^4(EI/Ml^3)$ = natural frequency of beam on rigid supports
$\omega_{22}^2 = k/M$ = natural frequency of rigid beam on springs

Solve these equations and show that

$$\omega^2 = \omega_{22}^2 \frac{\pi^2}{2}\left\{ \frac{(R + 1) \pm \sqrt{(R - 1)^2 + \dfrac{32}{\pi^2}R}}{\pi^2 - 8} \right\}$$

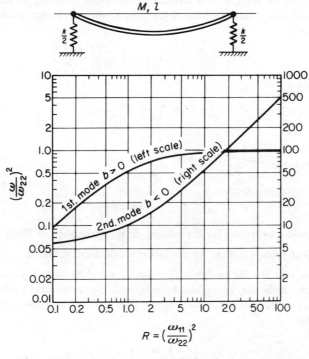

Figure P11-12.   First two natural frequencies of the system.

Let $y(x, t) = \left(b + \sin\dfrac{\pi x}{l}\right)q$ and use Rayleigh's method to obtain

$$\dfrac{q_2}{q_1} = b = \dfrac{\pi}{8}\left\{(R - 1) \mp \sqrt{(R - 1)^2 + \dfrac{32}{\pi^2}R}\right\}$$

$$R = \left(\dfrac{\omega_{11}}{\omega_{22}}\right)^2$$

A plot of the natural frequencies of the system is shown in Fig. P11-12b.

11-13   A uniform beam, clamped at both ends, is excited by a concentrated force $P_0 f(t)$ at midspan, as shown in Fig. P11-13. Determine the deflection under the load and the resulting bending moment at the clamped ends.

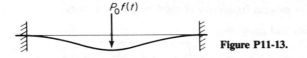

$P_0 f(t)$

Figure P11-13.

11-14   If a uniformly distributed load of arbitrary time variation is applied to a uniform cantilever beam, determine the participation factor for the first three modes.

11-15   A spring of stiffness k is attached to a uniform beam, as shown in Fig. P11-15. Show that the one-mode approximation results in the frequency equation

$$\left(\dfrac{\omega}{\omega_1}\right)^2 = 1 + 1.5\left(\dfrac{k}{M}\right)\left(\dfrac{Ml^3}{\pi^4 EI}\right)$$

where

$$\omega_1^2 = \dfrac{\pi^4 EI}{Ml^3}$$

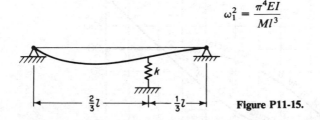

k

$\frac{2}{3}l$ $\frac{1}{3}l$ **Figure P11-15.**

11-16   Write the equations for the two-mode approximation of Prob. 11-15.

11-17   Repeat Prob. 11-16, using the mode acceleration method.

11-18   Show that for the problem of a spring attached to any point $x = a$ of a beam, both the constrained-mode and the mode-acceleration methods result in the same equation when only one mode is used, this equation being

$$\left(\dfrac{\omega}{\omega_1}\right)^2 = 1 + \dfrac{k}{M\omega_1^2}\varphi_1^2(a)$$

11-19 The beam shown in Fig. P11-19 has a spring of rotational stiffness K lb in./rad at the left end. Using two modes in Eq. (11.3-8), determine the fundamental frequency of the system as a function of $K/M\omega_1^2$ where ω_1 is the fundamental frequency of the simply supported beam.

Figure P11-19.

11-20 If both ends of the beam of Fig. P11-19 are restrained by springs of stiffness K, determine the fundamental frequency. As K approaches infinity, the result should approach that of the clamped ended beam.

11-21 An airplane is idealized to a simplified model of a uniform beam of length l and mass per unit length m with a lumped mass M_0 at its center, as shown in Fig. P11-21. Using the translation of M_0 as one of the generalized coordinates, write the equations of motion and establish the natural frequency of the symmetric mode. Use first cantilever mode for the wing.

Figure P11-21.

11-22 For the system of Prob. 11-21, determine the antisymmetric mode by using the rotation of the fuselage as one of the generalized coordinates.

11-23 If wing tip tanks of mass M_1 are added to the system of Prob. 11-21, determine the new frequency.

11-24 Using the method of constrained modes, show that the effect of adding a mass m_1 with moment of inertia J_1 to a point x_1 on the structure changes the first natural frequency ω_1 to

$$\omega_1' = \frac{\omega_1}{\sqrt{1 + \dfrac{m_1}{M_1}\varphi_1^2(x_1) + \dfrac{J_1}{M_1}\varphi_1'^2(x_1)}}$$

and the generalized mass and damping to

$$M_1' = M_1\left\{1 + \frac{m_1\varphi_1^2(x_1)}{M_1} + \frac{J_1}{M_1}\varphi_1'^2(x)\right\}$$

$$\zeta_1' = \frac{\zeta_1}{\sqrt{1 + \dfrac{m_1}{M_1}\varphi_1^2(x_1) + \dfrac{J_1}{M_1}\varphi_1'^2(x_1)}}$$

where a one-mode approximation is used for the inertia forces.

11-25 Formulate the vibration problem of the bent shown in Fig. P11-25 by the component mode synthesis. Assume the corners to remain at 90°.

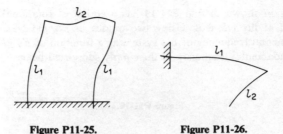

Figure P11-25.   **Figure P11-26.**

11-26 A rod of circular cross-section is bent at right angles in a horizontal plane as shown in Fig. P11-26. Using component mode synthesis, set up the equations for the vibration perpendicular to the plane of the rod. Note that member 1 is in flexure and torsion. Assume its bending only in the vertical plane.

NONLINEAR VIBRATIONS

Linear system analysis serves to explain much of the behavior of oscillatory systems. However, there are a number of oscillatory phenomena which cannot be predicted or explained by the linear theory.

In the linear systems which we have studied, cause and effect are related linearly; i.e., if we double the load, the response is doubled. In a nonlinear system this relationship between cause and effect is no longer proportional. For example, the center of an oil can may move proportionally to the force for small loads, but at a certain critical load it will snap over to a large displacement. The same phenomenon is also encountered in the buckling of columns, electrical oscillations of circuits containing inductance with an iron core, and vibration of mechanical systems with nonlinear restoring forces.

The differential equation describing a nonlinear oscillatory system may have the general form

$$\ddot{x} + f(\dot{x}, x, t) = 0$$

Such equations are distinguished from linear equations in that the principle of superposition does not hold for their solution.

Analytical procedures for the treatment of nonlinear differential equations are difficult and require extensive mathematical study. Exact solutions that are known are relatively few, and a large part of the progress in the knowledge of nonlinear systems comes from approximate and graphical solutions and from studies made on computing machines. Much can be learned about a nonlinear system, however, by using the state space approach and studying the motion presented in the phase plane.

12.1 PHASE PLANE

In an *autonomous* system, the time t does not appear explicitly in the differential equation of motion. Thus only the differential of time, dt, will appear in such an equation.

We will first study an automonous system with the differential equation

$$\ddot{x} + f(x, \dot{x}) = 0 \tag{12.1-1}$$

where $f(x, \dot{x})$ may be a nonlinear function of x and $\dot{x}$. In the method of *state space*, we express the above equation in terms of two first order equations as follows

$$\dot{x} = y \tag{12.1-2}$$
$$\dot{y} = -f(x, y)$$

If x and y are Cartesian coordinates, the xy plane is called the *phase plane*. The *state* of the system is defined by the coordinate x and $y = \dot{x}$, which represents a point on the phase plane. As the state of the system changes, the point on the phase plane moves, thereby generating a curve which is called the *trajectory*.

Another useful concept is the *state speed V* defined by the equation

$$V = \sqrt{\dot{x}^2 + \dot{y}^2} \tag{12.1-3}$$

When the state speed is zero, an *equilibrium state* is reached in that both the velocity of $\dot{x}$ and the acceleration $\dot{y} = \ddot{x}$ are zero.

Dividing the second of Eq. (12.1-2) by the first we obtain the relation

$$\frac{dy}{dx} = -\frac{f(x, y)}{y} = \phi(x, y) \tag{12.1-4}$$

Thus for every point x, y in the phase plane for which $\phi(xy)$ is not indeterminate, there is a unique slope of the trajectory.

If $y = 0$ (i.e., points along the x-axis) and $f(x, y) \neq 0$, the slope of the trajectory is infinite. Thus all trajectories corresponding to such points must cross the x-axis at right angles.

If $y = 0$ and $f(x, y) = 0$, the slope is indeterminate. We define such points as *singular points*. Singular points correspond to a state of equilibrium in that both the velocity $y = \dot{x}$ and the force $\ddot{x} = \dot{y} = -f(x, y)$ are zero. Further discussion is required to establish whether the equilibrium represented by the singular point is stable or unstable.

EXAMPLE 12.1-1

Determine the phase plane of a single degree of freedom oscillator

$$\ddot{x} + \omega^2 x = 0$$

Solution: With $y = \dot{x}$, the above equation is written in terms of two first order equations

$$\dot{y} = -\omega^2 x$$

$$\dot{x} = y$$

Dividing we obtain

$$\frac{dy}{dx} = -\frac{\omega^2 x}{y}$$

Separating variables and integrating

$$y^2 + \omega^2 x^2 = C$$

which is a series of ellipses, the size of which is determined by C. The above equation is also that of conservation of energy

$$\tfrac{1}{2} m\dot{x}^2 + \tfrac{1}{2} kx^2 = C'$$

Since the singular point is at $x = y = 0$, the phase plane plot appears as in Fig. 12.1-1. If y/ω is plotted in place of y, the ellipses of Fig. 12.1-1 reduce to circles.

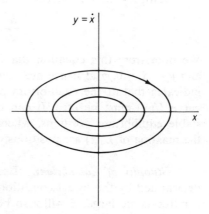

Figure 12.1-1.

12.2 CONSERVATIVE SYSTEMS

In a conservative system the total energy remains constant. Summing the kinetic and potential energies per unit mass, we have

$$\tfrac{1}{2}\dot{x}^2 + U(x) = E = \text{constant} \tag{12.2-1}$$

Solving for $y = \dot{x}$, the ordinate of the phase plane is given by the equation

$$y = \dot{x} = \pm\sqrt{2[E - U(x)]} \tag{12.2-2}$$

It is evident from this equation that the trajectories of a conservative system must be symmetric about the x-axis.

The differential equation of motion for a conservative system can be shown to have the form

$$\ddot{x} = f(x) \tag{12.2-3}$$

Since $\ddot{x} = \dot{x}(d\dot{x}/dx)$, the above equation can be written as

$$\dot{x}\,d\dot{x} - f(x)\,dx = 0 \tag{12.2-4}$$

Integrating, we have

$$\frac{\dot{x}^2}{2} - \int_0^x f(x)\,dx = E \tag{12.2-5}$$

and by comparison with Eq. (12.2-1) we find

$$U(x) = -\int_o^x f(x)\,dx$$
$$f(x) = -\frac{dU}{dx} \tag{12.2-6}$$

Thus for a conservative system the force is equal to the negative gradient of the potential energy.

With $y = \dot{x}$, Eq. (12.2-4) in the state-space becomes

$$\frac{dy}{dx} = \frac{f(x)}{y} \tag{12.2-7}$$

We note from this equation that singular points correspond to $f(x) = 0$ and $y = \dot{x} = 0$, and hence are equilibrium points. Equation (12.2-6) then indicates that at the equilibrium points the slope of the potential energy curve $U(x)$ must be zero. It can be shown that the minima of $U(x)$ are stable equilibrium positions, whereas the saddle points corresponding to the maxima of $U(x)$ are positions of unstable equilibrium.

Stability of Equilibrium. Examining Eq. (12.2-2) the value of E is determined by the initial conditions of $x(0)$ and $y(0) = \dot{x}(0)$. If the initial conditions are large, E will also be large. For every position x, there is a potential $U(x)$; for motion to take place, E must be greater than $U(x)$. Otherwise, Eq. (12.2-2) shows that the velocity $y = \dot{x}$ is imaginary.

Figure 12.2-1 shows a general plot of $U(x)$ and the trajectory y vs. x for various values of E computed from Eq. (12.2-2).

For $E = 7$, $U(x)$ lies below $E = 7$ only between $x = 0$ to 1.2, $x = 3.8$ to 5.9, and $x = 7$ to 8.7. The trajectories corresponding to $E = 7$ are closed

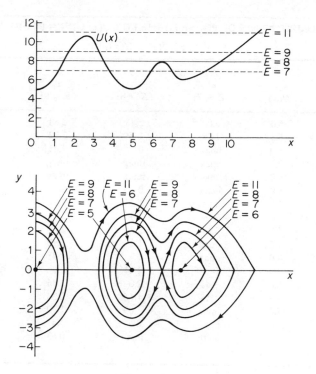

Figure 12.2-1.

curves and the period associated with them can be found from Eq. (12.2-2) by integration

$$\tau = 2\int_{x_1}^{x_2} \frac{dx}{\sqrt{2[E - U(x)]}}$$

where x_1 and x_2 are extreme points of the trajectory on the x-axis.

For smaller initial conditions, these closed trajectories become smaller. For $E = 6$ the trajectory about the equilibrium point $x = 7.5$ contracts to a point, while the trajectory about the equilibrium point $x = 5$ is a closed curve between $x = 4.2$ to 5.7.

For $E = 8$ one of the maxima of $U(x)$ at $x = 6.5$ is tangent to $E = 8$ and the trajectory at this point has four branches. The point $x = 6.5$ is a saddle point for $E = 8$ and the motion is unstable. The saddle point trajectories are called *separatrices*.

For $E > 8$ the trajectories may or may not be closed. $E = 9$ shows a closed trajectory between $x = 3.3$ to 10.2. Note that at $x = 6.5$, $dU/dx = -f(x) = 0$ and $y = \dot{x} \neq 0$ for $E = 9$, and hence equilibrium does not exist.

COMPUTATION OF PHASE PLANE FOR $U(x)$ GIVEN IN FIG. 12.2-1

$$y = \pm\sqrt{2[E - U(x)]}$$

x	$U(x)$	$\pm y$ at $(E = 7)$	$\pm y$ at $(E = 8)$	$\pm y$ at $(E = 9)$	$\pm y$ at $(E = 11)$
0	5.0	2.0	2.45	2.83	3.46
1.0	6.3	1.18	1.84	2.32	
1.5	8.0	imag	0	1.41	2.45
2.0	9.6	imag	imag	imag	
3.0	10.0	imag	imag	imag	1.41
3.5	8.0	imag	0	1.41	2.45
4.0	6.5	1.0	1.73	2.24	
5.0	5.0	2.0	2.45	2.83	3.46
5.5	5.7	1.61	2.24	2.57	
6.0	7.2	imag	1.26	1.90	
6.5	8.0	imag	0	1.41	2.45
7.0	7.0	0	1.41	2.0	
7.5	6.0	1.41	2.0	2.45	3.16
8.0	6.3	1.18	1.84	2.32	
9.0	7.4	imag	1.09	1.79	
9.5	8.0	imag	0	1.41	
10.0	8.8	imag	imag	0.63	
11.5					0

12.3   STABILITY OF EQUILIBRIUM

Expressed in the general form

$$\frac{dy}{dx} = \frac{P(x, y)}{Q(x, y)} \tag{12.3-1}$$

the singular points (x_s, y_s) of the equation are identified by

$$P(x_s, y_s) = Q(x_s, y_s) = 0 \tag{12.3-2}$$

Equation (12.3-1), of course, is equivalent to the two equations

$$\frac{dx}{dt} = Q(x, y)$$
$$\frac{dy}{dt} = P(x, y) \tag{12.3-3}$$

from which the time dt has been eliminated. A study of these equations in the neighborhood of the singular point provides us with answers as to the stability of equilibrium.

Recognize that the slope dy/dx of the trajectories does not vary with translation of the coordinate axes, we will translate the u, v axes to one of

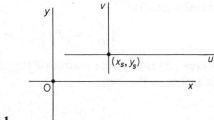

Figure 12.3-1.

the singular points to be studied, as shown in Fig. 12.3-1. We then have

$$x = x_s + u$$

$$y = y_s + v$$

$$\frac{dy}{dx} = \frac{dv}{du}$$

(12.3-4)

If $P(x, y)$ and $Q(x, y)$ are now expanded in terms of the Taylor series about the singular point (x_s, y_s), we obtain for $Q(x, y)$

$$Q(x, y) = Q(x_s, y_s) + \left(\frac{\partial Q}{\partial u}\right)_s u + \left(\frac{\partial Q}{\partial v}\right)_s v + \left(\frac{\partial^2 Q}{\partial u^2}\right)_s u^2 + \cdots$$

(12.3-5)

and a similar equation for $P(x, y)$. Since $Q(x_s, y_s)$ is zero and $(\partial Q/\partial u)_s$ and $(\partial Q/\partial v)_s$ are constants, Eq. (12.3-1) in the region of the singularity becomes

$$\frac{dv}{du} = \frac{cu + ev}{au + bv}$$

(12.3-6)

where the higher order derivatives of P and Q have been omitted. Thus a study of the singularity at (x_s, y_s) is possible by studying Eq. (12.3-6) for small u and v.

Returning to Eq. (12.3-3) and taking note of Eqs. (12.3-4) and (12.3-5), Eq. (12.3-6) is seen to be equivalent to

$$\frac{du}{dt} = au + bv$$

(12.3-7)

$$\frac{dv}{dt} = cu + ev$$

which may be rewritten in the matrix form

$$\begin{Bmatrix} \dot{u} \\ \dot{v} \end{Bmatrix} = \begin{bmatrix} a & b \\ c & e \end{bmatrix} \begin{Bmatrix} u \\ v \end{Bmatrix}$$

(12.3-8)

It was shown in Chapter 6, Sec. 6.7 that if the eigenvalues and eigenvectors of a matrix equation such as Eq. (12.3-8) are known, a transformation

$$\left\{ \begin{array}{c} u \\ v \end{array} \right\} = [P] \left\{ \begin{array}{c} \xi \\ \eta \end{array} \right\} = \left[\left\{ \begin{array}{c} u_1 \\ v_1 \end{array} \right\} \left\{ \begin{array}{c} u_2 \\ v_2 \end{array} \right\} \right] \left\{ \begin{array}{c} \xi \\ \eta \end{array} \right\} \qquad (12.3\text{-}9)$$

where $[P]$ is a modal matrix of the eigenvector columns will decouple the equation to the form

$$\left\{ \begin{array}{c} \dot{\xi} \\ \dot{\eta} \end{array} \right\} = [\Lambda] \left\{ \begin{array}{c} \xi \\ \eta \end{array} \right\} = \left[\begin{array}{cc} \lambda_1 & 0 \\ 0 & \lambda_2 \end{array} \right] \left\{ \begin{array}{c} \xi \\ \eta \end{array} \right\} \qquad (12.3\text{-}10)$$

Since Eq. (12.3-10) has the solution

$$\xi = e^{\lambda_1 t} \qquad (12.3\text{-}11)$$

$$\eta = e^{\lambda_2 t}$$

the solution for u and v are

$$u = u_1 e^{\lambda_1 t} + u_2 e^{\lambda_2 t}$$
$$v = v_1 e^{\lambda_1 t} + v_2 e^{\lambda_2 t} \qquad (12.3\text{-}12)$$

It is evident, then, that the stability of the singular point depends on the eigenvalues λ_1 and λ_2 determined from the characteristic equation

$$\left| \begin{array}{cc} (a - \lambda) & b \\ c & (e - \lambda) \end{array} \right| = 0$$

or

$$\lambda_{1,2} = \left(\frac{a + e}{2} \right) \pm \sqrt{\left(\frac{a + e}{2} \right)^2 - (ac - bc)} \qquad (12.3\text{-}13)$$

Thus

if $(ae - bc) > \left(\dfrac{a + e}{2} \right)^2$, the motion is oscillatory;

if $(ae - bc) < \left(\dfrac{a + e}{2} \right)^2$, the motion is aperiodic;

if $(a + e) > 0$, the system is unstable;

if $(a + e) < 0$, the system is stable.

The type of trajectories in the neighborhood of the singular point can be determined by first examining Eq. (12.3-10) in the form

$$\frac{d\xi}{d\eta} = \frac{\lambda_1}{\lambda_2} \frac{\xi}{\eta} \qquad (12.3\text{-}14)$$

which has the solution

$$\xi = (\eta)^{\lambda_1/\lambda_2}$$

and using the transformation of Eq. (12.3-9) to plot v vs. u.

12.4 METHOD OF ISOCLINES

Consider the autonomous system with the equation

$$\frac{dy}{dx} = -\frac{f(x, y)}{y} = \phi(x, y) \qquad (12.4\text{-}1)$$

that was discussed in Sec. 12.1, Eq. (12.1-4). In the method of isoclines we fix the slope dy/dx by giving it a definite number α, and solve for the curve

$$\phi(x, y) = \alpha \qquad (12.4\text{-}2)$$

With a family of such curves drawn, it is possible to sketch in a trajectory starting at any point x, y as shown in Fig. 12.4-1.

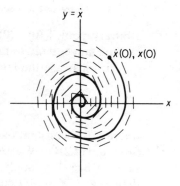

Figure 12.4-1.

EXAMPLE 12.4-1

Determine the isoclines for the simple pendulum.

Solution: The equation for the simple pendulum is

$$\ddot{\theta} + \frac{g}{l}\sin\theta = 0 \qquad (a)$$

Letting $x = \theta$ and $y = \dot{\theta} = \dot{x}$ we obtain

$$\frac{dy}{dx} = -\frac{g}{l}\frac{\sin x}{y} \qquad (b)$$

Thus for $dy/dx = \alpha$, a constant, the equation for the isocline, Eq. (12.4-2), becomes

$$y = -\left(\frac{g}{l\alpha}\right)\sin x \qquad (c)$$

It is evident from Eq. (b) that the singular points lie along the x-axis at $x = 0$, $\pm\pi$, $\pm 2\pi$ etc. Figure 12.4-2 shows isoclines in the first quadrant that correspond to negative values of α. Starting at an

375

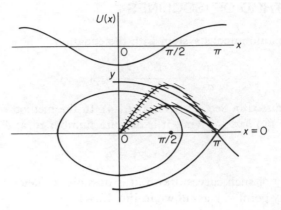

Figure 12.4-2. Isocline curves for the simple pendulum.

arbitrary point $x(0), y(0)$, the trajectory can be sketched in by proceeding tangentially to the slope segments.

In this case the integral of Eq. (a) is readily available as

$$\frac{y^2}{2} - \frac{g}{l} \cos x = E$$

where E is a constant of integration corresponding to the total energy (see Eq. (12.2-1)). We also have $U(x) = -g/l \cos x$ and the discussions of Sec. 12.2 apply. For the motion to exist, E must be greater than $-g/l$. $E = g/l$ corresponds to the separatrix and for $E > g/l$ the trajectory does not close. This means that the initial conditions are large enough to cause the pendulum to continue past $\theta = 2\pi$.

EXAMPLE 12.4-2

One of the interesting nonlinear equations which has been studied extensively is the *van der Pol* equation

$$\ddot{x} - \mu\dot{x}(1 - x^2) + x = 0$$

The equation somewhat resembles that of free vibration of a spring-mass system with viscous damping; however, the damping term of this equation is nonlinear in that it depends on both the velocity and the displacement. For small oscillations ($x < 1$) the damping is negative, and the amplitude will increase with time. For $x > 1$ the damping is positive, and the amplitude will diminish with time. If the system is initiated with $x(0)$ and $\dot{x}(0)$, the amplitude will increase or decrease, depending on whether x is small or large, and it will finally reach a stable state known as the *limit cycle*, graphically displayed by the phase plane plot of Fig. 12.4-3.

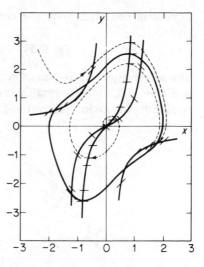

Figure 12.4-3. Isocline curves for van der Pol's equation with $\mu = 1.0$.

12.5 DELTA METHOD

The *delta method*, proposed by L. S. Jacobsen,* is a graphical method for the solution of the equation

$$\ddot{x} + f(\dot{x}, x, t) = 0 \qquad (12.5\text{-}1)$$

where $f(\dot{x}, x, t)$ must be continuous and single valued. The equation is first rewritten by adding and subtracting a term $\omega_0^2 x$

$$\ddot{x} + f(\dot{x}, x, t) - \omega_0^2 x + \omega_0^2 x = 0 \qquad (12.5\text{-}2)$$

Introducing new variables τ and y defined by

$$\tau = \omega_0 t \quad \text{and} \quad y = \frac{dx}{d\tau} = \frac{\dot{x}}{\omega_0} \qquad (12.5\text{-}3)$$

and letting

$$\delta(x, y, \tau) = \frac{1}{\omega_0^2}\left[f(\dot{x}, x, t) - \omega_0^2 x\right] \qquad (12.5\text{-}4)$$

Eq. (12.5-2) may be written as

$$\frac{dy}{dx} = \frac{-(x + \delta)}{y} \qquad (12.5\text{-}5)$$

The function $\delta(y, x, \tau)$ given in Eq. (12.5-4) depends on the variables y, x, and τ; however, for small changes in the variables, it may be assumed

*L. S. Jacobsen, "On a General Method of Solving Second Order Ordinary Differential Equations by Phase Plane Displacements," *J. Appl. Mech.* 19 (December 1952), pp. 543–53.

to remain constant. With δ a constant, Eq. (12.5-5) can be integrated to give

$$(x + \delta)^2 + y^2 = \rho^2 = \text{constant} \tag{12.5-6}$$

The above equation is that of a circle of radius ρ with its center at $y = 0$ and $x = -\delta$. Thus for small increments of τ, the solution corresponds to a small arc of a circle as shown in Fig. 12.5-1.

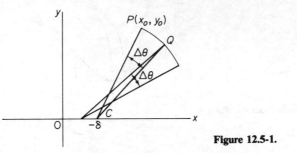

Figure 12.5-1.

The procedure for the drawing of the trajectory is as follows

1   Locate the initial point $P(x_0 y_0)$ in the xy phase plane.
2   From Eq. (12.5-4), calculate the initial value of $\delta(x_0, y_0, 0)$ and locate the point $-\delta$ on the x-axis.
3   With the center $(-\delta, 0)$ draw a short arc through point $P(x_0 y_0)$. The length of the arc over which the solution is valid depends on the variation of δ, which is assumed to be constant.
4   For an assumed value of $\Delta\theta$, locate the next point Q on the trajectory, and using the new values of x and y corresponding to point Q, compute a new δ.
5   Draw a line through the new $-\delta$ and the point Q and measure $\Delta\theta$ from this line for the location of the third point as shown in Fig. 12.5-1. The procedure is then repeated.

The relationship between the angular rotation $d\theta$ of the line CP and the time increment $d\tau$ can be found as follows

$$d\theta = \frac{ds}{\rho} = \frac{\sqrt{1 + \left(\frac{dy}{dx}\right)^2}\, dx}{\sqrt{y^2 + (x + \delta)^2}}$$

Substituting for dy/dx from Eq. (12.5-5) and noting from Eq. (12.5-3) that $dx = yd\tau$, we obtain

$$d\theta = \frac{dx}{y} = d\tau \tag{12.5-7}$$

It is also evident from Eq. (12.5-5) that the slope of the trajectory is negative in the first quadrant and that $d\theta$ proceeds in the clockwise direction.

EXAMPLE 12.5-1

Determine the phase plane trajectory of the equation

$$\ddot{x} + \mu|\dot{x}|\dot{x} + \omega^2 x = 0$$

with $\omega = 10$ and $x(0) = 4$, $\dot{x}(0) = 0$.

Solution: We first make the substitution $\tau = \omega t$ and $dx/d\tau = y$ to reduce the equation to the form

$$\frac{dy}{d\tau} + \mu|y|y + x = 0$$

or

$$\frac{dy}{dx} = \frac{-(\mu|y|y + x)}{y}$$

Thus δ for the delta method is

$$\delta = \mu|y|y$$

which for a given value of μ is a parabola as plotted in Fig. 12.5-2. If the delta method is applied to the problem, the center $-\delta$ of the trajectory arc for any point P is M. Thus in a half cycle, the point M

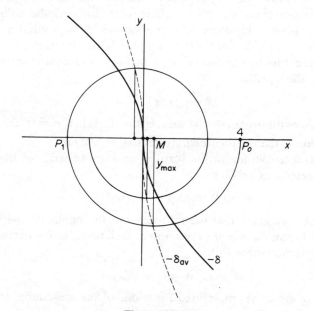

Figure 12.5-2.

moves from the origin to some extreme point corresponding to the maximum value of $-\delta$ and back again to 0.

Instead of using the step-by-step delta method we will consider here the average δ method, which will enable one to draw the trajectory for the half cycle as a circle with a center at the average value of $-\delta$.

The average $-\delta$ for this problem can be found for any y by integration

$$\delta_{av} = \frac{1}{y} \int_0^y \mu y^2 \, dy = \mu \frac{y^2}{3}$$

Thus the average δ curve of $\frac{1}{3}$ the original delta can be drawn on the phase plane, and the trajectory with the center at $-\delta_{av}$ must cut this curve with the same $x = -\delta_{av}$. This is easily done with a compass, employing trial and error, such that $MP_0 = MP_1 = My_{max}$.

12.6 PERTURBATION METHOD

The *perturbation method* is applicable to problems where a small parameter μ is associated with the nonlinear term of the differential equation. The solution is formed in terms of a series of the perturbation parameter μ, the result being a development in the neighborhood of the solution of the linearized problem. If the solution of the linearized problem is periodic, and if μ is small, we can expect the perturbed solution to be periodic also. We can reason from the phase plane that the periodic solution must represent a closed trajectory. The period which depends on the initial conditions is then a function of the amplitude of vibration.

Consider the free oscillation of a mass on a nonlinear spring which is defined by the equation

$$\ddot{x} + \omega_n^2 x + \mu x^3 = 0 \tag{12.6-1}$$

with initial conditions $x(0) = A$ and $\dot{x}(0) = 0$. When $\mu = 0$, the frequency of oscillation is that of the linear system $\omega_n = \sqrt{k/m}$.

We seek a solution in the form of an infinite series of the perturbation parameter μ as follows

$$x = x_0(t) + \mu x_1(t) + \mu^2 x_2(t) + \cdots \tag{12.6-2}$$

Furthermore, we know that the frequency of the nonlinear oscillation will depend on the amplitude of oscillation as well as on μ. We express this fact also in terms of a series in μ.

$$\omega^2 = \omega_n^2 + \mu \alpha_1 + \mu^2 \alpha_2 + \cdots \tag{12.6-3}$$

where the α_i are as yet undefined functions of the amplitude, and ω is the frequency of the nonlinear oscillations.

We will consider only the first two terms of Eqs. (12.6-2) and (12.6-3), which will adequately illustrate the procedure. Substituting these into Eq. (12.6-1), we obtain

$$\ddot{x}_0 + \mu\ddot{x}_1 + (\omega^2 - \mu\alpha_1)(x_0 + \mu x_1) + \mu(x_0^3 + 3\mu x_0^2 x_1 + \cdots) = 0$$
(12.6-4)

Since the perturbation parameter μ could have been chosen arbitrarily, the coefficients of the various powers of μ must be equated to zero. This leads to a system of equations which can be solved successively

$$\ddot{x}_0 + \omega^2 x_0 = 0$$
$$\ddot{x}_1 + \omega^2 x_1 = \alpha_1 x_0 - x_0^3$$
(12.6-5)
$$\vdots$$

The solution to the first equation, subject to the initial conditions $x(0) = A$, $\dot{x}(0) = 0$, is

$$x_0 = A \cos \omega t$$
(12.6-6)

which is called the *generating solution*. Substituting this into the right side of the second equation in Eq. (12.6-5) we obtain

$$\ddot{x}_1 + \omega^2 x_1 = \alpha_1 A \cos \omega t - A^3 \cos^3 \omega t$$
(12.6-7)
$$= \left(\alpha_1 - \frac{3}{4}A^2\right)A \cos \omega t - \frac{A^3}{4}\cos 3\omega t$$

where $\cos^3 \omega t = \frac{3}{4}\cos \omega t + \frac{1}{4}\cos 3\omega t$ has been used. We note here that the forcing term $\cos \omega t$ would lead to a secular term $t \cos \omega t$ in the solution for x_1(i.e., we have a condition of resonance). Such terms violate the initial stipulation that the motion is to be periodic; hence, we impose the condition

$$\left(\alpha_1 - \frac{3}{4}A^2\right) = 0$$

Thus α_1, which we stated earlier to be some function of the amplitude A, is evaluated to be equal to

$$\alpha_1 = \frac{3}{4}A^2$$
(12.6-8)

With the forcing term $\cos \omega t$ eliminated from the right side of the equation, the general solution for x_1 is

$$x_1 = C_1 \sin \omega t + C_2 \cos \omega t + \frac{A^3}{32\omega^2}\cos 3\omega t$$
(12.6-9)
$$\omega^2 = \omega_n^2 + \frac{3}{4}\mu A^2$$

Imposing the initial conditions $x_1(0) = \dot{x}_1(0) = 0$, the constants C_1 and C_2

are evaluated as

$$C_1 = 0 \quad C_2 = -\frac{A^3}{32\omega^2}$$

Thus

$$x_1 = \frac{A^3}{32\omega^2}(\cos 3\omega t - \cos \omega t) \tag{12.6-10}$$

and the solution at this point from Eq. (12.6-2) becomes

$$x = A \cos \omega t + \mu\frac{A^3}{32\omega^2}(\cos 3\omega t - \cos \omega t)$$

$$\tag{12.6-11}$$

$$\omega = \omega_n\sqrt{1 + \frac{3}{4}\frac{\mu A^2}{\omega_0^2}}$$

The solution is thus found to be periodic, and the fundamental frequency ω is found to increase with the amplitude, as expected for a hardening spring.

Mathieu Equation: Consider the nonlinear equation

$$\ddot{x} + \omega_n^2 x + \mu x^3 = F \cos \omega t \tag{a}$$

and assume a perturbation solution

$$x = x_1(t) + \xi(t) \tag{b}*$$

Substituting Eq. (b) into (a), we obtain the following two equations

$$\ddot{x}_1 + \omega_n^2 x_1 + \mu x_1^3 = F \cos \omega t \tag{c}$$

$$\ddot{\xi} + (\omega_n^2 + \mu 3 x_1^2)\xi = 0 \tag{d}$$

If μ is assumed to be small, we can let

$$x_1 \cong A \sin \omega t \tag{e}$$

and substitute it into Eq. (d), which becomes

$$\ddot{\xi} + \left[\left(\omega_n^2 + \frac{3\mu}{2}A^2\right) - \frac{3\mu}{2}A^2 \cos 2\omega t\right]\xi = 0 \tag{f}$$

This equation is of the form

$$\frac{d^2 y}{dz^2} + (a - 2b \cos 2z)y = 0 \tag{g}$$

which is known as the *Mathieu equation*. The stable and unstable regions of the Mathieu equation depend on the parameters a and b, and are shown in Fig. 12.6-1.

*See Ref. 5, pp. 259–73.

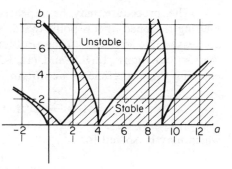

Figure 12.6-1. Stable region of Mathieu equation indicated by shaded area, which is symmetric about the horizontal axis.

12.7 METHOD OF ITERATION

Duffing[*] made an exhaustive study of the equation

$$m\ddot{x} + c\dot{x} + kx \pm \mu x^3 = F \cos \omega t$$

which represents a mass on a cubic spring, excited harmonically. The $\pm$ sign signifies a hardening or softening spring. The equation is nonautonomous in that the time t appears explicitly in the forcing term.

In this section we wish to examine a simpler equation where damping is zero, written in the form

$$\ddot{x} + \omega_n^2 x \pm \mu x^3 = F \cos \omega t \qquad (12.7\text{-}1)$$

We seek only the steady state harmonic solution by the *method of iteration*, which is essentially a process of *successive approximation*. An assumed solution is substituted into the differential equation, which is integrated to obtain a solution of improved accuracy. The procedure may be repeated any number of times to achieve the desired accuracy.

For the first assumed solution, let

$$x_0 = A \cos \omega t \qquad (12.7\text{-}2)$$

and substitute into the differential equation

$$\ddot{x} = -\omega_n^2 A \cos \omega t \mp \mu A^3\left(\tfrac{3}{4}\cos \omega t + \tfrac{1}{4}\cos 3\omega t\right) + F \cos \omega t$$

$$= \left(-\omega_n^2 A \mp \tfrac{3}{4}\mu A^3 + F\right)\cos \omega t \mp \tfrac{1}{4}\mu A^3 \cos 3\omega t$$

In integrating this equation it is necessary to set the constants of integration to zero if the solution is to be harmonic with period $\tau = 2\pi/\omega$. Thus, we obtain for the improved solution

$$x_1 = \frac{1}{\omega^2}\left(\omega_n^2 A \pm \frac{3}{4}\mu A^3 - F\right)\cos \omega t \mp \cdots \qquad (12.7\text{-}3)$$

where the higher harmonic term is ignored.

*See Ref. 7.

The procedure may be repeated but we will not go any further. Duffing reasoned at this point that if the first and second approximations are reasonable solutions to the problem, then the coefficients of cos ωt in the two equations (12.7-2) and (12.7-3) must not differ greatly. Thus, by equating these coefficients we obtain

$$A = \frac{1}{\omega^2}\left(\omega_n^2 A \pm \frac{3}{4}\mu A^3 - F\right) \qquad (12.7\text{-}4)$$

which may be solved for ω^2

$$\omega^2 = \omega_n^2 \pm \frac{3}{4}\mu A^2 - \frac{F}{A} \qquad (12.7\text{-}5)$$

It is evident from this equation that if the nonlinear parameter is zero, we obtain the exact result for the linear system

$$A = \frac{F}{\omega_n^2 - \omega^2}$$

For $\mu \neq 0$, the frequency ω is a function of μ, F and A. It is evident that when $F = 0$, we obtain the frequency equation for free vibration

$$\frac{\omega^2}{\omega_n^2} = 1 \pm \frac{3}{4}\mu\frac{A^2}{\omega_n^2}$$

discussed in the previous section. Here we see that the frequency increases with the amplitude for the hardening spring ($+$), and decreased for the softening spring ($-$).

For $\mu \neq 0$ and $F \neq 0$, it is convenient to hold both μ and F constant and plot $|A|$ against ω/ω_n. In the construction of these curves, it is helpful

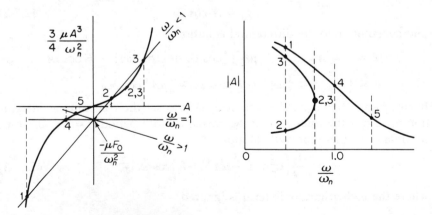

Figure 12.7-1.   Solution of Equation 12.7-6.

to rearrange Eq. (11.10-5) to

$$\frac{3}{4}\mu\frac{A^3}{\omega_n^2} = \left(1 - \frac{\omega^2}{\omega_n^2}\right)A - \frac{F}{\omega_n^2} \qquad (12.7\text{-}6)$$

each side of which can be plotted against A as shown in Fig. 12.7-1. The left side of this equation is a cubic, whereas the right side is a straight line of slope $(1 - \omega^2/\omega_n^2)$ and intercept $-F/\omega_n^2$. For $\omega/\omega_n < 1$, the two curves intersect at three points 1, 2, 3, which are also shown in the amplitude-frequency plot. As ω/ω_n increases towards unity, points 2 and 3 approach each other, after which only one value of the amplitude will satisfy Eq. (12.7-6). When $\omega/\omega_n = 1$, or when $\omega/\omega_n > 1$, these points are 4 or 5.

The Jump Phenomenon. In problems of this type, it is found that the amplitude A undergoes a sudden discontinuous jump near resonance. The *jump phenomenon* can be described as follows. For the softening spring, with increasing frequency of excitation, the amplitude gradually increases until point "a" in Fig. 12.7-2 is reached. It then suddenly jumps to a larger value indicated by the point b, and diminishes along the curve to its right. In decreasing the frequency from some point c, the amplitude continues to increase beyond b to point d, and suddenly drops to a smaller value at e. The shaded region in the amplitude-frequency plot is unstable; the extent of unstableness depends on a number of factors such as the amount of damping present, the rate of change of the exciting frequency, etc. if a hardening spring had been chosen instead of the softening spring, the same type of analysis would be applicable and the result would be a curve of the type shown in Fig. 12.7-3.

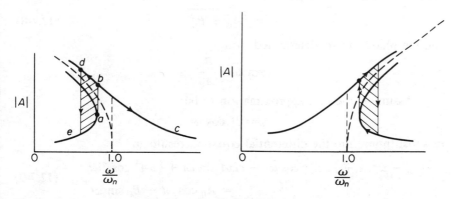

Figure 12.7-2. The jump phenomenon for the softening spring.

Figure 12.7-3. The jump phenomenon for the hardening spring.

Effect of Damping. In the undamped case the amplitude-frequency curves approach the backbone curve (shown dotted) asymptotically. This is also the case for the linear system where the backbone curve is the vertical line at $\omega/\omega_n = 1.0$.

With a small amount of damping present, the behavior of the system cannot differ appreciably from that of the undamped system. The upper end of the curve, instead of approaching the backbone curve asymptotically, will cross over in a continuous curve as shown in Fig. 12.7-4. The jump phenomenon is also present here but damping generally tends to reduce the size of the unstable region.

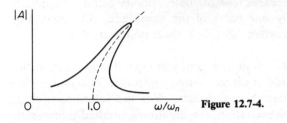

Figure 12.7-4.

The method of successive approximation is also applicable to the damped vibration case. The major difference in its treatment lies in the phase angle between the force and the displacement, which is no longer $0°$ or $180°$ as in the undamped problem. It is found that by introducing the phase in the force term rather than the displacement, the algebraic work is somewhat simplified. The differential equation can then be written as

$$\ddot{x} + c\dot{x} + \omega_n^2 x + \mu x^3 = F\cos(\omega t + \phi) \tag{12.7-7}$$
$$= A_0 \cos \omega t - B_0 \sin \omega t$$

where the magnitude of the force is

$$F = \sqrt{A_0^2 + B_0^2} \tag{12.7-8}$$

and the phase can be determined from

$$\tan \phi = \frac{B_0}{A_0}$$

Assuming the first approximation to be

$$x_0 = A \cos \omega t$$

its substitution into the differential equation results in

$$\left[(\omega_n^2 - \omega^2)A + \tfrac{3}{4}\mu A^3\right]\cos \omega t - c\omega A \sin \omega t + \tfrac{1}{4}\mu A^3 \cos 3\omega t \tag{12.7-9}$$
$$= A_0 \cos \omega t - B_0 \sin \omega t$$

We again ignore the $\cos 3\omega t$ term and equate coefficients of $\cos \omega t$ and $\sin \omega t$ to obtain

$$(\omega_n^2 - \omega^2)A + \tfrac{3}{4}\mu A^3 = A_0 \tag{12.7-10}$$
$$c\omega A = B_0$$

Squaring and adding these results, the relationship between the frequency, amplitude and force becomes

$$F^2 = \left[(\omega_n^2 - \omega^2)A + \tfrac{3}{4}A^3 \right]^2 + \left[c\omega A \right]^2 \qquad (12.7\text{-}11)$$

By fixing μ, c and F, the frequency ratio ω/ω_n can be computed for assigned values of A.

EXAMPLE 12.7-1

Using the iteration method, solve for the period of the linear equation

$$\ddot{x} + \omega_n^2 x = 0 \qquad (a)$$

with the initial conditions $x(0) = A$ and $\dot{x}(0) = 0$.

Solution: *Assume for the first solution $x = 1$ and substitute into the differential equation

$$\ddot{x} = -\omega_n^2 1$$

Integrate to obtain

$$\dot{x} = -\omega_n^2 \int_0^\xi 1 \, d\xi = -\omega_n^2 \xi$$

and

$$\int_A^x dx = -\omega_n^2 \int_0^t \xi \, d\xi$$

$$x(t) = A - \omega_n^2 \frac{t^2}{2} \qquad (b)$$

Letting $t = t_1$ at a quarter cycle and noting that $x(t_1) = 0$, the above equation may be written as

$$x = A \left(1 - \frac{t^2}{t_1^2} \right) \qquad (c)$$

We now substitute Eq. (c) into Eq. (a) and repeat the process

$$\dot{x}(t) = -\omega_n^2 A \int_0^\xi \left(1 - \frac{\xi^2}{t_1^2} \right) d\xi$$

$$= -\omega_n^2 A \left(\xi - \frac{\xi^3}{3t_1^2} \right)$$

$$x(t) = A - \omega_n^2 A \int_0^t \left(\xi - \frac{\xi^3}{3t_1^2} \right) d\xi \qquad (d)$$

$$= A - \omega_n^2 A \left(\frac{t^2}{2} - \frac{t^4}{12t_1^2} \right)$$

*See Ref. 2.

Next let $t = t_1$ and $x(t_1) = 0$.

$$0 = A\left[1 - \omega_n^2\left(\frac{t_1^2}{2} - \frac{t_1^2}{12}\right)\right]$$

Solving for t_1 we obtain

$$t_1 = \frac{1}{\omega_n}\sqrt{\frac{12}{5}} = \frac{\tau}{2\pi}\sqrt{\frac{12}{5}} = \frac{\tau}{4.05}$$

and we find that after two iterations, the value of t_1 is found to be nearly equal to the exact value of $\tau/4$.

12.8 SELF-EXCITED OSCILLATIONS

Oscillations which depend on the motion itself are called self-excited. The shimmy of automobile wheels, the flutter of airplane wings, and the oscillations of the van der Pol equation are some examples.

Self-excited oscillations may occur in a linear or a nonlinear system. The motion is induced by an excitation that is some function of the velocity or of the velocity and the displacement. If the motion of the system tends to increase the energy of the system, the amplitude will increase, and the system may become unstable.

As an example, consider a viscously damped single degree of freedom linear system excited by a force which is some function of the velocity. Its equation of motion is

$$m\ddot{x} + c\dot{x} + kx = F(\dot{x}) \qquad (12.8\text{-}1)$$

Rearranging the equation to the form

$$m\ddot{x} + (c\dot{x} - F(\dot{x})) + kx = 0 \qquad (12.8\text{-}2)$$

we can recognize the possibility of negative damping if $F(\dot{x})$ becomes greater than $c\dot{x}$.

Suppose that $\phi(\dot{x}) = c\dot{x} - F(\dot{x})$ in the above equations varies as in Fig. 12.8-1. For small velocities the apparent damping $\phi(\dot{x})$ is negative, and the amplitude of oscillation will increase. For large velocities the opposite is true, and hence the oscillations will tend to a limit cycle.

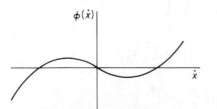

Figure 12.8-1. System with apparent damping $\phi(\dot{x}) = c\dot{x} - F(\dot{x})$.

EXAMPLE 12.8-1

The coefficient of kinetic friction μ_k is generally less than the coefficient of static friction μ_s, this difference increasing somewhat with the velocity. Thus if the belt of Fig. 12.8-2 is started, the mass will move with the belt until the spring force is balanced by the static friction.

$$kx_0 = \mu_s mg \tag{a}$$

At this point the mass will start to move back to the left, and the forces will again be balanced on the basis of kinetic friction when

$$k(x_0 - x) = \mu_{kl} mg$$

From these two equations, the amplitude of oscillation is found to be

$$x = x_0 - \mu_{kl} \frac{mg}{k} = \frac{(\mu_s - \mu_{kl})g}{\omega_n^2} \tag{b}$$

While the mass is moving to the left, the relative velocity between it and the belt is greater than when it is moving to the right, thus μ_{kl} is less than μ_{kr} where the subscripts l and r refer to left and right. It is evident then that the work done by the friction force while moving to the right is greater than that while moving to the left, so that more energy is put into the spring-mass system than taken out. This then represents one type of self-excited oscillation and the amplitude will continue to increase.

The work done by the spring from 2 to 3 is

$$-\tfrac{1}{2}k\big[(x_0 + \Delta x) + (x_0 - 2x)\big](2x + \Delta x)$$

The work done by friction from 2 to 3 is

$$\mu_{kr} mg(2x + \Delta x)$$

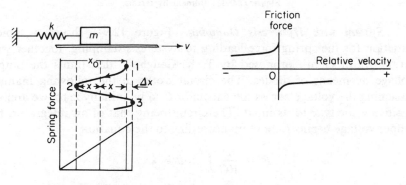

Figure 12.8-2. Coulomb friction between belt and mass.

Equating the net work done between 2 and 3 to the change in kinetic energy which is zero,

$$-\tfrac{1}{2}k(2x_0 - 2x + \Delta x) + \mu_{kr} mg = 0 \tag{c}$$

Substituting (a) and (b) into (c), the increase in amplitude per cycle of oscillation is found to be

$$\Delta x = \frac{2g(\mu_{kr} - \mu_{kl})}{\omega_n^2} \tag{d}$$

12.9   ANALOG COMPUTER CIRCUITS FOR NONLINEAR SYSTEMS

Many nonlinear systems can be studied by the use of the electronic analog computer. Presented in this section are some of the circuit diagrams associated with nonlinear systems with a brief discussion as to their working principles.

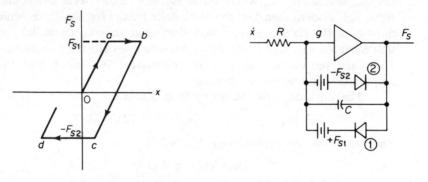

Figure 12.9-1.   Bilinear-hysteresis.

System with Hysteresis Damping. Figure 12.9-1 shows a typical variation for the spring force leading to hysteresis damping, together with an integrator circuit proposed by T. K. Caughey* that limits the output voltage by means of diodes: The circuit works in the following manner. Assuming the voltage across the capacitor C to be initially zero, we apply a positive velocity $\dot{x}$ to its input. The circuit being that of an integrator, the output voltage begins to built up according to the equation

$$F_s = \frac{1}{RC} \int_0^t \dot{x}\, dt = kx$$

Noting that the potential of the grid g is essentially zero, this voltage

*See Ref. 4.

appears across C. During this time diode ② cannot conduct and hence circuit ② appears as open. Diode ① can conduct only when its voltage from the output side exceeds the bias voltage $+F_{s1}$ which is set to the level indicated by a in the force displacement curve. When diode ① conducts, the voltage across C is limited to $+F_{s1}$ until $\dot{x}$ becomes negative at b, at which time diode ① becomes nonconducting and the circuit appears as if only C is present across the amplifier. When the output voltage reaches some negative value set by $-F_{s2}$ at c, the diode ② conducts and limits the negative voltage on C until again the velocity $\dot{x}$ becomes positive at d.

Figure 12.9-2 shows how the limited integrator is incorporated into the circuit to solve the equation

$$\frac{d^2x}{d\tau^2} + F_s = F(t) \tag{12.9-1}$$

By inserting an additional capacitor C_2 (shown dotted), it is possible to give the line ab and cd of the stiffness curve a slope other than zero.

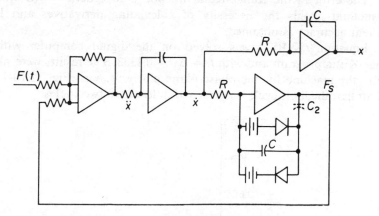

Figure 12.9-2.   Circuit for Equation 12.9-1.

12.10   RUNGE-KUTTA METHOD

The Runge-Kutta method discussed in Chapter 4 can be used to solve nonlinear differential equations. We will consider the nonlinear equation

$$\frac{d^2x}{d\tau^2} + 0.4\frac{dx}{d\tau} + x + 0.5x^3 = 0.5 \cos{(0.5\tau)} \tag{12.10-1}$$

and rewrite it in the first order form by letting $y = dx/d\tau$ as follows

$$\frac{dy}{d\tau} = 0.5 \cos{(0.5\tau)} - x - 0.5x^3 - 0.4y = F(\tau, x, y)$$

The computational equations to be used are programmed for the digital computer in the following order. From these results the values of x and y

τ	x	y	F
$t_1 = \tau_1$	$k_1 = x_1$	$g_1 = y_1$	$f_1 = F(t_1, k_1, g_1)$
$t_2 = \tau_1 + h/2$	$k_2 = x_1 + g_1 h/2$	$g_2 = y_1 + f_1 h/2$	$f_2 = F(t_2, k_2, g_2)$
$t_3 = \tau_1 + h/2$	$k_3 = x_1 + g_1 h/2$	$g_3 = y_1 + f_2 h/2$	$f_3 = F(t_3, k_3, g_3)$
$t_4 = \tau + h$	$k_4 = x_1 + g_3 h$	$g_4 = y_1 + f_3 h$	$f_4 = F(t_4, k_4, g_4)$

are determined from the following recurrence equations, where $h = \Delta t$.

$$x_{i+1} = x_i + \frac{h}{6}(g_1 + 2g_2 + 2g_3 + g_4) \qquad (12.10\text{-}2)$$

$$y_{i+1} = y_i + \frac{h}{6}(f_1 + 2f_2 + 2f_3 + f_4) \qquad (12.10\text{-}3)$$

Thus with $i = 1$, x_2 and y_2 are found, and with $\tau_2 = \tau_1 + \Delta\tau$, the previous table of t, k, g, and f is computed and again substituted into the recurrence equations to find x_3 and y_3.

The error in the Runge-Kutta method is of order $h^5 = (\Delta\tau)^5$. Also, the method avoids the necessity of calculating derivatives and hence excellent accuracy is obtained.

Equation (12.10-1) was solved on the digital computer with the Runge-Kutta program and with $h = \Delta\tau = 0.1333$. The results were plotted out by the machine for the phase plane plot y vs. x in Fig. 12.10-1. It is evident that the limit cycle was reached in less than two cycles.

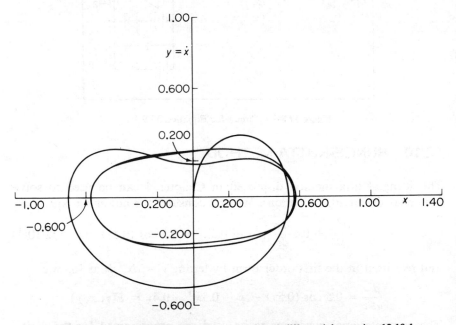

Figure 12.10-1.   Runge-Kutta solution for nonlinear differential equation 12.10-1.

Using the digital computer, the van der Pol equation
$$\ddot{x} - \mu\dot{x}(1 - x^2) + x = 0$$
was solved by the Runge-Kutta method for $\mu = 0.2, 0.7, 1.5, 3$ and 4 with a small initial displacement. Both the phase plane and the time plots were automatically plotted.

For the case $\mu = 0.2$ the response is practically sinusoidal and the phase plane plot is nearly an elliptic spiral. The effect of the nonlinearity is quite evident for $\mu = 1.5$ which is shown in Figs. 12.10-2 and 12.10-3.

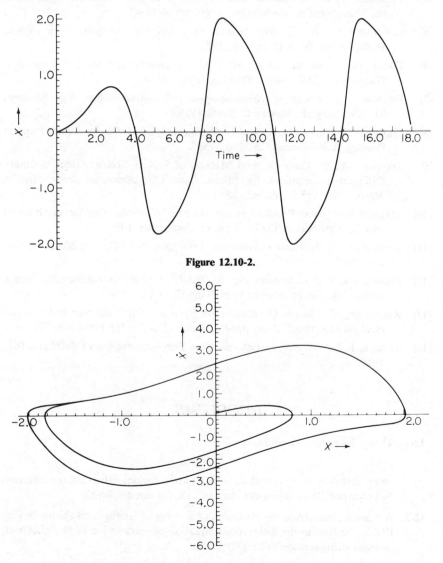

Figure 12.10-2.

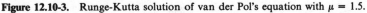

Figure 12.10-3.   Runge-Kutta solution of van der Pol's equation with $\mu = 1.5$.

REFERENCES

[1] BELLMAN, R. *Perturbation Techniques in Mathematics, Physics and Engineering*. New York: Holt, Rinehart & Winston, Inc., 1964.

[2] BROCK, J. E. "An Iterative Numerical Method for Nonlinear Vibrations," *Jour. Appl'd. Mech.* (March 1951), pp. 1–11.

[3] BUTENIN, N. V. *Elements of the Theory of Nonlinear Oscillations*. New York: Blaisdell Publishing Co., 1965.

[4] CAUGHEY, T. K. "Sinusoidal Excitation of a System with Bilinear Hysteresis," *Jour. Appl'd. Mech.* (Dec. 1960), pp. 640–43.

[5] CUNNINGHAM, W. J. *Introduction to Nonlinear Analysis*. New York: McGraw-Hill Book Company, 1958.

[6] DAVIS, H. T. *Introduction to Nonlinear Differential and Integral Equations*. Washington, D.C.: Govt. Printing Office, 1956.

[7] DUFFING, G. *Erwugene Schwingungen bei veranderlicher Eigenfrequenz*. Braunschweig: F. Vieweg u. Sohn, 1918.

[8] HAYASHI, C. *Forced Oscillations in Nonlinear Systems*. Osaka, Japan: Nippon Printing & Publishing Co., Ltd., 1953.

[9] JACOBSEN, L. S. "On a General Method of Solving Second Order Ordinary Differential Equations by Phase Plane Displacements," *Jour. Appl'd. Mech.* (Dec. 1952), pp. 543–53.

[10] MALKIN, I. G. *Some Problems in the Theory of Nonlinear Oscillations*, Books I and II. Washington, D.C.: Dept. of Commerce, 1959.

[11] MINORSKY, N. *Nonlinear Oscillations*. Princeton, N.J.: D. Van Nostrand Co., Inc., 1962.

[12] NISHIKAWA, Y. *A Contribution of the Theory of Nonlinear Oscillations*. Osaka, Japan: Nippon Printing & Publishing Co. Ltd., 1964.

[13] RAUSCHER, M. "Steady Oscillations of Systems with Nonlinear and Unsymmetrical Elasticity," *Jour. Appl'd. Mech.* (Dec. 1938), pp. A169–77.

[14] STOKER, J. J. *Nonlinear Vibrations*. New York: Interscience Publishers, Inc., 1950.

PROBLEMS

12-1 Using the nonlinear equation

$$\ddot{x} + x^3 = 0$$

show that if $x_1 = \varphi_1(t)$ and $x_2 = \varphi_2(t)$ are solutions satisfying the differential equation, their superposition $(x_1 + x_2)$ is not a solution.

12-2 A mass is attached to the midpoint of a string of length $2l$ as shown in Fig. P12-2. Determine the differential equation of motion for large deflection. Assume string tension to be T.

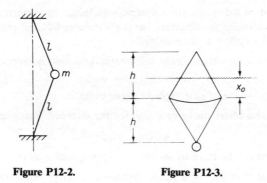

Figure P12-2.   Figure P12-3.

12-3 A buoy is composed of two cones of diameter $2r$ and height h as shown in Fig. P12-3. A weight attached to the bottom allows it to float in the equilibrium position x_0. Establish the differential equation of motion for vertical oscillation.

12-4 Determine the differential equation of motion for the spring-mass system with the discontinuous stiffness resulting from the free gaps of Fig. P12-4.

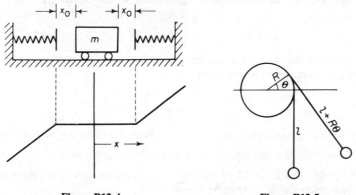

Figure P12-4.   Figure P12-5.

12-5 The cord of a simple pendulum is wrapped around a fixed cylinder of radius R such that its length is l when in the vertical position as shown in Fig. P12-5. Determine the differential equation of motion.

12-6 Plot the phase plane trajectory for the undamped spring-mass system including the potential energy curve $U(x)$. Discuss the initial conditions associated with the plot.

12-7 From the plot of $U(x)$ vs. x of Prob. 12-6, determine the period from the equation

$$\tau = 4 \int_0^{x_{max}} \frac{dx}{\sqrt{2[E - U(x)]}}$$

(Remember that E in the text was for a unit mass.)

12-8   For the undamped spring-mass system with initial conditions $x(0) = A$ and $\dot{x}(0) = 0$ determine the equation for the state speed V and state under what condition the system is in equilibrium.

12-9   The solution to a certain linear differential equation is given as
$$x = \cos \pi t + \sin 2\pi t$$
Determine $y = \dot{x}$ and plot a phase plane diagram.

12-10   Determine the phase plane equation for the damped spring-mass system
$$\ddot{x} + 2\zeta\omega_n\dot{x} + \omega_n^2 x = 0$$
and plot one of the trajectories with $v = y/\omega_n$ and x as coordinates.

12-11   If the potential energy of a simple pendulum is given with the positive sign
$$U(\theta) = +\frac{g}{l}\cos\theta$$
determine which of the singular points are stable or unstable and explain their physical implications. Compare the phase plane with Fig. 12.4-2.

12-12   Given the potential $U(x) = 8 - 2\cos\pi x/4$, plot the phase plane trajectories for $E = 6, 7, 8, 10, 12$, and discuss the curves.

12-13   Determine the eigenvalues and eigenvectors of the equations
$$\dot{x} = 5x - y$$
$$\dot{y} = 2x + 2y$$

12-14   Determine the modal transformation of the equations of Prob. 12-13 which will decouple them to the form
$$\dot{\xi} = \lambda_1\xi$$
$$\dot{\eta} = \lambda_2\eta$$

12-15   Plot the ξ, η phase plane trajectories of Prob. 12-14 for $\lambda_1/\lambda_2 = 0.5$ and 2.0.

12-16   For $\lambda_1/\lambda_2 = 2.0$ in Prob. 12-15, plot the trajectory y vs. x.

12-17   If λ_1 and λ_2 of Prob. 12-14 are complex conjugates $-\alpha \pm i\beta$, show that the equation in the u, v plane becomes
$$\frac{dv}{du} = \frac{\beta u + \alpha v}{\alpha u - \beta v}$$

12-18   Using the transformation $u = \rho\cos\theta$ and $v = \rho\sin\theta$ show that the phase plane equation for Prob. 12-17 becomes
$$\frac{d\rho}{\rho} = \frac{\alpha}{\beta}d\theta$$
with the trajectories identified as logarithmic spirals
$$\rho = e^{(\alpha/\beta)\theta}$$

12-19   Near a singular point in the x, y plane, the trajectories appear as shown in Fig. P12-19.
Determine the form of the phase plane equation and the corresponding trajectories in the $\xi - \eta$ plane.

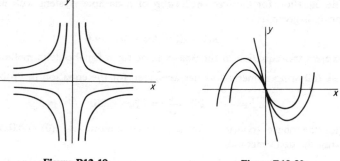

Figure P12-19.   Figure P12-20.

12-20   The phase plane trajectories in the vicinity of a singularity of an over-damped system ($\zeta > 1$) is shown in Fig. P12-20.

Identify the phase plane equation and plot the corresponding trajectories in the $\xi - \eta$ plane.

12-21   Show that the solution of the equation

$$\frac{dy}{dx} = \frac{-x - y}{x + 3y}$$

is $x^2 + 2xy + 3y^2 = C$, which is a family of ellipses with axes rotated from the x, y coordinates. Determine the rotation of the semimajor axis and plot one of the ellipses.

12-22   Show that the isoclines of the linear differential equation of second order are straight lines.

12-23   Draw the isoclines for the equation

$$\frac{dy}{dx} = xy(y - 2)$$

12-24   Consider the nonlinear equation

$$\ddot{x} + \omega_n^2 x + \mu x^3 = 0$$

Replacing $\ddot{x}$ by $y(dy/dx)$ where $y = \dot{x}$, its integral becomes

$$y^2 + \omega_n^2 x^2 + \tfrac{1}{2} \mu x^4 = 2E$$

With $y = 0$ when $x = A$, show that the period is available from

$$\tau = 4 \int_0^A \frac{dx}{\sqrt{2[E - U(x)]}}$$

12-25   What do the isoclines of Prob. 12-24 look like?

12-26   Plot of the isoclines of the van der Pol's equation

$$\ddot{x} - \mu\dot{x}(1 - x^2) + x = 0$$

for $\mu = 2.0$ and $dy/dx = 0, -1$ and $+1$

12-27   The equation for the free oscillation of a damped system with hardening spring is given as

$$m\ddot{x} + c\dot{x} + kx + \mu x^3 = 0$$

Express this equation in the phase plane form for the delta method.

12-28   The following numerical values are given for the equation in Prob. 12-27

$$\omega_n^2 = \frac{k}{m} = 25, \quad \frac{c}{m} = 2\zeta\omega_n = 2.0, \quad \frac{\mu}{m} = 5$$

Plot the phase trajectory for the initial conditions $x(0) = 4.0$, $\dot{x}(0) = 0$, using the delta method.

12-29   Using the delta method, plot the phase plane trajectory for the simple pendulum with the initial conditions $\theta(0) = 60°$ and $\dot{\theta}(0) = 0$.

12-30   Determine the period of the pendulum of Prob. 12-29 and compare with that of the linear system.

12-31   The equation of motion for a spring-mass system with constant Coulomb damping can be written as

$$\ddot{x} + \omega_n^2 x + C\,\text{sgn}(\dot{x}) = 0$$

where $\text{sgn}(\dot{x})$ signifies either a positive or negative sign equal to that of the sign of $\dot{x}$. Express this equation in a form suitable for the Delta method.

12-32   A system with Coulomb damping has the following numerical values: $k = 3.60$ lb/in., $m = 0.10$ lb sec^2 in.$^{-1}$ $\mu = 0.20$. Using the Delta method, plot the trajectory for $x(0) = 20''$, $\dot{x}(0) = 0$.

12-33   Consider the motion of a simple pendulum with viscous damping and determine the singular points. With the aid of Fig. 12.4-2, and the knowledge that the trajectories must spiral into the origin, sketch some approximate trajectories.

12-34   Apply the perturbation method to the simple pendulum with $\sin\theta$ replaced by $\theta - \frac{1}{6}\theta^3$. Use only the first two terms of the series for x and ω.

12-35   From the perturbation method, what is the equation for the period of the simple pendulum as a function of the amplitude.

12-36   For a given system the numerical values of Eq. (12.7-7) are given as

$$\ddot{x} + 0.15\dot{x} + 10x + x^3 = 5\cos(\omega t + \varphi)$$

Plot A vs. ω from Eq. (12.7-11) by first assuming a value of A and solving for ω^2.

12-37   Determine the phase angle φ vs. ω for Prob. 12-36.

12-38   The supporting end of a simple pendulum is given a motion as shown in Fig. P12-38. Show that the equation of motion is

$$\ddot{\theta} + \left(\frac{g}{l} - \frac{\omega^2 y_0}{l}\cos 2\omega t\right)\sin\theta = 0$$

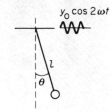

Figure P12-38.

12-39   For a given value of g/l, determine the frequencies of the excitation for which the simple pendulum of Prob. 12-38 with a stiff arm l will be stable in the inverted position.

12-40   Determine the perturbation solution for the system shown in Fig. P12-40 leading to a Mathieu equation. Use initial conditions $\dot{x}(0) = 0$, $x(0) = A$.

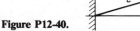

Figure P12-40.

12-41   A circuit which simulates a dead zone in the spring stiffness is shown in Fig. P12-41. Complete the analogue circuit to solve Prob. 12-4.

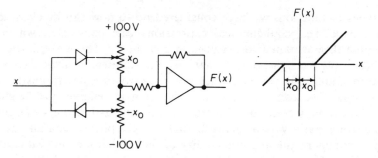

Figure P12-41.

12-42   Using the Runge-Kutta routine and $g/l = 1.0$, calculate the angle θ for the simple pendulum of Prob. 12-29.

12-43   With damping added to Prob. 12-42, the equation of motion is given as

$$\ddot{\theta} + 0.30\dot{\theta} + \sin\theta = 0$$

Solve by the Runge-Kutta method for the initial conditions $\theta(0) = 60°$, $\dot{\theta}(0) = 0$.

12-44   Obtain a numerical solution for the system of Prob. 12-40 by using (a) the central difference method and (b) the Runge-Kutta method.

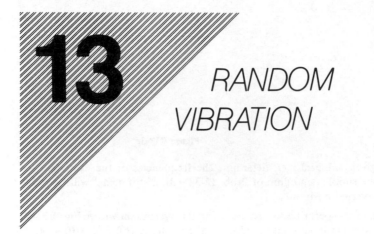

RANDOM VIBRATION

The types of functions we have considered up to now can be classified as deterministic, i.e., mathematical expressions can be written which will determine their instantaneous values at any time t. There are, however, a number of physical phenomena that result in nondeterministic data where future instantaneous values cannot be predicted in a deterministic sense. As examples, we can mention the noise of a jet engine, the heights of waves in a choppy sea, ground motion during an earthquake, and pressure gusts encountered by an airplane in flight. These phenomena all have one thing in common: the unpredictability of their instantaneous value at any future time. Nondeterministic data of this type are referred to as *random time functions*.

13.1   RANDOM PHENOMENA

A sample of a typical random time function is shown in Fig. 13.1-1. In spite of the irregular character of the function, many random phenomena exhibit some degree of statistical regularity, and certain averaging procedures can be applied to establish gross characteristics useful in engineering design.

In any statistical method, a large amount of data is necessary to establish reliability. For example, to establish the statistics of the pressure fluctuation due to air turbulance over a certain air route, an airplane may collect hundreds of records of the type shown in Fig. 13.1-2.

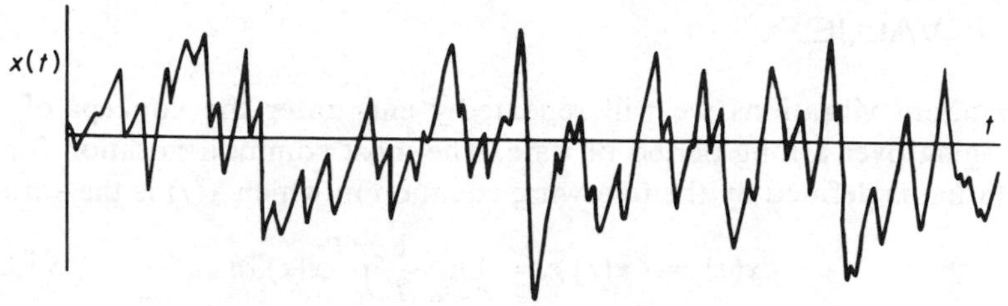

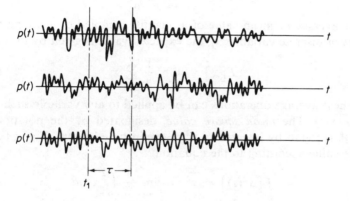

Figure 13.1-1.   A record of random time function.

Figure 13.1-2.   An ensemble of random time functions.

Each record is called a *sample*, and the total collection of samples is called the *ensemble*. We can compute the ensemble average of the instantaneous pressures at time t_1. We can also multiply the instantaneous pressures in each sample at times t_1 and $t_1 + \tau$, and average these results for the ensemble. If such averages do not differ as we choose different values of t_1, then the random process described by the above ensemble is said to be *stationary*.

If the ensemble averages are replaced next by time averages, and if the results computed from each sample are the same as those of any other sample and equal to the ensemble average, then the random process is said to be *ergodic*.

Thus for a stationary ergodic random phenomena its statistical properties are available from a single time function of sufficiently long time period. Although such random phenomena may exist only theoretically, its assumption greatly simplifies the task of dealing with random variables. This chapter will treat only this class of stationary ergodic random functions.

13.2 TIME AVERAGING AND EXPECTED VALUE

In random vibrations we will repeatedly encounter the concept of time averaging over a long period of time. The most common notation for this operation is defined by the following equation in which $x(t)$ is the variable.

$$\overline{x(t)} = \langle x(t) \rangle = \lim_{T \to \infty} \frac{1}{T} \int_0^T x(t)\, dt \qquad (13.2\text{-}1)$$

The above number is also equal to the *expected value* of $x(t)$ which is written as

$$E[x(t)] = \lim_{T \to \infty} \frac{1}{T} \int_0^T x(t)\, dt \qquad (13.2\text{-}2)$$

It is the average or mean value of a quantity sampled over a long time. In the case of discrete variables x_i, the expected value is given by the equation

$$E[x] = \lim_{n \to \infty} \frac{1}{n} \sum_{i=1}^n x_i \qquad (13.2\text{-}3)$$

These average operations can be applied to any variable such as $x^2(t)$ or $x(t) \cdot y(t)$. The *mean square value*, designated by the notation $\overline{x^2}$ or $E[x^2(t)]$, is found by integrating $x^2(t)$ over a time interval T and taking its average value according to the equation

$$E[x^2(t)] = \overline{x^2} = \lim_{T \to \infty} \frac{1}{T} \int_0^T x^2\, dt \qquad (13.2\text{-}4)$$

It is often desirable to consider the time series in terms of the mean and its fluctuation from the mean. A property of importance describing the fluctuation is the *variance* σ^2, which is the mean square value about the mean, given by the equation

$$\sigma^2 = \lim_{T \to \infty} \frac{1}{T} \int_0^T (x - \bar{x})^2\, dt \qquad (13.2\text{-}5)$$

By expanding the above equation, it is easily seen that

$$\sigma^2 = \overline{x^2} - (\bar{x})^2 \qquad (13.2\text{-}6)$$

so that the variance is equal to the mean square value minus the square of the mean. The positive square root of the variance is the *standard deviation* σ.

In the discussions to follow in this chapter, we will often represent a time function by a Fourier series in the exponential form. In Chapter 1 the exponential Fourier series was shown to be

$$x(t) = \sum_{-\infty}^{\infty} c_n e^{in\omega_1 t} = c_0 + \sum_{n=1}^{\infty} \left(c_n e^{in\omega_1 t} + c_n^* e^{-in\omega_1 t} \right) \qquad (13.2\text{-}7)$$

This series, which is a real function, involves a summation over negative and positive frequencies, and it also contains a constant term c_0. The constant term c_0 is the average value of $x(t)$ and since it can be dealt with separately, we will exclude it in future considerations. Moreover, actual measurements are made in terms of positive frequencies, and it would be more desirable to work with the equation

$$x(t) = Re \sum_{n=1}^{\infty} C_n e^{in\omega_1 t} \tag{13.2-8}$$

The one-sided summation in the above equation is complex and hence the real part of the series must be stipulated for $x(t)$ real. Since the real part of a vector is one-half the sum of the vector and its conjugate [see Eq. (1.1-9)],

$$x(t) = Re \sum_{n=1}^{\infty} C_n e^{in\omega_1 t} = \frac{1}{2} \sum_{n=1}^{\infty} (C_n e^{in\omega_1 t} + C_n^* e^{-in\omega_1 t})$$

By comparison with Eq.(1.2-6), we find

$$C_n = 2c_n = \frac{2}{T} \int_{-T/2}^{T/2} x(t) e^{-in\omega_1 t}\, dt$$

$$= a_n - ib_n \tag{13.2-9}$$

EXAMPLE 13.2-1

Determine the mean square value of a record of random vibration $x(t)$ containing many discrete frequencies.

Solution: The record being periodic, we can represent it by the real part of the Fourier series

$$x(t) = Re \sum_{1}^{\infty} C_n e^{in\omega_0 t}$$

$$= \frac{1}{2} \sum_{1}^{\infty} (C_n e^{in\omega_0 t} + C_n^* e^{-in\omega_0 t})$$

where C_n is a complex number and C_n^* is its complex conjugate. [See Eq. (13.2-9).] Its mean square value is

$$\overline{x^2} = \lim_{T \to \infty} \frac{1}{T} \int_0^T \frac{1}{4} \sum_{n=1}^{\infty} (C_n e^{in\omega_0 t} + C_n^* e^{-in\omega_0 t})^2\, dt$$

$$= \lim_{T \to \infty} \sum_{n=1}^{\infty} \frac{1}{4} \left[\frac{C_n^2 e^{i2n\omega_0 t}}{i2n\omega_0 T} + 2C_n C_n^* + \frac{C_n^{*2} e^{-i2n\omega_0 t}}{-i2n\omega_0 T} \right]_0^T$$

$$= \sum_{n=1}^{\infty} \frac{1}{2} C_n C_n^* = \sum_{n=1}^{\infty} \frac{1}{2} |C_n|^2 = \sum_{n=1}^{\infty} \overline{C_n^2}$$

In the above equation, $e^{\pm i2n\omega_0 t}$ for any t, is bounded between ± 1, and due to $T \to \infty$ in the denominator, the first and last terms become zero. The middle term, however, is independent of T. Thus the mean square value of the periodic function is simply the sum of the mean square value of each harmonic component present.

13.3   PROBABILITY DISTRIBUTION

Referring to the random time function of Fig. 13.3-1, what is the probability of its instantaneous value being less than (more negative than) some specified value x_1? To answer this question, we draw a horizontal line at the specified value x_1 and sum the time intervals Δt_i during which $x(t)$ is less than x_1. This sum divided by the total time then represents the fraction of the total time that $x(t)$ is less than x_1, which is the probability that $x(t)$ will be found less than x_1.

$$P(x_1) = \text{Prob.}\left[x(t) < x_1 \right]$$

$$= \lim_{t \to \infty} \frac{1}{t} \sum \Delta t_i$$

(13.3-1)

If a large negative number is chosen for x_1, none of the curve will extend negatively beyond x_1, and hence $P(x_1 \to -\infty) = 0$. As the horizontal line corresponding to x_1 is moved up, more of $x(t)$ will extend negatively beyond x_1, and the fraction of the total time in which $x(t)$ extends below x_1 must increase as shown in Fig. 13.3-2a. As $x \to \infty$, all $x(t)$ will lie in the region less than $x = \infty$, and hence the probability of $x(t)$ being less than $x = \infty$ is certain, or $P(x = \infty) = 1.0$. Thus the curve of Fig. 13.3-2a which is cumulative towards positive x must increase monotonically from zero at $x = -\infty$ to 1.0 at $x = +\infty$. The curve is called the cumulative probability distribution function $P(x)$.

If next we wish to determine the probability of $x(t)$ lying between the values x_1 and $x_1 + \Delta x$, all we need to do is subtract $P(x_1)$ from $P(x_1 + \Delta x)$, which is also proportional to the time occupied by $x(t)$ in the zone x_1 to $x_1 + \Delta x$.

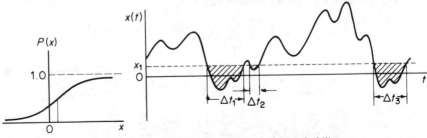

Figure 13.3-1.   Calculation of cumulative probability.

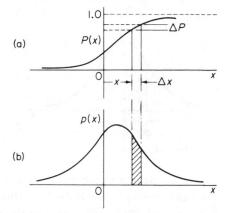

(a)

(b)

Figure 13.3-2. (a) Cumulative probability, (b) Probability density.

We now define the *probability density function p(x)* as

$$p(x) = \lim_{\Delta x \to 0} \frac{P(x + \Delta x) - P(x)}{\Delta x} = \frac{dP(x)}{dx} \qquad (13.3\text{-}2)$$

and it is evident from Fig. 13.3-2b that $p(x)$ is the slope of the cumulative probability distribution $P(x)$. From the above equation we can also write

$$P(x_1) = \int_{-\infty}^{x_1} p(x)\,dx \qquad (13.3\text{-}3)$$

The area under the probability density curve of Fig. 13.3-2b between two values of x represents the probability of the variable being in this interval. Since the probability of $x(t)$ being between $x = \pm \infty$ is certain

$$P(\infty) = \int_{-\infty}^{+\infty} p(x)\,dx = 1.0 \qquad (13.3\text{-}4)$$

and the total area under the $p(x)$ curve must be unity. Figure 13.3-3 again illustrates the probability density $p(x)$ which is the fraction of the time occupied by $x(t)$ in the interval x to $x + dx$.

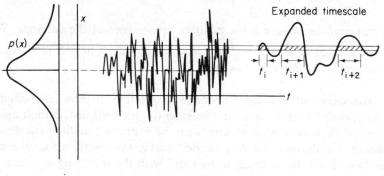

Figure 13.3-3.

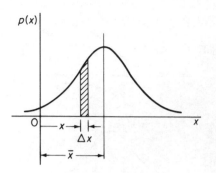

Figure 13.3-4. First and second moments of $p(x)$.

The mean and the mean square value, previously defined in terms of the time average, are related to the probability density function in the following manner. The mean value $\bar{x}$ coincides with the centroid of the area under the probability density curve $p(x)$, as shown in Fig. 13.3-4. It can therefore be determined by the first moment

$$\bar{x} = \int_{-\infty}^{\infty} xp(x)dx \qquad (13.3\text{-}5)$$

Likewise the mean square value is determined from the second moment

$$\overline{x^2} = \int_{-\infty}^{\infty} x^2 p(x)dx \qquad (13.3\text{-}6)$$

which is analogous to the moment of inertia of the area under the probability density curve about $x = 0$.

The *variance* σ^2, previously defined as the mean square value about the mean, is

$$
\begin{aligned}
\sigma^2 &= \int_{-\infty}^{\infty} (x - \bar{x})^2 p(x)dx \\
&= \int_{-\infty}^{\infty} x^2 p(x)dx - 2\bar{x}\int_{-\infty}^{\infty} xp(x)dx + (\bar{x})^2 \int_{-\infty}^{\infty} p(x)dx \\
&= \overline{x^2} \qquad\quad -2(\bar{x})^2 \qquad\qquad +(\bar{x})^2 \\
&= \overline{x^2} \qquad\quad -(\bar{x})^2
\end{aligned}
\qquad (13.3\text{-}7)
$$

The *standard deviation* σ is the positive square root of the variance. When the mean value is zero, $\sigma = \sqrt{\overline{x^2}}$ and the standard deviation is equal to the root mean square (rms) value.

Gaussian and Rayleigh Distribution. Certain distributions which occur frequently in nature are the *Gaussian* (or normal) distribution and the *Rayleigh* distribution, both of which can be expressed mathematically. The Gaussian distribution is a bell shaped curve, symmetric about the mean value (which will be assumed to be zero) with the following equation.

$$p(x) = \frac{1}{\sigma\sqrt{2\pi}} e^{-x^2/2\sigma^2} \qquad (13.3\text{-}8)$$

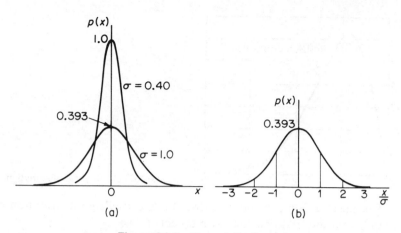

Figure 13.3-5. Normal distribution.

The standard deviation σ is a measure of the spread about the mean value; the smaller the value of σ, the narrower the $p(x)$ curve (remember that the total area $= 1.0$), as shown in Fig. 13.3-5a.

In Fig. 13.3-5b the Gaussian distribution is plotted nondimensionally in terms of x/σ. The probability of $x(t)$ being between $\pm\lambda\sigma$ where λ is any positive number is found from the equation

$$\text{Prob}\big[-\lambda\sigma \le x(t) \le \lambda\sigma\big] = \frac{1}{\sigma\sqrt{2\pi}} \int_{-\lambda\sigma}^{\lambda\sigma} e^{-x^2/2\sigma^2}\, dx \qquad (13.3\text{-}9)$$

The following table presents numerical values associated with $\lambda = 1, 2,$ and 3.

λ	$Prob[-\lambda\sigma \le x(t) \le \lambda\sigma]$	$Prob[|x| > \lambda\sigma]$
1	68.3%	31.7%
2	95.4%	4.6%
3	99.7%	0.3%

The probability of $x(t)$ lying outside $\pm\lambda\sigma$ is the probability of $|x|$ exceeding $\lambda\sigma$, which is 1.0 minus the above values, or the equation

$$\text{Prob}\big[|x| > \lambda\sigma\big] = \frac{2}{\sigma\sqrt{2\pi}} \int_{\lambda\sigma}^{\infty} e^{-x^2/2\sigma^2}\, dx = \text{erfc}\left(\frac{\lambda}{\sqrt{2}}\right) \qquad (13.3\text{-}10)$$

Random variables restricted to positive values, such as the absolute value A of the amplitude often tend to follow the Rayleigh distribution, which is defined by the equation

$$p(A) = \frac{A}{\sigma^2} e^{-A^2/2\sigma^2} \qquad A > 0 \qquad (13.3\text{-}11)$$

The probability density $p(A)$ is zero here for $A < 0$ and has the shape shown in Fig. 13.3-6.

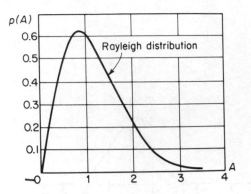

Figure 13.3-6. Rayleigh distribution.

The mean and mean square values for the Rayleigh distribution can be found from the first and second moments to be

$$\bar{A} = \int_0^\infty Ap(A)dA = \int_0^\infty \frac{A^2}{\sigma^2}e^{-A^2/2\sigma^2}\,dA = \sqrt{\frac{\pi}{2}}\,\sigma$$

(13.3-12)

$$\overline{A^2} = \int_0^\infty A^2 p(A)dA = \int_0^\infty \frac{A^3}{\sigma^2}e^{-A^2/2\sigma^2}\,dA = 2\sigma^2$$

The variance associated with the Rayleigh distribution is

$$\sigma_A^2 = \overline{A^2} - (\bar{A})^2 = \left(\frac{4-\pi}{2}\right)\sigma^2$$

(13.3-13)

$$\therefore \sigma_A \cong \frac{2}{3}\sigma$$

Also, the probability of A exceeding a specified value $\lambda\sigma$ is

$$\text{Prob}[A > \lambda\sigma] = \int_{\lambda\sigma}^\infty \frac{A}{\sigma^2}e^{-A^2/2\sigma^2}\,dA$$

(13.3-14)

which has the following numerical values

λ	$P[A > \lambda\sigma]$
0	100%
1	60.7%
2	13.5%
3	1.2%

Three important examples of time records frequently encountered in practice are shown in Fig. 13.3-7 where the mean value is arbitrarily chosen to be zero. The cumulative probability distribution for the sine wave is easily shown to be

$$P(x) = \tfrac{1}{2} + \frac{1}{\pi}\sin^{-1}\frac{x}{A}$$

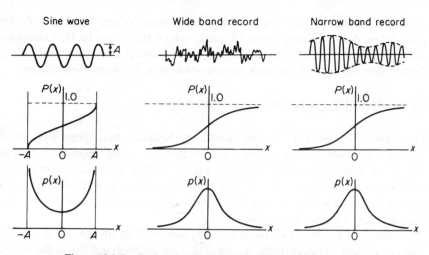

Figure 13.3-7. Probability functions for three types of records.

and its probability density, by differentiation, is

$$p(x) = \frac{1}{\pi\sqrt{A^2 - x^2}} \qquad |x| < A$$
$$= 0 \qquad |x| > A$$

For the wide-band record, the amplitude, phase, and frequency all vary randomly and an analytical expression is not possible for its instantaneous value. Such functions are encountered in radio noise, jet engine pressure fluctuation, atmospheric turbulence, etc., and a most likely probability distribution for such records is the *Gaussian distribution*.

When a wide-band record is put through a narrow-band filter, or a resonance system where the filter bandwidth is small compared to its central frequency f_0, we obtain the third type of wave which is essentially a constant frequency oscillation with slowly varying amplitude and phase. The probability distribution for its instantaneous values is the same as that for the wide band random function. However, the absolute values of its peaks, corresponding to the envelope, will have a Rayleigh distribution.

Another quantity of great interest is the distribution of the peak values. Rice* shows that the distribution of the peak values depends on a quantity $N_0/2M$ where N_0 is the number of zero crossings and $2M$ is the number of positive and negative peaks. For a sine wave or a narrow band, N_0 is equal to $2M$ so that the ratio $N_0/2M = 1$. For a wide-band random record, the number of peaks will greatly exceed the number of zero

*See Ref. 8, at end of chapter.

crossings so that $N_0/2M$ tends to approach zero. When $N_0/2M = 0$, the probability density distribution of peak values turns out to be Gaussian, whereas when $N_0/2M = 1$, as in the narrow band case, the probability density distribution of the peak values tends to a *Rayleigh distribution*.

13.4 CORRELATION

Correlation is a measure of the similarity between two quantities. As it applies to vibration waveforms, correlation is a time-domain analysis useful for detecting hidden periodic signals buried in measurement noise and propagation time through the structure and for determining other information related to its spectral characteristics, which are better discussed under Fourier transforms.

Suppose we have two records, $x_1(t)$ and $x_2(t)$ as shown in Fig. 13.4-1. The *correlation* between them is computed by multiplying the ordinates of the two records at each time t and determining the average value $\langle x_1(t)x_2(t) \rangle$ by dividing the sum of the products by the number of products. It is evident that the correlation so found will be largest when the two records are similar or identical. For dissimilar records, some of the products will be positive and others will be negative, so their sum will be smaller.

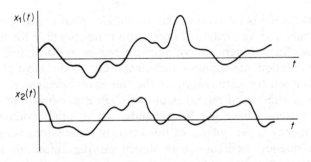

Figure 13.4-1.   Correlation between $x_1(t)$ and $x_2(t)$.

Consider next the case where $x_2(t)$ is identical to $x_1(t)$ but shifted to the left by a time τ as shown in Fig. 13.4-2. Then at time t, when x_1 is $x(t)$, the value of x_2 is $x(t + \tau)$, and the correlation will be given by $\langle x(t)x(t + \tau) \rangle$. Here, if $\tau = 0$, we have complete correlation. As τ increases the correlation will decrease.

It is evident that the above result can be computed from a single record by multiplying the ordinates at times t and $t + \tau$ and determining the average. We then call this result the *autocorrelation* and designate it by

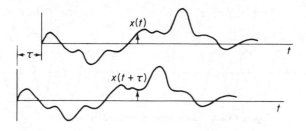

Figure 13.4-2. Function $x(t)$ shifted by τ.

$R(\tau)$. It is also the expected value of the product $x(t)x(t + \tau)$, or

$$R(\tau) = E[x(t)x(t + \tau)] = \langle x(t)x(t + \tau)\rangle$$

$$= \lim_{T \to \infty} \frac{1}{T} \int_{-T/2}^{T/2} x(t)x(t + \tau)\, dt \qquad (13.4\text{-}1)$$

When $\tau = 0$, the above definition reduces to the mean square value

$$R(0) = \overline{x^2} = \sigma^2 \qquad (13.4\text{-}2)$$

Since the second record of Fig. 13.4-2 can be considered to be delayed with respect to the first record, or the first record advanced with respect to the second record, it is evident that $R(\tau) = R(-\tau)$ is symmetric about the origin $\tau = 0$ and is always less than $R(0)$.

Highly random functions, such as the wide-band noise shown in Fig. 13.4-3, soon lose their similarity within a short time shift. Its autocorrelation, therefore, is a sharp spike at $\tau = 0$ that drops off rapidly with $\pm\tau$ as shown. It implies that wide-band random records have little or no correlation except near $\tau = 0$.

For the special case of a periodic wave, the autocorrelation must be periodic of the same period since shifting the wave one period brings the wave into coincidence again. Fig. 13.4-4 shows a sine wave and its autocorrelation.

For the narrow-band record shown in Fig. 13.4-5, the autocorrelation has some of the characteristics found for the sine wave in that it is again an even function with a maximum at $\tau = 0$ and frequency ω_0 corresponding

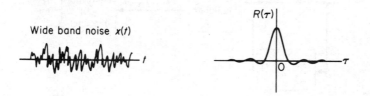

Wide band noise $x(t)$

$R(\tau)$

Figure 13.4-3. Highly random function and its autocorrelation.

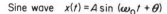

Type of record
Autocorrelation

Sine wave $x(t) = A \sin(\omega_0 t + \theta)$

$R(\tau) = \dfrac{A^2}{2} \cos \omega_0 \tau$

Figure 13.4-4. Sine wave and its autocorrelation.

$R(\tau) = c e^{-k|\tau|} \cos \omega_0 t$

Narrow band response

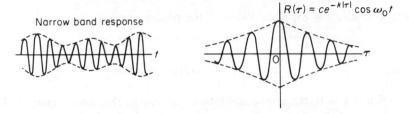

Figure 13.4-5. Autocorrelation for the narrow-band record.

to the dominant or central frequency. The difference appears in the fact that $R(\tau)$ approaches zero for large τ for the narrow band record. It is evident from this discussion that hidden periodicities in a noisy random record can be detected by correlating the record with a sinusoid. There will be almost no correlation between the sinusoid and the noise that will be suppressed. By exploring with sinusoids of differing frequencies the hidden periodic signal can be detected. Figure 13.4-6 shows a block diagram for the determination of the autocorrelation. The signal $x(t)$ is delayed by τ and multiplied, after which it is integrated and averaged. The delay time τ is fixed during each run and is changed in steps or is continuously changed by a slow sweeping technique. If the record is on magnetic tape, the time delay τ can be accomplished by passing the tape between two identical pickup units as shown in Fig. 13.4-7.

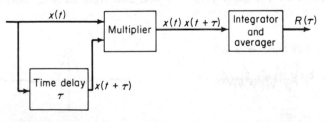

Figure 13.4-6. Block diagram of the autocorrelation analyzer.

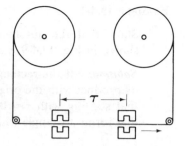

Figure 13.4-7.   Time delay for autocorrelation.

Cross Correlation.   Consider two random quantities $x(t)$ and $y(t)$. The correlation between these two quantities is defined by the equation

$$R_{xy}(\tau) = E[x(t)y(t + \tau)] = \langle x(t)y(t + \tau)\rangle$$

$$= \lim_{T\to\infty} \frac{1}{T} \int_{-T/2}^{T/2} x(t)y(t + \tau)\, dt \qquad (13.4\text{-}3)$$

that can also be called the *cross correlation* between the quantities x and y.

Such quantities often arise in dynamical problems. For example, let $x(t)$ be the deflection at the end of a beam due to a load $F_1(t)$ at some specified point. $y(t)$ is the deflection at the same point, due to a second load $F_2(t)$ at a different point than the first, as illustrated in Fig. 13.4-8. The deflection due to both loads is then $z(t) = x(t) + y(t)$, and the autocorrelation of $z(t)$ as a result of the two loads is

$$R_z(\tau) = \langle z(t)z(t + \tau)\rangle$$

$$= \langle [x(t) + y(t)][x(t + \tau) + y(t + \tau)]\rangle$$

$$= \langle x(t)x(t + \tau)\rangle + \langle x(t)y(t + \tau)\rangle \qquad (13.4\text{-}4)$$

$$+\langle y(t)x(t + \tau)\rangle + \langle y(t)y(t + \tau)\rangle$$

$$= R_x(\tau) + R_{xy}(\tau) + R_{yx}(\tau) + R_y(\tau)$$

Thus the autocorrelation of a deflection at a given point due to separate loads $F_1(t)$ and $F_2(t)$ cannot be determined simply by adding the autocorrelations $R_x(\tau)$ and $R_y(\tau)$ resulting from each load acting separately. $R_{xy}(\tau)$ and $R_{yx}(\tau)$ are here referred to as *cross correlation*, and, in general, they are not equal.

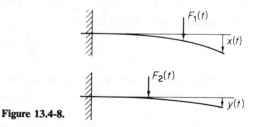

Figure 13.4-8.

EXAMPLE 13.4-1

Show that the autocorrelation of the rectangular gating function shown in Fig. 13.4-9 is a triangle.

Solution: If the rectangular pulse is shifted in either direction by τ, its product with the original pulse is $A^2(T - \tau)$. It is easily seen then that starting with $\tau = 0$, the autocorrelation curve is a straight line that forms a triangle with height A^2 and base equal to $2T$.

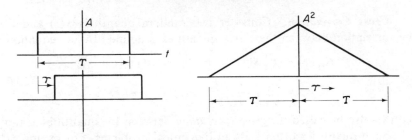

Figure 13.4-9. Autocorrelation of a rectangle is a triangle.

13.5 POWER SPECTRUM AND POWER SPECTRAL DENSITY

The frequency composition of a random function can be described in terms of the spectral density of the mean square value. We found earlier in Example 13.2-1 that the mean square value of a periodic time function is the sum of the mean square value of the individual harmonic component present.

$$\overline{x^2} = \sum_{n=1}^{\infty} \tfrac{1}{2} C_n C_n^*$$

Thus $\overline{x^2}$ is made up of discrete contributions in each frequency interval Δf.

We will first define the contribution to the mean square in the frequency interval Δf as the *power spectrum* $G(f_n)$.

$$G(f_n) = \tfrac{1}{2} C_n C_n^* \tag{13.5-1}$$

The mean square value is then

$$\overline{x^2} = \sum_{n=1}^{\infty} G(f_n) \tag{13.5-2}$$

We will now define the discrete *power spectral density* $S(f_n)$ as power spectrum divided by the frequency interval Δf.

$$S(f_n) = \frac{G(f_n)}{\Delta f} = \frac{C_n C_n^*}{2\Delta f} \tag{13.5-3}$$

The mean square value can then be written as

$$\overline{x^2} = \sum_{n=1}^{\infty} S(f_n)\Delta f \tag{13.5-4}$$

The power spectrum and the power spectral density will hereafter be abbreviated as PS and PSD respectively.

An example of discrete PSD is shown in Fig. 13.5-1. When $x(t)$ contains a very large number of frequency components, the lines of the discrete spectrum become closer together and they more nearly resemble a continuous spectrum, as shown in Fig. 13.5-2. We now define the PSD, $S(f)$, for a continuous spectrum as the limiting case of $S(f_n)$ as $\Delta f \to 0$.

$$\lim_{\Delta f \to 0} S(f_n) = S(f) \tag{13.5-5}$$

The mean square value is then

$$\overline{x^2} = \int_0^{\infty} S(f)df \tag{13.5-6}$$

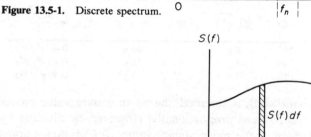

Figure 13.5-1. Discrete spectrum.

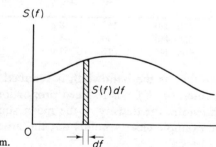

Figure 13.5-2. Continuous spectrum.

To illustrate the meaning of PS and PSD, the following experiment is described. A Xtal accelerometer is attached to a shaker, and its output is amplified, filtered, and read by a rms voltmeter as shown by the block diagram of Fig. 13.5-3. The rms voltmeter should have a long time constant, which corresponds to a long averaging time.

We will excite the shaker by a wide-band random input that is constant over the frequency range 0 Hz to 2000 Hz. If the filter is

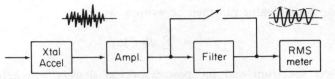

Figure 13.5-3. Measurement of random data.

bypassed, the rms voltmeter will read the rms vibration in the entire frequency spectrum. Assuming an ideal filter which will pass all vibrations of frequencies within the pass band, the output of the filter represents a narrow-band vibration.

We will consider a central frequency of 500 Hz and first set the upper and lower cut-off frequencies at 580 Hz and 420 Hz. The rms meter will now read only the vibration within this 160-Hz band. Let us say that the reading is 8 g. The mean square value is then $G(f_n) = 64$ g^2, and its spectral density is $S(f_n) = 64$ g$^2/160 = 0.40$ g^2/Hz.

We will next reduce the pass band to 40 Hz by setting the upper and lower filter frequencies to 520 Hz and 480 Hz. The mean square value passed by the filter is now one-quarter of the previous value, or 16 g^2 and the rms meter will read 4 g.

Reducing the pass band further to 10 Hz, between 505 Hz and 495 Hz, the rms meter reading becomes 2 g, as shown in the following tabulation.

Frequencies f	Band width Δf	rms *Meter* Reading $\sqrt{\Delta(\overline{x^2})}$	*Filtered* Mean Square $G(f_n) = \Delta(\overline{x^2})$	*Spectral* Density $S(f_n) = \dfrac{\Delta(\overline{x^2})}{\Delta f}$
580–420	160	8 g	64 g^2	0.40 g^2/Hz
520–480	40	4 g	16 g^2	0.40 g^2/Hz
505–495	10	2 g	4 g^2	0.40 g^2/Hz

Note that as the bandwidth is reduced, the mean square value passed by the filter, or $G(f_n)$, is reduced proportionally. However, by dividing by the bandwidth, the density of the mean square value, $S(f_n)$ remains constant. The example clearly points out the advantage of plotting $S(f_n)$ instead of $G(f_n)$.

The PSD can also be expressed in terms of the delta function. As seen from Fig. 13.5-4, the area of a rectangular pulse of height $1/\Delta f$ and width Δf is always unity, and in the limiting case when $\Delta f \to 0$, it becomes a delta function. Thus $S(f)$ becomes

$$S(f) = \lim_{\Delta f \to 0} S(f_n) = \lim_{\Delta f \to 0} \frac{G(f_n)}{\Delta f} = G(f)\delta(f - f_n)$$

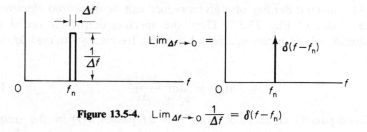

Figure 13.5-4. $\text{Lim}_{\Delta f \to 0} \dfrac{1}{\Delta f} = \delta(f - f_n)$

Typical spectral density functions for two common types of random records are shown in Figs. 13.5-5 and 13.5-6. The first is a wide-band noise-type of record which has a broad spectral density function. The second is a narrow-band random record which is typical of a response of a sharply resonant system to a wide-band input. Its spectral density function is concentrated around the frequency of the instantaneous variation within the envelope.

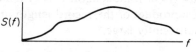

Figure 13.5-5. Wide band record and its spectral density.

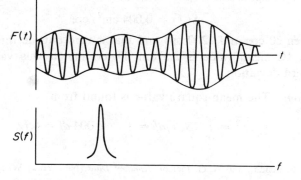

Figure 13.5-6. Narrow band record and its spectral density.

The spectral density of a given record can be measured electronically by the circuit of Fig. 13.5-7. Here the spectral density is noted as the contribution of the mean square value in the frequency interval Δf divided by Δf.

$$S(f) = \lim_{\Delta f \to 0} \frac{\Delta(\overline{x^2})}{\Delta f} \tag{13.5-7}$$

The band-pass filter of pass band $B = \Delta f$ passes $x(t)$ in the frequency interval f to $f + \Delta f$, and the output is squared, averaged, and divided by Δf.

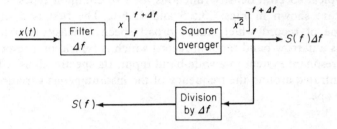

Figure 13.5-7. Power spectral density analyzer.

For high resolution, Δf should be made as narrow as possible; however, the pass band of the filter cannot be reduced indefinitely without losing the reliability of the measurement. Also, a long record is required for the true estimate of the mean square value, but actual records are always of finite length. It is evident now that a parameter of importance is the product of the record length and the band width, $2BT$, which must be sufficiently large.*

EXAMPLE 13.5-1

A random signal has a spectral density that is a constant

$$S(f) = 0.004 \text{ cm}^2/\text{cps}$$

between 20 cps and 1200 cps and that is zero outside this frequency range. Its mean value is 2.0 cm. Determine its rms value and its standard deviation.

Solution: The mean square value is found from

$$\overline{x^2} = \int_0^\infty S(f)\,df = \int_{20}^{1200} 0.004\,df = 4.72$$

*See J. S. Bendat, and A. G. Piersol, *Random Data* (New York: Wiley Interscience, 1971), p. 96.

and the rms value is

$$\text{rms} = \sqrt{\overline{x^2}} = \sqrt{4.72} = 2.17 \text{ cm}$$

The variance σ^2 is defined by Eq. (13.2-6)

$$\sigma^2 = \overline{x^2} - (\bar{x})^2$$

$$= 4.72 - (2)^2 = 0.72$$

and the standard deviation becomes

$$\sigma = \sqrt{0.72} = 0.85 \text{ cm}$$

The problem is graphically displayed by Fig. 13.5-8, which shows the time variation of the signal and its probability distribution.

Figure 13.5-8.

EXAMPLE 13.5-2

Determine the Fourier coefficients C_n and the power spectral density of the periodic function shown in Fig. 13.5-9.

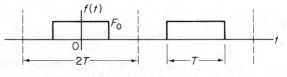

Figure 13.5-9.

Solution: The period is $2T$ and C_n are

$$C_0 = \frac{2}{2T} \int_{-T/2}^{T/2} F_0 d\xi = F_0$$

$$C_n = \frac{2}{2T} \int_{-T/2}^{T/2} F_0 e^{-in\omega_0 \xi} d\xi = F_0 \left(\frac{\sin(n\pi/2)}{n\pi/2} \right)$$

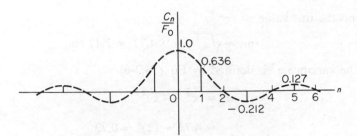

Figure 13.5-10.   Fourier coefficients versus n.

Numerical values of C_n are computed as follows and plotted in Fig. 13.5-10.

n	$\dfrac{n\pi}{2}$	$\sin\dfrac{n\pi}{2}$	$\dfrac{1}{2}C_n$
0	0	0	$\dfrac{F_0}{2} = 1.0\dfrac{F_0}{2}$
1	$\dfrac{\pi}{2}$	1	$\left(\dfrac{2}{\pi}\right)\dfrac{F_0}{2} = 0.636\dfrac{F_0}{2}$
2	π	0	0
3	$3\dfrac{\pi}{2}$	-1	$\left(-\dfrac{2}{3\pi}\right)\dfrac{F_0}{2} = -0.212\dfrac{F_0}{2}$
4	2π	0	0
5	$5\dfrac{\pi}{2}$	1	$\left(\dfrac{2}{5\pi}\right)\dfrac{F_0}{2} = 0.127\dfrac{F_0}{2}$

The mean square value is determined from the equation

$$\overline{x^2} = \lim_{T\to\infty}\frac{1}{2T}\int_{-T}^{T}x^2(t)\,dt = \lim_{T\to\infty}\frac{1}{2T}\int_{T}^{-T}\frac{1}{4}\left\{\sum_n\left(C_n e^{in\omega_0 t}\right.\right.$$
$$\left.\left. + C_n^* e^{-in\omega_0 t}\right)\right\}^2 dt$$

$$= \sum_{n=1}^{\infty}\frac{C_n C_n^*}{2}$$

and since $\overline{x^2} = \int_{\infty}^{0}S_f(\omega)\,d\omega$, the spectral density function can be represented by a series of delta functions as

$$S_f(\omega) = \sum_{n=1}^{\infty}\frac{C_n C_n^*}{2}\delta(\omega - n\omega_0)$$

13.6   FOURIER TRANSFORMS

The discrete frequency spectrum of periodic functions becomes a continuous one when the period T is extended to infinity. Random vibrations are generally not periodic and the determination of its continuous frequency spectrum requires the use of the Fourier integral which can be regarded as a limiting case of the Fourier series as the period approaches infinity.

The Fourier transform has become the underlying operation for the modern time series analysis. In many of the modern instruments for spectral analysis, the calculation performed is that of determining the amplitude and phase of a given record.

The *Fourier integral* is defined by the equation

$$x(t) = \int_{-\infty}^{\infty} X(f)e^{i2\pi ft}\, df \tag{13.6-1}$$

In contrast to the summation of the discrete specturm of sinusoids in the Fourier series, the Fourier integral may be regarded as a summation of the continous spectrum of sinusoids. The quantity $X(f)$ in the above equation is called the *Fourier transform* of $x(t)$, which may be evaluated from the equation

$$X(f) = \int_{-\infty}^{\infty} x(t)e^{-i2\pi ft}\, dt \tag{13.6-2}$$

Like the Fourier coefficient C_n, $X(f)$ is a complex quantity which is a continuous function of f from $-\infty$ to $+\infty$. Equation (13.6-2) resolves the function $x(t)$ into harmonic components $X(f)$, whereas Eq. (13.6-1) synthesizes these harmonic components to the original time function $x(t)$. The two equations above are referred to as the *Fourier transform pair*.

Fourier transform (FT) of basic functions. To demonstrate the spectral character of the FT, we will consider the FT of some basic functions.

EXAMPLE 13.6-1

$$x(t) = Ae^{i2\pi f_n t} \tag{a}$$

From Eq. (13.6-1) we have

$$Ae^{i2\pi f_n t} = \int_{-\infty}^{\infty} X(f)e^{i2\pi ft}df$$

Recognizing the properties of a delta function, the above equation is satisfied if

$$X(f) = A\delta(f - f_n) \tag{b}$$

By substituting into Eq. (13.6-2), we obtain

$$\delta(f - f_n) = \int_{-\infty}^{\infty} e^{-i2\pi(f-f_n)t}\, dt \tag{c}$$

The FT of $x(t)$ is displayed in Fig. 13.6-1, which demonstrates its spectral character.

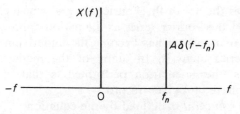

Figure 13.6-1.   FT of $Ae^{i2\pi f_n t}$.

EXAMPLE 13.6-2

$$x(t) = a_n \cos 2\pi f_n t \tag{a}$$

Since

$$\cos 2\pi f_n t = \frac{1}{2}(e^{i2\pi f_n t} + e^{-i2\pi f_n t})$$

the result of Example 13.6-1 immediately gives

$$X(f) = \frac{a_n}{2}\left[\delta(f - f_n) + \delta(f + f_n)\right] \tag{b}$$

Figure 13.6-2 shows that $X(f)$ is a two-sided function of f.

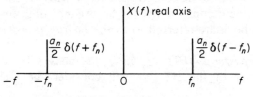

Figure 13.6-2.   FT of $A_n \cos 2\pi f_n t$.

In a similar manner, the FT of $b_n \sin 2\pi f_n t$ is

$$X(f) = -i\frac{b}{2}n\left[\delta(f - f_n) - \delta(f + f_n)\right]$$

which is shown on the imaginary plane of Fig. 13.6-3.

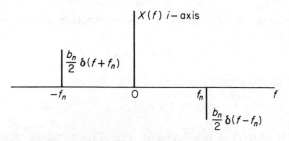

Figure 13.6-3.   FT of $b_n \sin 2\pi f_n t$.

If we put the two FTS together in perpendicular planes, as shown in Fig. 13.6-4, we obtain the complex conjugate coefficients $C_n = a_n - ib_n$ and $C_n^* = a_n + ib_n$. Thus the product

$$\frac{C_n C_n^*}{4} = \frac{1}{4}(a_n^2 + b_n^2) = c_n c_n^*$$

is the square of the magnitude of the Fourier series which is generally plotted at $\pm f$.

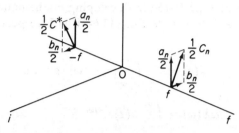

Figure 13.6-4. FT of $(a_n \cos 2\pi f_n t + b_n \sin 2\pi f_n t)$.

EXAMPLE 13.6-3

We will next determine the FT of a rectangular pulse, which is an example of an aperiodic function. (See Fig. 13.6-5.) Its FT is

$$X(f) = \int_{-\infty}^{\infty} x(t) e^{-i2\pi ft} \, dt = \int_{-T/2}^{T/2} A e^{-i2\pi ft} \, dt = AT\left(\frac{\sin \pi fT}{\pi fT}\right)$$

Note that the FT is now a continuous function instead of a discontinuous function. The product XX^* which is a real number is also plotted here. Later it will be shown to be equal to the spectral density function.

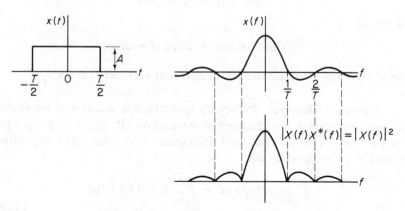

Figure 13.6-5. Rectangular pulse and its spectra.

FT of derivatives. When the FT is expressed in terms of ω instead of f, a factor $1/2\pi$ is introduced in the equation for $x(t)$.

$$x(t) = \frac{1}{2\pi} \int_{-\infty}^{\infty} X(\omega)e^{i\omega t} \, d\omega \qquad (13.6\text{-}3)$$

$$X(\omega) = \int_{-\infty}^{\infty} x(t)e^{-i\omega t} \, dt \qquad (13.6\text{-}4)$$

This form is sometimes preferred in developing mathematical relationships. For example, if we differentiate Eq. (13.6-3) with respect to t, we obtain the FT pair

$$\dot{x}(t) = \frac{1}{2\pi} \int_{-\infty}^{\infty} \left[i\omega X(\omega) \right] e^{i\omega t} \, d\omega$$

$$i\omega X(\omega) = \int_{-\infty}^{\infty} \dot{x}(t)e^{-i\omega t} \, dt$$

Thus the FT of a derivative is simply the FT of the function multiplied by $i\omega$.

$$\text{FT}\left[\dot{x}(t) \right] = i\omega \text{FT}\left[x(t) \right] \qquad (13.6\text{-}5)$$

Differentiating again, we obtain

$$\text{FT}\left[\ddot{x}(t) \right] = -\omega^2 \text{FT}\left[x(t) \right] \qquad (13.6\text{-}6)$$

These equations enable one to conveniently take the FT of differential equations. For example, if we take the FT of the differential equation

$$m\ddot{y} + c\dot{y} + ky = x(t)$$

we obtain

$$(-m\omega^2 + i\omega c + k)Y(\omega) = X(\omega)$$

where $X(\omega)$ and $Y(\omega)$ are the FT of $x(t)$ and $y(t)$ respectively.

Parseval's Theorem. Parseval's theorem is a useful tool for converting time integration into frequency integration. If $X_1(f)$ and $X_2(f)$ are Fourier transforms of real time functions $x_1(t)$ and $x_2(t)$ respectively, Parseval's theorem states that

$$\int_{-\infty}^{\infty} x_1(t)x_2(t) \, dt = \int_{-\infty}^{\infty} X_1(f)X_2^*(f) \, df$$

$$= \int_{-\infty}^{\infty} X_1^*(f)X_2(f) \, df \qquad (13.6\text{-}7)$$

This relationship may be proved using the Fourier transform as follows

$$x_1(t)x_2(t) = x_2(t)\int_{-\infty}^{\infty} X_1(f)e^{i2\pi ft}\,df$$

$$\int_{-\infty}^{\infty} x_1(t)x_2(t)\,dt = \int_{-\infty}^{\infty} x_2(t)\int_{-\infty}^{\infty} X_1(f)e^{i2\pi ft}\,df\,dt$$

$$= \int_{-\infty}^{\infty} X_1(f)\left[\int_{-\infty}^{\infty} x_2(t)e^{i2\pi ft}\,dt\right]df$$

$$= \int_{-\infty}^{\infty} X_1(f)X_2^*(f)\,df$$

All of the previous formulas for the mean square value, autocorrelation, and cross correlation can now be expressed in terms of the Fourier transform by Parseval's theorem.

EXAMPLE 13.6-4

Express the mean square value in terms of the Fourier transform. Letting $x_1(t) = x_2(t) = x(t)$, and averaging over T, which is allowed to go to ∞, we obtain

$$\overline{x^2} = \lim_{T\to\infty}\frac{1}{T}\int_{-T/2}^{T/2} x^2(t)\,dt = \int_{-\infty}^{\infty}\lim_{T\to\infty}\frac{1}{T}X(f)X^*(f)\,df$$

Comparing this with Eq. (13.3-6), we obtain the relationship

$$S(f_\pm) = \lim_{T\to\infty}\frac{1}{T}X(f)X^*(f) \tag{13.6-8}$$

where $S(f_\pm)$ is the spectral density function over positive and negative frequencies.

EXAMPLE 13.6-5

Express the auto correlation in terms of the Fourier transform. We begin with the Fourier transform of $x(t + \tau)$

$$x(t + \tau) = \int_{-\infty}^{\infty} X(f)e^{i2\pi f(t+\tau)}\,df$$

Substituting this into the expression for the autocorrelation, we obtain

$$R(\tau) = \lim_{T\to\infty}\frac{1}{T}\int_{-\infty}^{\infty} x(t)x(t+\tau)\,dt$$

$$= \lim_{T\to\infty}\frac{1}{T}\int_{-\infty}^{\infty} x(t)\int_{-\infty}^{\infty} X(f)e^{i2\pi ft}e^{i2\pi f\tau}\,df\,dt$$

$$= \int_{-\infty}^{\infty}\lim_{T\to\infty}\frac{1}{T}\left\{\int_{-\infty}^{\infty} x(t)e^{i2\pi ft}\,dt\right\}X(f)e^{i2\pi f\tau}\,df$$

$$= \int_{-\infty}^{\infty}\left\{\lim_{T\to\infty}\frac{1}{T}X^*(f)X(f)\right\}e^{i2\pi f\tau}\,df$$

Substituting from Eq. (13.6-8) the above equation becomes

$$R(\tau) = \int_{-\infty}^{\infty} S(f)e^{i2\pi f\tau}\, df \qquad (13.6\text{-}9)$$

The inverse of the above equation is also available from the Fourier transform

$$S(f) = \int_{-\infty}^{\infty} R(\tau)e^{-i2\pi f\tau}\, d\tau \qquad (13.6\text{-}10)$$

Since $R(\tau)$ is symmetric about $\tau = 0$, the last equation can also be written as

$$S(f) = 2\int_{0}^{\infty} R(\tau)\cos 2\pi f\tau\, d\tau \qquad (13.6\text{-}11)$$

These are the *Wiener-Khintchine* equations, and they state that the spectral density function is the FT of the autocorrelation function.

As a parallel to the Wiener-Khintchine equations, we can define the cross correlation between two quantities $x(t)$ and $y(t)$ as

$$R_{xy}(\tau) = \langle x(t)y(t+\tau) \rangle = \lim_{T\to\infty} \frac{1}{T}\int_{-T/2}^{T/2} x(t)y(t+\tau)\, dt$$

$$= \int_{-\infty}^{\infty} \lim_{T\to\infty} \frac{1}{T} X^*(f)Y(f)e^{i2\pi f\tau}\, df \qquad (13.6\text{-}12)$$

$$R_{xy}(\tau) = \int_{-\infty}^{\infty} S_{xy}(f)e^{i2\pi f\tau}\, df$$

where the cross spectral density is defined as

$$S_{xy}(f) = \lim_{T\to\infty} \frac{1}{T} X^*(f)Y(f) \qquad -\infty \le f \le \infty$$

$$= \lim_{T\to\infty} \frac{1}{T} X(f)Y^*(f) \qquad (13.6\text{-}13)$$

$$= S_{xy}^*(f) = S_{xy}(-f)$$

Its inverse from the Fourier transform is

$$S_{xy}(f) = \int_{-\infty}^{\infty} R_{xy}(\tau)e^{-i2\pi f\tau} \qquad (13.6\text{-}14)$$

which is the parallel to Eq. (13.6-10). Unlike the autocorrelation, the cross correlation and the cross spectral density functions are, in general, not even functions; hence, the limits $-\infty$ to $+\infty$ are retained.

EXAMPLE 13.6-6

Using the relationship

$$S(f) = 2 \int_0^\infty R(\tau) \cos 2\pi f \tau \, d\tau$$

and the results of Example 13.4-1,

$$R(\tau) = A^2(T - \tau)$$

find $S(f)$ for the rectangular pulse.

Solution: Since $R(\tau) = 0$ for τ outside $\pm T$, we have

$$S(f) = 2 \int_0^T A^2(T - \tau) \cos 2\pi f \tau \, d\tau$$

$$= 2A^2 T \int_0^T \cos 2\pi f \tau \, d\tau - 2A^2 \int_0^T \tau \cos 2\pi f \tau \, d\tau$$

$$= 2A^2 T \frac{\sin 2\pi f \tau}{2\pi f} \bigg|_0^T - 2A^2 \left[\frac{\cos 2\pi f \tau}{(2\pi f)^2} + \frac{\tau}{2\pi f} \sin 2\pi f \tau \right] \bigg|_0^T$$

$$= \frac{2A^2}{(2\pi f)^2} (1 - \cos 2\pi f T) = A^2 T^2 \left(\frac{\sin \pi f T}{\pi f T} \right)^2$$

Thus the power spectral density of a rectangular pulse using Eq. (13.6-11), is

$$S(f) = A^2 T^2 \left(\frac{\sin \pi f T}{\pi f T} \right)^2$$

Note from Example 13.6-3 that this is also equal to $X(f)X^*(f) = |X(f)|^2$.

13.7   FREQUENCY RESPONSE FUNCTION

In any linear system there is a direct linear relationship between the input and the output. This relationship, which also holds for random functions, is represented by the block diagram of Fig. 13.7-1.

In the time domain the system behavior may be determined in terms of the system impulse response $h(t)$ used in the convolution integral of Eq. (4.2-1).

$$y(t) = \int_0^t x(\xi) h(t - \xi) d\xi \tag{13.7-1}$$

A much simpler relationship is available for the frequency domain in terms

Input $x(t)$ → [System $h(t)$ $H(\omega)$] → Output $y(t)$ **Figure 13.7-1.** Block diagram of a linear
$X(\omega)$ $Y(\omega)$ system.

of the *frequency response function* $H(\omega)$, which is actually the FT of the
impulse response function $h(t)$. We can also define the frequency response
function as the ratio of the output to the input under steady-state conditions, with the input equal to a harmonic time function of unit amplitude.
The transient solution is thus excluded in this consideration.

Applying this definition to a single degree of freedom system,

$$m\ddot{y} + c\dot{y} + ky = x(t)$$

let the input be $x(t) = e^{i\omega t}$. The steady-state output will then be $y = H(\omega)e^{i\omega t}$ where $H(\omega)$ is a complex function. Substituting these into the
differential equation and cancelling $e^{i\omega t}$ from each side, we obtain

$$(-m\omega^2 + ic\omega + k)H(\omega) = 1$$

The frequency response function is then

$$
H(\omega) = \frac{1}{k - m\omega^2 + ic\omega}
$$

$$
= \frac{1}{k} \frac{1}{1 - \left(\dfrac{\omega}{\omega_n}\right)^2 + i2\zeta\left(\dfrac{\omega}{\omega_n}\right)}
$$

(13.7-2)*

Note that $H(\omega)$ is a complex function of (ω/ω_n) and the damping factor ζ
and that it has the dimension of displacement divided by force. It can also
be expressed in terms of its absolute value and phase angle.

$$H(\omega) = |H(\omega)| \underline{/\phi(\omega)}$$

(13.7-3)

as in Eqs. (3.1-3) and (3.1-4) and its variation with frequency plotted as in
Fig. 3.1-3.

EXAMPLE 13.7-1

Show that the frequency response function $H(\omega)$ is the Fourier
transform of the impulse response function $h(t)$.

Solution: From the convolution integral, Eq. (4.2-1), the response

---

*Often the dimensional factor [$1/k$ in Eq. (13.7-2)] is considered together with the
force, leaving the frequency response function a nondimensional quantity

$$
H(\omega) = \frac{1}{1 - \left(\dfrac{\omega}{\omega_n}\right)^2 + i2\zeta\left(\dfrac{\omega}{\omega_n}\right)}
$$

equation in terms of the impulse response function is

$$x(t) = \int_{-\infty}^{t} f(\xi)h(t - \xi)d\xi$$

where the lower limit has been extended to $-\infty$ to account for all past excitations. By letting $\tau = (t - \xi)$, the above integral becomes

$$x(t) = \int_{0}^{\infty} f(t - \tau)h(\tau)d\tau$$

For a harmonic excitation $f(t) = e^{i\omega t}$, the above equation becomes

$$x(t) = \int_{0}^{\infty} e^{i\omega(t-\tau)}h(\tau)d\tau$$

$$= e^{i\omega t} \int_{0}^{\infty} h(\tau)e^{-i\omega\tau} d\tau$$

Since the steady-state output for the input $y(t) = e^{i\omega t}$ is $x = H(\omega)e^{i\omega t}$, the frequency response function is

$$H(\omega) = \int_{0}^{\infty} h(\tau)e^{-i\omega\tau} d\tau = \int_{-\infty}^{\infty} h(\tau)e^{-i\omega\tau} d\tau$$

which is the FT of the impulse response function $h(t)$. The lower limit in the above integral has been changed from 0 to $-\infty$ since $h(t) = 0$ for negative t.

In engineering design, we often need to know the relationship between different points in the system. For example, how much of the roughness of a typical road is transmitted through the suspension system to the body of an automobile? (Here the term transfer function[†] is often used for the frequency response function.) Furthermore, it is often not possible to introduce a harmonic excitation to the input point of the system. It may be necessary to accept measurements $x(t)$ and $y(t)$ at two different points in the system for which the frequency response function is desired. The frequency response function for these points can be obtained by taking the FT of the input and output. The quantity $H(\omega)$ is then available from

$$H(\omega) = \frac{Y(\omega)}{X(\omega)} = \frac{\text{FT of output}}{\text{FT of input}} \tag{13.7-4}$$

where $X(\omega)$ and $Y(\omega)$ are the FT of $x(t)$ and $y(t)$.

[†]Strictly speaking, the term transfer function is the ratio of the Laplace transform of the output to the Laplace transform of the input. In the frequency domain, however, the real part of $s = \alpha + i\omega$ is zero, and the LT becomes the FT.

If we multiply and divide the above equation by the complex conjugate $X^*(\omega)$, the result is

$$H(\omega) = \frac{Y(\omega)X^*(\omega)}{X(\omega)X^*(\omega)} \qquad (13.7\text{-}5)$$

The denominator $X(\omega)X^*(\omega)$ is now a real quantity. The numerator is the cross spectrum $Y(\omega)X^*(\omega)$ between the input and the output and is a complex quantity. The phase of $H(\omega)$ is then found from the real and imaginary parts of the cross spectrum which is simply

$$|Y(\omega)| \, \underline{/\phi_y} \cdot |X^*(\omega)| \, \underline{/\phi_x} = |Y(\omega)X^*(\omega)| \, \underline{/(\phi_y - \phi_x)} \qquad (13.7\text{-}6)$$

Another useful relationship can be found by multiplying $H(\omega)$ by its conjugate $H^*(\omega)$. The result is

$$H(\omega)H^*(\omega) = \frac{Y(\omega)Y^*(\omega)}{X(\omega)X^*(\omega)}$$

or

$$Y(\omega)Y^*(\omega) = |H(\omega)|^2 X(\omega)X^*(\omega) \qquad (13.7\text{-}7)$$

Thus the output power spectrum is equal to the square of the system transfer function multiplied by the input power spectrum. Obviously, each side of the above equation is real and the phase does not enter in.

We wish now to examine the mean square value of the response. From Eq. (13.6-8) the mean square value of the input $x(t)$ is

$$\overline{x^2} = \int_{-\infty}^{\infty} S_x(f_\pm)\,df = \int_{-\infty}^{\infty} \lim_{T\to\infty} \frac{1}{T} X(f)X^*(f)\,df$$

The mean square value of the output $y(t)$ is

$$\overline{y^2} = \int_{-\infty}^{\infty} S_y(f_\pm)\,df = \int_{-\infty}^{\infty} \lim_{T\to\infty} \frac{1}{T} Y(f)Y^*(f)\,df$$

Substituting $YY^* = |H(f)|^2 XX^*$, we obtain

$$\begin{aligned}
\overline{y^2} &= \int_{-\infty}^{\infty} |H(f)|^2 \left[\lim_{T\to\infty} \frac{1}{T} X(f)X^*(f) \right] df \\
&= \int_{-\infty}^{\infty} |H(f)|^2 S_x(f_\pm)\,df
\end{aligned} \qquad (13.7\text{-}8)$$

which is the mean square value of the response in terms of the system response function and the spectral density of the input.

In these expressions $S(f_\pm)$ are the two-sided spectral density functions over both the positive and negative frequencies. Also, $S(f_\pm)$ are even functions. In actual practice, it is desirable to work with spectral densities over only the positive frequencies. Equation (13.7-8) can then be written as

$$\overline{y^2} = \int_0^\infty |H(f)|^2 S_x(f_+)\,df \tag{13.7-9}$$

and since the two expressions must result in the same value for the mean square value, the relationship between the two must be

$$S(f_+) = S(f) = 2S(f_\pm) \tag{13.7-10}$$

Some authors also use the expression

$$\overline{y^2} = \int_0^\infty |H(\omega)|^2 S_x(\omega)\,d\omega \tag{13.7-11}$$

Again, the equations must result in the same mean square value so that

$$2\pi S(\omega) = S(f) \tag{13.7-12}$$

For a single degree of freedom system we have

$$H(f) = \frac{\dfrac{1}{k}}{\left[1 - (f/f_n)^2\right] + i\left[2\zeta(f/f_n)\right]} \tag{13.7-13}$$

If the system is lightly damped, the response function $H(f)$ is peaked steeply at resonance, and the system acts like a narrow-band filter. If the spectral density of the excitation is broad, as in Fig. 13.7-2, the mean square response for the single degree of freedom system can be approximated by the equation

$$\overline{y^2} \cong \frac{f_n}{k^2} S_x(f_n) \frac{\pi}{4\zeta} \tag{13.7-14}$$

where

$$\frac{\pi}{4\zeta} = \int_0^\infty \frac{d\left(\dfrac{f}{f_n}\right)}{\left[1 - \left(\dfrac{f}{f_n}\right)^2\right]^2 + \left(2\zeta \dfrac{f}{f_n}\right)^2}$$

and $S_x(f_n)$ is the spectral density of the excitation at frequency f_n.

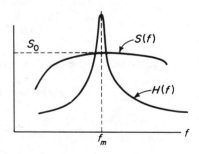

Figure 13.7-2. $S(f)$ and $H(f)$ leading to $\overline{y^2}$ of
Equation 13.7-9.

EXAMPLE 13.7-2

A single degree of freedom system with natural frequency $\omega_n = \sqrt{k/m}$ and damping $\zeta = 0.20$ is excited by the force

$$F(t) = F \cos \frac{1}{2}\omega_n t + F \cos \omega_n t + F \cos \frac{3}{2}\omega_n t$$

$$= \sum_{m=1/2, \, 1, \, 3/2} F \cos m\omega_n t$$

Determine the mean square response and compare the output spectrum with that of the input.

Solution: The response of the system is simply the sum of the response of the single degree of freedom system to each of the harmonic components of the exciting force.

$$x(t) = \sum_{m=1/2, \, 1, \, 3/2} |H(m\omega)| F \cos(m\omega_n t - \phi_m)$$

where

$$\left|H\left(\tfrac{1}{2}\omega_n\right)\right| = \frac{\dfrac{1}{k}}{\sqrt{\tfrac{9}{16} + (0.20)^2}} = \frac{1.29}{k}$$

$$\left|H(\omega_n)\right| = \frac{\dfrac{1}{k}}{\sqrt{4(0.20)^2}} = \frac{2.50}{k}$$

$$\left|H\left(\tfrac{3}{2}\omega_n\right)\right| = \frac{\dfrac{1}{k}}{\sqrt{\tfrac{25}{16} + 9(0.20)^2}} = \frac{0.72}{k}$$

$$\phi_{1/2} = \tan^{-1} \frac{4\zeta}{3} = 0.083\pi$$

$$\phi_1 = \tan^{-1} \infty = 0.50\pi$$

$$\phi_{3/2} = \tan^{-1} \frac{-12\zeta}{5} = -0.142\pi$$

Substituting these values into $x(t)$, we obtain the equation

$$x(t) = \frac{F}{k}\left[1.29 \cos(0.5\omega_n - 0.083\pi)\right.$$

$$+ 2.50 \cos(\omega_n t - 0.50\pi)$$

$$\left. + 0.72 \cos(1.5\omega_n t + 0.142\pi)\right]$$

The mean square response is then

$$\overline{x^2} = \frac{F^2}{2k^2}\left[(1.29)^2 + (2.50)^2 + (0.72)^2\right]$$

Figure 13.7-3 shows the input and output spectra for the problem. The components of the mean square input are the same for each frequency and equal to $F^2/2$. The output spectrum is modified by the system frequency response function.

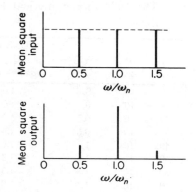

Figure 13.7-3. Input and output spectra with discrete frequencies.

EXAMPLE 13.7-3

The response of any structure to a single point random excitation can be computed by a simple numerical procedure, provided the spectral density of the excitation and the frequency response curve of the structure are known. For example, consider the structure of Fig. 13.7-4a whose base is subjected to a random acceleration input with the power spectral density function shown in Fig. 13.7-4b. It is desired to compute the response of the point p and establish the probability of exceeding any specified acceleration.

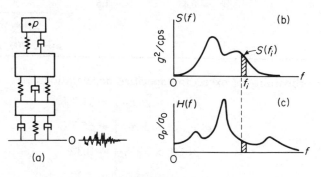

Figure 13.7-4.

The frequency response function $H(f)$ for the point p may be obtained experimentally by applying to the base a variable frequency sinusoidal shaker with a constant acceleration input a_0, and measuring the acceleration response at p. Dividing the measured acceleration by a_0, $H(f)$ may appear as in Fig. 13.7-4c.

The mean-square response $\overline{a_p^2}$ at p is calculated numerically from the equation

$$\overline{a_p^2} = \sum_i S(f_i)|H(f_i)|^2 \Delta f_i$$

The following numerical table illustrates the computational procedure.

NUMERICAL EXAMPLE

f cps	Δf cps	$S(f_i)$ g^2/cps	$\|H(f_i)\|$ Nondimensional	$\|H(f_i)\|^2\Delta f$ cps	$S(f_i)\|H(f_i)\|^2\Delta f$ g^2 units
0	10	0.	1.0	10.	0.
10	10	0.	1.0	10.	0.
20	10	0.2	1.1	12.1	2.4
30	10	0.6	1.4	19.6	11.8
40	10	1.2	2.0	40.	48.0
50	10	1.8	1.3	16.9	30.5
60	10	1.8	1.3	16.9	30.5
70	10	1.1	2.0	40.	44.0
80	10	0.9	3.7	137.	123.
90	10	1.1	5.4	291.	320.
100	10	1.2	2.2	48.4	57.7
110	10	1.1	1.3	16.9	18.6
120	10	0.8	0.8	6.4	5.1
130	10	0.6	0.6	3.6	2.2
140	10	0.3	0.5	2.5	0.8
150	10	0.2	0.6	3.6	0.7
160	10	0.2	0.7	4.9	0.1
170	10	0.1	1.3	16.9	1.7
180	10	0.1	1.1	12.1	1.2
190	10	0.5	0.7	4.9	2.3
200	10	0.	0.5	2.5	0.
210	10	0.	0.4	1.6	0.

$$\overline{a^2} = 700.6g^2$$
$$\sigma = \sqrt{700.6g^2} = 26.6g$$

The probability of exceeding specified accelerations are

$$p[\,|a| > 26.6g\,] = 31.7\%$$
$$p[\,a_{\text{peak}} > 26.6g\,] = 60.7\%$$
$$p[\,|a| > 79.8g\,] = 0.3\%$$
$$p[\,a_{\text{peak}} > 79.8g\,] = 1.2\%$$

REFERENCES

[1] BENDAT, J. S. *Principles and Applications of Random Noise Theory*. New York: John Wiley & Sons, Inc., 1958.

[2] BENDAT, J. S., and PIERSOL, A. G. *Measurement and Analysis of Random Data*. New York: John Wiley & Sons, Inc., 1966.

[3] BLACKMAN, R. B., and TUKEY, J. W. *The Measurement of Power Spectra*. New York: Dover Publications, Inc., 1958.

[4] CLARKSON, B. L. "The Effect of Jet Noise on Aircraft Structures." *Aeronautical Quarterly*, Vol. 10, Part 2, May 1959.

[5] CRAMER, H. *The Elements of Probability Theory*. New York: John Wiley & Sons, Inc., 1955.

[6] CRANDALL, S. H. *Random Vibration*. Cambridge, Mass.: The Technology Press of M.I.T., 1948.

[7] CRANDALL, S. H. *Random Vibration*, Vol. 2. Cambridge, Mass.: The Technology Press of M.I.T., 1963.

[8] RICE, S. O. *Mathematical Analysis of Random Noise*. New York: Dover Publications, Inc., 1954.

[9] ROBSON, J. D. *Random Vibration*. Edinburgh University Press 1964.

[10] THOMSON, W. T., and BARTON, M. V. "The Response of Mechanical Systems to Random Excitation." *J. Appld. Mech.*, June 1957, pp. 248–51.

PROBLEMS

13-1 Give examples of random data and indicate classifications for each example.

13-2 Discuss the differences between nonstationary, stationary, and ergodic data.

13-3 Discuss what we mean by the expected value. What is the expected number of heads when 8 coins are thrown 100 times; 1000 times? What is the probability for tails?

13-4 Throw a coin 50 times, recording 1 for head and 0 for tail. Determine the probability of obtaining heads by dividing the cumulative heads by the number of throws and plot this number as a function of the number of throws. The curve should approach 0.5.

13-5 For the series of triangular waves shown in Fig. P13-5, determine the mean and the mean square values.

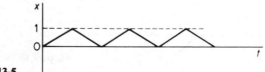

Figure P13-5.

13-6   A sine wave with a steady component has the equation

$$x = A_0 + A_1 \sin \omega t$$

Determine the expected values $E(x)$ and $E(x^2)$.

13-7   Determine the mean and mean square values for the rectified sine wave.

13-8   Discuss why the probability distribution of the peak values of a random function should follow the Rayleigh distribution or one similar in shape to it.

13-9   Show that for the Gaussian probability distribution $p(x)$ the central moments are given by

$$E(x^n) = \int_{-\infty}^{\infty} x^n p(x)\,dx = \begin{cases} 0 \text{ for } n \text{ odd} \\ 1 \cdot 3 \cdot 5 \cdots (n-1)\sigma^n \text{ for } n \text{ even} \end{cases}$$

13-10   Derive the equations for the cumulative probability and the probability density functions of the sine wave. Plot these results.

13-11   What would the cumulative probability and the probability density curves look like for the rectangular wave shown in Fig. P13-11.

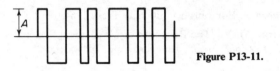

Figure P13-11.

13-12   Determine the autocorrelation of a cosine wave $x(t) = A \cos t$, and plot it against τ.

13-13   Determine the autocorrelation of the rectangular wave shown in Fig. P13-26.

13-14   Determine the autocorrelation of the rectangular pulse and plot it against τ.

13-15   Determine the autocorrelation of the binary sequence shown in Fig. P13-15. Suggestion: Trace the above wave on transparent graph paper and shift it through τ.

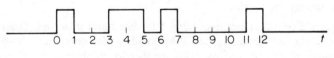

Figure P13-15.

13-16   Determine the autocorrelation of the triangular wave shown in Fig. P13-16.

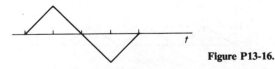

Figure P13-16.

13-17   Figure P13-17 shows the acceleration spectral density plot of a random vibration. Approximate the area by a rectangle and determine the rms value in m/sec ^{2}.

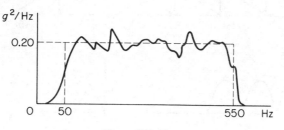

Figure P13-17.

13-18   Determine the rms value of the spectral density plot shown in Fig. P13-18.

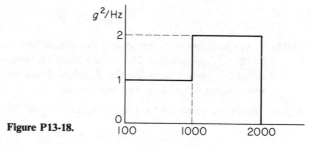

Figure P13-18.

13-19   The power spectral density plot of a random vibration is shown in Fig. P13-19. The slopes represent a 6-db/octave. Replot the result on a linear scale and estimate the rms value.

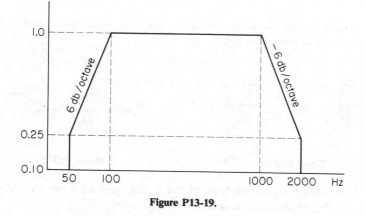

Figure P13-19.

13-20   Determine the spectral density function for the waves in Fig. P13-20.

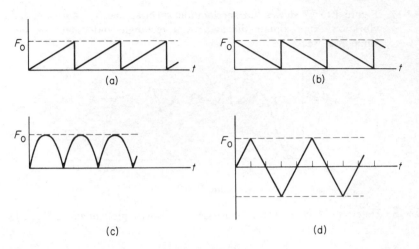

Figure P13-20.

13-21 A random signal is found to have a constant spectral density of $S(f) = 0.002$ in.^{2}/cps between 20 cps and 2000 cps. Outside this range, the spectral density is zero. Determine the standard deviation and the rms value if the mean value is 1.732 in. Plot this result.

13-22 Derive the equation for the coefficients C_n of the periodic function

$$f(t) = Re \sum_{n=0}^{\infty} C_n e^{in\omega_0 t}$$

13-23 Show that for Prob. 13-22, $C_{-n} = C_n^*$, and that $f(t)$ can be written as

$$f(t) = \sum_{n=-\infty}^{\infty} C_n e^{in\omega_0 t}$$

13-24 Determine the Fourier series for the saw tooth wave shown in Fig. P13-24 and plot its spectral density.

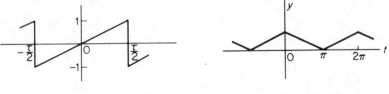

Figure P13-24.   **Figure P13-25.**

13-25 Determine the complex form of the Fourier series for the wave shown in Fig. P13-25, and plot its spectral density.

13-26 Determine the complex form of the Fourier series for the rectangular wave shown in Fig. P13-26, and plot its spectral density.

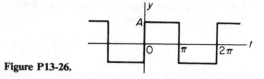

Figure P13-26.

13-27 The sharpness of the frequency response curve near resonance is often expressed in terms of $Q = \frac{1}{2}\zeta$. Points on either side of resonance where the response falls to a value $1/\sqrt{2}$ are called half-power points. Determine the respective frequencies of the half-power points in terms of ω_n and Q.

13-28 Show that

$$\int_0^\infty \frac{d\eta}{[1 - \eta^2]^2 + [2\zeta\eta]^2} = \frac{\pi}{4\zeta} \text{ for } \zeta \ll 1$$

13-29 The differential equation of a system with structural damping is given as

$$m\ddot{x} + k(1 + i\gamma)x = F(t)$$

Determine the frequency response function.

13-30 A single degree of freedom system with natural frequency ω_n and damping factor $\zeta = 0.10$ is excited by the force

$$F(t) = F\cos(0.5\omega_n t - \theta_1) + F\cos(\omega_n t - \theta_2) + F\cos(2\omega_n t - \theta_3)$$

Show that the mean square response is

$$\overline{y^2} = (1.74 + 25.0 + 0.110)\frac{1}{2}\left(\frac{F}{k}\right)^2 = 13.43\left(\frac{F}{k}\right)^2$$

13-31 In Example 13.7-3 what is the probability of the instantaneous acceleration exceeding a value $53.2g$? Of the peak value exceeding this value?

13-32 A large hydraulic press stamping out metal parts is operating under a series of forces approximated by Fig. P13-32. The mass of the press on its foundation is 40 kg and its natural frequency is 2.20 Hz. Determine the Fourier spectrum of the excitation and the mean square value of the response.

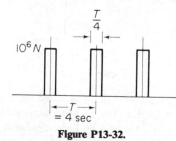

Figure P13-32.

13-33   For a single degree of freedom system, the substitution of Eq. (13.7-13) into
Eq. (13.7-9) results in

$$\overline{y^2} = \int_0^\infty S_x(f_+)\frac{1}{k^2}\frac{df}{\left[1-\left(\frac{f}{f_n}\right)^2\right]^2 + \left(2\zeta\frac{f}{f_n}\right)^2}$$

where $S_x(f_+)$ is the spectral density of the excitation force. When the
damping ζ is small and the variation of $S_x(f_+)$ is gradual, the above
equation becomes

$$\overline{y^2} \cong S_x(f_n)\frac{f_n}{k^2}\int_0^\infty \frac{d\left(\frac{f}{f_n}\right)}{\left[1-\left(\frac{f}{f_n}\right)^2\right]^2 + \left(2\zeta\frac{f}{f_n}\right)^2} = S_x(f_n)\frac{f_n}{k^2}\frac{\pi}{4\zeta}$$

which is Eq. (13.7-14). Derive a similar equation for the mean square value
of the relative motion z of a single degree of freedom system excited by the
base motion, in terms of the spectral density $S_y(f_+)$ of the base acceleration. (See Sec. 3.5.) If the spectral density of the base acceleration is
constant over a given frequency range, what must be the expression for $\overline{z^2}$.

13-34   Referring to Sec. 3.5 we can write the equation for the absolute acceleration
of the mass undergoing base excitation as

$$\ddot{x} = \frac{k + i\omega c}{k - m\omega^2 + i\omega c}\cdot\ddot{y}$$

Determine the equation for the mean square acceleration $\overline{\ddot{x}^2}$. Establish a
numerical integration technique for the computer evaluation of $\overline{\ddot{x}^2}$.

13-35   A radar dish with a mass of 60 kg is subject to wind loads with the spectral
density shown in Fig. P13-35. The dish-support system has a natural
frequency of 4 Hz. Determine the mean square response and the probability of the dish exceeding a vibration amplitude of 0.132 m. Assume $\zeta = .05$.

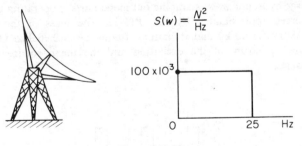

Figure P13-35.

13-36   A jet engine with a mass of 272 kg is tested on a stand which results in a
natural frequency of 26. Hz. The spectral density of the jet force under test
is shown in Fig. P13-36. Determine the probability of the vibration amplitude in the axial direction of the jet thrust exceeding .012 m. Assume
$\zeta = 0.10$.

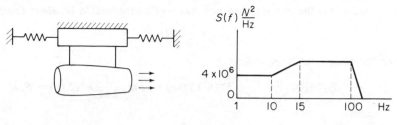

Figure P13-36.

13-37 An SDF system with viscous damping $\zeta = 0.03$ is excited by white noise excitation $F(t)$ having a constant power spectral density of 5×10^6 N ^{2}/Hz. The system has a natural frequency of $\omega_n = 30$ rad/sec and a mass of 1500 kg. Determine σ. Assuming Rayleigh distribution for peaks, determine the probability that the maximum peak response will exceed 0.037 meter.

13-38 Starting with the relationship

$$x(t) = \int_0^\infty f(t - \xi)h(\xi)d\xi$$

and using the Fourier transform technique, show that

$$X(i\omega) = F(i\omega)H(i\omega)$$

and

$$\overline{x^2} = \int_0^\infty S_F(\omega)|H(i\omega)|^2 \, d\omega$$

where

$$S_F(\omega) = \lim_{T\to\infty} \frac{1}{2\pi T} F(i\omega)F^*(i\omega)$$

13-39 Starting with the relationship

$$H(i\omega) = |H(i\omega)|e^{i\phi(\omega)}$$

show that

$$\frac{H(i\omega)}{H^*(i\omega)} = e^{i2\phi(\omega)}$$

13-40 Find the frequency spectrum of the rectangular pulse shown in Fig. P13-40.

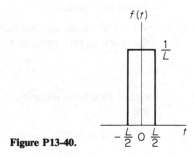

Figure P13-40.

13-41 Show that the unit step function has no Fourier transform. *Hint*: Examine

$$\int_{-\infty}^{\infty} |f(t)|\, dt$$

13-42 Starting with the equations

$$S_{FX}(\omega) = \lim_{T\to\infty} \frac{1}{2\pi T} F^*(i\omega)X(i\omega) = \lim_{T\to\infty} \frac{1}{2\pi T} F^*(FH) = S_F H$$

and

$$S_{XF}(\omega) = \lim_{T\to\infty} \frac{1}{2\pi T} X^* F = \lim_{T\to\infty} \frac{1}{2\pi T}(F^* H^*)F = S_F H^*$$

show that

$$\frac{S_{FX}(\omega)}{S_{XF}(\omega)} = e^{i2\phi(\omega)}$$

and

$$\frac{S_F(\omega)}{S_{XF}(\omega)} = \frac{S_{FX}(\omega)}{S_F(\omega)} = H(i\omega)$$

13-43 The differential equation for the longitudinal motion of a uniform slender rod is

$$\frac{\partial^2 u}{\partial t^2} = c^2 \frac{\partial^2 u}{\partial x^2}$$

Show that for an arbitrary axial force at the end $x = 0$, with the other end $x = l$ free, the Laplace transform of the response is

$$\bar{u}(x, s) = \frac{-c\bar{F}(s)e^{-s(l/c)}}{sAE(1 - e^{-2s(l/c)})} \left\{ e^{(s/c)(x-l)} + e^{-(s/c)(x-l)} \right\}$$

13-44 If the force in Prob. 13-43 is harmonic and equal to $F(t) = F_0 e^{i\omega t}$, show that

$$u(x, t) = \frac{cF_0 e^{i\omega t} \cos[(\omega l/c)(x/l - 1)]}{\omega AE \sin(\omega l/c)}$$

and

$$\sigma(x, t) = \frac{-\sin[(\omega l/c)(x/l - 1)]}{\sin(\omega l/c)} \frac{F_0}{A} e^{i\omega t}$$

where σ is the stress.

13-45 With $S(\omega)$ as the spectral density of the excitation stress at $x = 0$, show that the mean square stress in Prob. 13-43 is

$$\overline{\sigma^2} \cong \frac{2\pi}{\gamma} \sum_n \frac{c}{n\pi l} S(\omega_n) \sin^2 n\pi \frac{x}{l}$$

where structural damping is assumed. The normal modes of the problem are

$$\varphi_n(x) = \sqrt{2}\ \cos n\pi(x/l - 1), \quad \omega_n = n\pi(c/l), \quad c = \sqrt{AE/m}.$$

13-46 Determine the FT of $x(t - t_0)$ and show that it is equal to $e^{-i2\pi f t_0}X(f)$ where $X(f) = \text{FT}[x(t)]$.

13-47 Prove that the FT of a convolution is the product of the separate FT.

$$\text{FT}[x(t)*y(t)] = X(f)Y(f)$$

13-48 Using the derivative theorem, show that the FT of the derivative of a rectangular pulse is a sine wave.

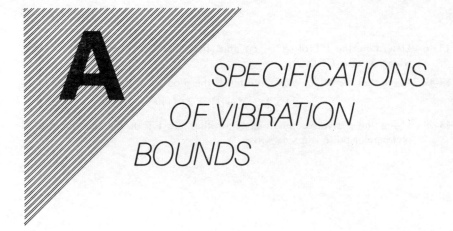

A SPECIFICATIONS OF VIBRATION BOUNDS

Specifications for vibrations are often based on harmonic motion.

$$x = x_0 \sin \omega t$$

The velocity and acceleration are then available from differentiation and the following relationships for the peak values can be written.

$$\dot{x}_0 = 2\pi f x_0$$

$$\ddot{x}_0 = -4\pi^2 f^2 x_0 = -2\pi f \dot{x}_0$$

These equations can be represented on the log-log paper by rewriting them in the form

$$\ln \dot{x}_0 = \ln x_0 + \ln 2\pi f$$

$$\ln \dot{x}_0 = -\ln \ddot{x}_0 - \ln 2\pi f$$

By letting x_0 = constant, the plot of $\ln \dot{x}_0$ against $\ln 2\pi f$ is a straight line of slope equal to $+1$. By letting $\ddot{x}_0$ = constant, the plot of $\ln \dot{x}_0$ versus $\ln 2\pi f$ is again a straight line of slope -1. These lines are shown graphically in Fig. A-1. The graph is often used to specify bounds for the vibration. Shown in heavy lines are bounds for a maximum acceleration of 10 g, minimum and maximum frequencies of 5 and 500 c.p.s., and an upper limit for the displacement of 0.30 inch.

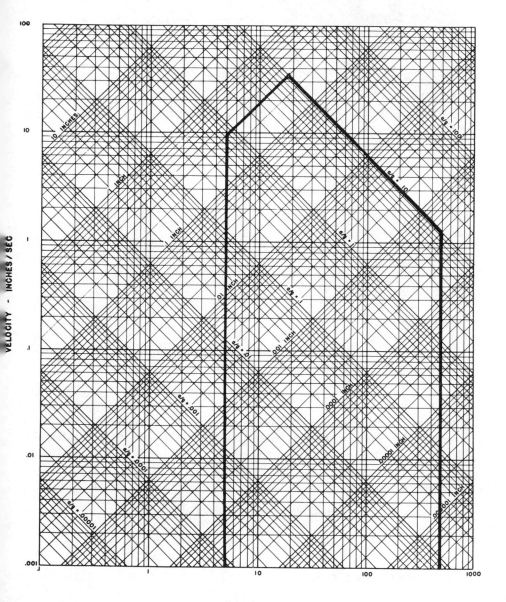

FREQUENCY - CPS

Figure A-1.

445

INTRODUCTION TO LAPLACE TRANSFORMATION

B

Definition

If $f(t)$ is a known function of t for values of $t > 0$, its Laplace transform $\bar{f}(s)$ is defined by the equation

$$\bar{f}(s) = \int_0^\infty e^{-st} f(t)\, dt = \mathcal{L}f(t) \tag{1}$$

where s is a complex variable. The integral exists for the real part of $s > 0$ provided $f(t)$ is an absolutely integrable function of t in the time interval 0 to ∞.

EXAMPLE 1

Let $f(t)$ be a constant c for $t > 0$. Its L.T. is

$$\mathcal{L}c = \int_0^\infty ce^{-st}\, dt = -\left. \frac{ce^{-st}}{s} \right|_0^\infty = \frac{c}{s}$$

which exists for $R(s) > 0$.

EXAMPLE 2.

Let $f(t) = t$. Its L.T. is found by integration by parts, letting

$$u = t \qquad\qquad du = dt$$

$$dv = e^{-st}\, dt \qquad v = -\frac{e^{-st}}{s}$$

The result is

$$\mathcal{L}t = -\left. \frac{te^{-st}}{s} \right]_0^\infty + \frac{1}{s} \int_0^\infty e^{-st} \, dt = \frac{1}{s^2} \qquad R(s) > 0$$

Laplace Transform of Derivatives. If $\mathcal{L}f(t) = \bar{f}(s)$ exists, where $f(t)$ is continuous, then $f(t)$ tends to $f(0)$ as $t \to 0$ and the L.T. of its derivative $f'(t) = df(t)/dt$ is equal to

$$\mathcal{L}f'(t) = s\bar{f}(s) - f(0) \qquad (2)$$

The above relation is found by integration by parts

$$\int_0^\infty e^{-st} f'(t) \, dt = \left. e^{-st} f(t) \right]_0^\infty + \frac{1}{s} \int_0^s e^{-st} f(t) \, dt$$

$$= -f(0) + s\bar{f}(s)$$

Similarly the L.T. of the second derivative can be shown to be

$$\mathcal{L}f''(t) = s^2 \bar{f}(s) - sf(0) - f'(0) \qquad (3)$$

Shifting Theorem

Consider the L.T. of the function $e^{at}x(t)$.

$$\mathcal{L}e^{at}x(t) = \int_0^\infty e^{-st} \left[e^{at}x(t) \right] dt = \int_0^\infty e^{-(s-a)t}x(t) \, dt$$

We conclude from this expression that

$$\mathcal{L}e^{at}x(t) = \bar{x}(s - a) \qquad (4)$$

where $\mathcal{L}x(t) = \bar{x}(s)$. Thus, the multiplication of $x(t)$ by e^{at} shifts the transform by a, where a may be any number, real or complex.

Transformation of Ordinary Differential Equations

Consider the differential equation

$$m\ddot{x} + c\dot{x} + kx = F(t) \qquad (5)$$

Its L.T. is

$$m\left[s^2\bar{x}(s) - sx(0) - \dot{x}(0) \right] + c\left[s\bar{x}(s) - x(0) \right] + k\bar{x}(s) = \bar{F}(s)$$

which can be rearranged to

$$\bar{x}(s) = \frac{\bar{F}(s)}{ms^2 + cs + k} + \frac{(ms + c)x(0) + m\dot{x}(0)}{ms^2 + cs + k} \qquad (6)$$

The above equation is called the subsidiary equation of the differential equation. The response $x(t)$ is found from the inverse transformation, the first term representing the forced response and the second term the response due to the initial conditions.

For the more general case, the subsidiary equation can be written in the form

$$\bar{x}(s) = \frac{A(s)}{B(s)} \tag{7}$$

where $A(s)$ and $B(s)$ are polynomials. $B(s)$ is in general of higher order than $A(s)$.

Transforms Having Simple Poles

Considering the subsidiary equation

$$\bar{x}(s) = \frac{A(s)}{B(s)}$$

we examine the case where $B(s)$ is factorable in terms of n roots a_k which are distinct (simple poles).

$$B(s) = (s - a_1)(s - a_2) \cdots (s - a_n)$$

The subsidiary equation can then be expanded in the following partial fractions

$$\bar{x}(s) = \frac{A(s)}{B(s)} = \frac{C_1}{s - a_1} + \frac{C_2}{s - a_2} + \cdots + \frac{C_n}{s - a_n} \tag{8}$$

To determine the constants C_k, we multiply both sides of the above equation by $(s - a_k)$ and let $s = a_k$. Every term on the right will then be zero except C_k and we arrive at the result

$$C_k = \lim_{s \to a_k} (s - a_k) \frac{A(s)}{B(s)} \tag{9}$$

Since $\mathcal{L}^{-1} C_k / (s - a_k) = C_k e^{a_k t}$ the inverse transform of $\bar{x}(s)$ becomes

$$x(t) = \sum_{k=1}^{n} \lim_{s \to a_k} (s - a_k) \frac{A(s)}{B(s)} e^{a_k t} \tag{10}$$

Another expression for the above equation becomes apparent by noting that

$$B(s) = (s - a_k) B_1(s)$$
$$B'(s) = (s - a_k) B_1'(s) + B_1(s)$$
$$\lim_{s \to a_k} B'(s) = B_1(a_k)$$

Since $(s - a_k) A(s) / B(s) = A(s) / B_1(s)$, it is evident that

$$x(t) = \sum_{k=1}^{n} \frac{A(a_k)}{B'(a_k)} e^{a_k t} \tag{11}$$

Transforms Having Poles of Higher Order

If in the subsidiary equation

$$\bar{x}(s) = \frac{A(s)}{B(s)}$$

a factor in $B(s)$ is repeated m times, we say that $\bar{x}(s)$ has an mth order pole. Assuming that there is an mth order pole at a_1, $B(s)$ will have the form

$$B(s) = (s - a_1)^m(s - a_2)(s - a_3) \cdots$$

The partial fraction expansion of $\bar{x}(s)$ then becomes

$$
\begin{aligned}
\bar{x}(s) = \frac{C_{11}}{(s - a_1)^m} &+ \frac{C_{12}}{(s - a_1)^{m-1}} + \cdots \\
&+ \frac{C_{1m}}{(s - a_1)} + \frac{C_2}{(s - a_2)} + \frac{C_3}{(s - a_3)} + \cdots
\end{aligned}
\tag{12}
$$

The coefficient C_{11} is determined by multiplying both sides of the equation by $(s - a_1)^m$ and letting $s = a_1$

$$
\begin{aligned}
(s - a_1)^m \bar{x}(s) = C_{11} &+ (s - a_1)C_{12} + \cdots \\
&+ (s - a_1)^{m-1}C_{1m} + \frac{(s - a_1)^m}{(s - a_2)}C_2 + \cdots
\end{aligned}
$$

$$\therefore C_{11} = \left[(s - a_1)^m \bar{x}(s)\right]_{s=a_1} \tag{13}$$

The coefficient C_{12} is determined by differentiating the equation for $(s - a_1)^m \bar{x}(s)$ with respect to s and then letting $s = a_1$

$$C_{12} = \left[\frac{d}{ds}(s - a_1)^m \bar{x}(s)\right]_{s=a_1} \tag{14}$$

It is evident then that

$$C_{1n} = \frac{1}{(n-1)!}\left[\frac{d^{n-1}}{ds^{n-1}}(s - a_1)^m \bar{x}(s)\right]_{s=a_1} \tag{15}$$

The remaining coefficients C_2, C_3, etc., are evaluated as in the previous section for simple poles.

Since by the shifting theorem

$$\mathcal{L}^{-1}\frac{1}{(s - a_1)^n} = \frac{t^{n-1}}{(n-1)!}a_1t$$

the inverse transform of $\bar{x}(s)$ becomes

$$
\begin{aligned}
x(t) = \left[C_{11}\frac{t^{m-1}}{(m-1)!} + C_{12}\frac{t^{m-2}}{(m-2)!} + \cdots\right]e^{a_1t} \\
+ C_2 e^{a_2t} + C_3 e^{a_3t} + \cdots
\end{aligned}
\tag{16}
$$

Most ordinary differential equations can be solved by the elementary theory of L.T. The tables here give the L.T. of simple functions. The table is also used to establish the inverse L.T., since if

$$\mathcal{L}f(t) = \bar{f}(s)$$

then

$$f(t) = \mathcal{L}^{-1}\bar{f}(s).$$

SHORT TABLE OF LAPLACE TRANSFORMS

	$f(s)$	$f(t)$
(1)	1	$\delta(t)$ = unit impulse at $t = 0$
(2)	$\dfrac{1}{s}$	$\mathcal{U}(t)$ = unit step function at $t = 0$
(3)	$\dfrac{1}{s^n}(n = 1, 2, \cdots)$	$\dfrac{t^{n-1}}{(n-1)!}$
(4)	$\dfrac{1}{s + a}$	e^{-at}
(5)	$\dfrac{1}{(s + a)^2}$	te^{-at}
(6)	$\dfrac{1}{(s + a)^n}(n = 1, 2, \cdots)$	$\dfrac{1}{(n-1)!}t^{n-1}e^{-at}$
(7)	$\dfrac{1}{s(s + a)}$	$\dfrac{1}{a}(1 - e^{-at})$
(8)	$\dfrac{1}{s^2(s + a)}$	$\dfrac{1}{a^2}(e^{-at} + at - 1)$
(9)	$\dfrac{s}{s^2 + a^2}$	$\cos at$
(10)	$\dfrac{s}{s^2 - a^2}$	$\cosh at$
(11)	$\dfrac{1}{s^2 + a^2}$	$\dfrac{1}{a}\sin at$
(12)	$\dfrac{1}{s^2 - a^2}$	$\dfrac{1}{a}\sinh at$
(13)	$\dfrac{1}{s(s^2 + a^2)}$	$\dfrac{1}{a^2}(1 - \cos at)$
(14)	$\dfrac{1}{s^2(s^2 + a^2)}$	$\dfrac{1}{a^3}(at - \sin at)$
(15)	$\dfrac{1}{(s^2 + a^2)^2}$	$\dfrac{1}{2a^3}(\sin at - at \cos at)$
(16)	$\dfrac{s}{(s^2 + a^2)^2}$	$\dfrac{t}{2a}\sin at$

SHORT TABLE OF LAPLACE TRANSFORMS (Continued)

	$f(s)$	$f(t)$
(17)	$\dfrac{s^2 - a^2}{(s^2 + a^2)^2}$	$t \cos at$
(18)	$\dfrac{1}{s^2 + 2\zeta\omega_0 s + \omega_0^2}$	$\dfrac{1}{\omega_0\sqrt{1 - \zeta^2}} e^{-\zeta\omega_0 t} \sin \omega_0\sqrt{1 - \zeta^2}\, t$

REFERENCE

1. Thomson, W. T. *Laplace Transformation*, 2nd Ed., Englewood Cliffs, N.J.: Prentice-Hall, Inc., 1960.

DETERMINANT AND MATRICES

I DETERMINANT

A determinant of the second order and its numerical evaluation are defined by the following notation and operation

$$D = \begin{vmatrix} a & b \\ c & d \end{vmatrix} = ad - bc$$

An nth order determinant has n rows and n columns, and in order to identify the position of its elements, the following notation is adopted

$$\begin{vmatrix} a_{11} & a_{12} & a_{13} \cdots a_{1n} \\ a_{21} & a_{22} & a_{23} \cdots a_{2n} \\ \vdots \\ a_{n1} & a_{n2} & a_{n3} \cdots a_{nn} \end{vmatrix}$$

Minors

A minor M_{ij} of the element a_{ij} is a determinant formed by deleting the ith row and the jth column from the original determinant.

Cofactor

The cofactor C_{ij} of the element a_{ij} is defined by the equation

$$C_{ij} = (-1)^{i+j} M_{ij}$$

EXAMPLE

Given the third order determinant

$$\begin{vmatrix} 2 & 1 & 5 \\ 4 & 2 & 1 \\ 2 & 0 & 3 \end{vmatrix}$$

The minor of the term $a_{21} = 4$ is

$$M_{21} \text{ of } \begin{vmatrix} 2 & 1 & 5 \\ 4 & 2 & 1 \\ 2 & 0 & 3 \end{vmatrix} = \begin{vmatrix} 1 & 5 \\ 0 & 3 \end{vmatrix} = 3$$

and its cofactor is

$$C_{21} = (-1)^{2+1} 3 = -3$$

Expansion of a Determinant

The order of a determinant can be reduced by one by expanding any row or column in terms of its cofactors.

EXAMPLE

The determinant of the previous example is expanded in terms of the second column as

$$D = \begin{vmatrix} 2 & 1 & 5 \\ 4 & 2 & 1 \\ 2 & 0 & 3 \end{vmatrix} = 1(-1)^{1+2} \begin{vmatrix} 4 & 1 \\ 2 & 3 \end{vmatrix} + 2(-1)^{2+2} \begin{vmatrix} 2 & 5 \\ 2 & 3 \end{vmatrix}$$

$$+ 0(-1)^{3+2} \begin{vmatrix} 2 & 5 \\ 4 & 1 \end{vmatrix}$$

$$= -10 - 8 = -18$$

Properties of Determinants

The following properties of determinants are stated without proof.

(1) Interchange of any two columns or rows changes the sign of the determinant.
(2) If two rows or two columns are identical, the determinant is zero.
(3) Any row or column may be multiplied by a constant and added to another row or column without changing the value of the determinant.

II MATRICES

Matrix. A rectangular array of terms arranged in m rows and n columns is called a matrix. For example

$$A = \begin{bmatrix} a_{11} & a_{12} & a_{13} & a_{14} \\ a_{21} & a_{22} & a_{23} & a_{24} \\ a_{31} & a_{32} & a_{33} & a_{34} \end{bmatrix}$$

is a 3×4 matrix.

Square Matrix. A square matrix is one in which the number of rows is equal to the number of columns. It is referred to as an $n \times n$ matrix or a matrix of order n.

Symmetric Matrix. A square matrix is said to be symmetric if the elements on the upper right half can be obtained by flipping the matrix about the diagonal

$$A = \begin{bmatrix} 2 & 1 & 3 \\ 1 & 5 & 0 \\ 3 & 0 & 1 \end{bmatrix} = \text{symmetric matrix}$$

Trace. The sum of the diagonal elements of a square matrix is called the trace. For the matrix above

$$\text{Trace } A = 2 + 5 + 1 = 8$$

Singular Matrix. If the determinant of a matrix is zero, the matrix is said to be singular.

Row Matrix. A row matrix has $m = 1$.

$$B = \begin{bmatrix} b_1 & b_2 & b_3 \end{bmatrix}$$

Column Matrix. A column matrix has $n = 1$.

$$C = \begin{Bmatrix} C_1 \\ C_2 \\ C_3 \end{Bmatrix}$$

Zero Matrix. The zero matrix is defined as one in which all elements are zero.

$$0 = \begin{bmatrix} 0 & 0 & 0 \\ 0 & 0 & 0 \end{bmatrix}$$

Unit Matrix.   The unit matrix

$$I = \begin{bmatrix} 1 & 0 & 0 \\ 0 & 1 & 0 \\ 0 & 0 & 1 \end{bmatrix}$$

is a square matrix in which the diagonal elements from the top left to the bottom right are unity with all other elements equal to zero.

Diagonal Matrix.   A square matrix having elements a_{ii} along the diagonal with all other elements equal to zero is a diagonal matrix

$$[a_{ii}] = \begin{bmatrix} a_{11} & 0 & 0 \\ 0 & a_{22} & 0 \\ 0 & 0 & a_{33} \end{bmatrix}$$

Transpose.   The transpose A' of a matrix A is one in which the rows and columns are interchanged. For example

$$A = \begin{bmatrix} a_{11} & a_{12} & a_{13} \\ a_{21} & a_{22} & a_{23} \end{bmatrix} \qquad A' = \begin{bmatrix} a_{11} & a_{21} \\ a_{12} & a_{22} \\ a_{13} & a_{23} \end{bmatrix}$$

The transpose of a column matrix is a row matrix

$$X = \begin{Bmatrix} x_1 \\ x_2 \\ x_3 \end{Bmatrix} \qquad X' = [x_1 x_2 x_3]$$

Minor.   A minor M_{ij} of a matrix A is formed by deleting the ith row and the jth column from the determinant of the original matrix

$$\text{Let } A = \begin{bmatrix} a_{11} & a_{12} & a_{13} \\ a_{21} & a_{22} & a_{23} \\ a_{31} & a_{32} & a_{33} \end{bmatrix}$$

$$M_{12} \begin{vmatrix} a_{11} & a_{12} & a_{13} \\ a_{21} & a_{22} & a_{23} \\ a_{31} & a_{32} & a_{33} \end{vmatrix} = \begin{vmatrix} a_{21} & a_{23} \\ a_{31} & a_{33} \end{vmatrix}$$

Cofactor.   The cofactor C_{ij} is equal to the signed minor $(-1)^{i+j}M_{ij}$. From the previous example

$$C_{12} = (-1)^{1+2}M_{12} = -M_{12}$$

Adjoint Matrix. An adjoint matrix of a square matrix A is a transpose of the matrix of cofactors of A.

Let cofactor matrix of A be

$$[C_{ij}] = \begin{bmatrix} C_{11} & C_{12} & C_{13} \\ C_{21} & C_{22} & C_{23} \\ C_{31} & C_{32} & C_{33} \end{bmatrix}$$

$$adj\ A = [C_{ij}]' = [C_{ji}] = \begin{bmatrix} C_{11} & C_{21} & C_{31} \\ C_{12} & C_{22} & C_{32} \\ C_{13} & C_{23} & C_{33} \end{bmatrix}$$

Inverse Matrix. The inverse A^{-1} of a matrix A satisfies the relationship

$$A^{-1}A = AA^{-1} = I$$

Orthogonal Matrix. An orthogonal matrix A satisfies the relationship

$$A'A = AA' = I$$

From the definition of an inverse matrix it is evident that for an orthogonal matrix $A' = A^{-1}$.

III   RULES OF MATRIX OPERATIONS

Addition. Two matrices having the same number of rows and columns may be added by summing the corresponding elements.

EXAMPLE

$$\begin{bmatrix} 1 & 3 & 2 \\ 4 & 1 & 1 \end{bmatrix} + \begin{bmatrix} 2 & 0 & 4 \\ 1 & -2 & -3 \end{bmatrix} = \begin{bmatrix} 3 & 3 & 6 \\ 5 & -1 & -2 \end{bmatrix}$$

Multiplication. The product of two matrices A and B is another matrix C.

$$AB = C$$

The element C_{ij} of C is determined by multiplying the elements of the ith row in A by the elements of the jth column in B according to the rule

$$c_{ij} = \sum_k a_{ik} b_{kj}$$

EXAMPLE

$$\text{Let } A = \begin{bmatrix} 1 & 1 & 1 \\ 1 & 2 & 2 \\ 1 & 2 & 3 \end{bmatrix} \qquad B = \begin{bmatrix} 2 & 0 \\ 0 & 1 \\ 3 & -1 \end{bmatrix}$$

$$AB = \begin{bmatrix} 1 & 1 & 1 \\ 1 & 2 & 2 \\ 1 & 2 & 3 \end{bmatrix} \begin{bmatrix} 2 & 0 \\ 0 & 1 \\ 3 & -1 \end{bmatrix} = \begin{bmatrix} 5 & 0 \\ 8 & 0 \\ 11 & -1 \end{bmatrix} = C$$

i.e., $\qquad c_{21} = 1 \times 2 + 2 \times 0 + 2 \times 3 = 8$

It is evident that the number of columns in A must equal the number of rows in B, or that the matrices must be conformable. We also note that $AB \neq BA$.

The post-multiplication of a matrix by a column matrix results in a column matrix

EXAMPLE

$$\begin{bmatrix} 1 & 1 & 1 \\ 1 & 5 & 2 \\ 2 & 1 & 3 \end{bmatrix} \begin{Bmatrix} 1 \\ 3 \\ 2 \end{Bmatrix} = \begin{Bmatrix} 6 \\ 20 \\ 11 \end{Bmatrix}$$

Pre-multiplication of a matrix by a row matrix (or transpose of a column matrix) results in a row matrix

EXAMPLE

$$\begin{bmatrix} 1 & 3 & 2 \end{bmatrix} \begin{bmatrix} 1 & 1 & 1 \\ 1 & 5 & 2 \\ 2 & 1 & 3 \end{bmatrix} = \begin{bmatrix} 8 & 18 & 13 \end{bmatrix}$$

The transpose of a product $AB = C$ is $C' = B'A'$

EXAMPLE

$$\text{Let } A = \begin{bmatrix} 1 & 1 \\ 2 & 3 \end{bmatrix} \qquad B = \begin{bmatrix} 2 & 1 \\ 1 & 1 \end{bmatrix}$$

$$C = AB = \begin{bmatrix} 3 & 2 \\ 7 & 5 \end{bmatrix} \qquad C' = B'A' = \begin{bmatrix} 2 & 1 \\ 1 & 1 \end{bmatrix} \begin{bmatrix} 1 & 2 \\ 1 & 3 \end{bmatrix} = \begin{bmatrix} 3 & 7 \\ 2 & 5 \end{bmatrix}$$

Inversion of a Matrix.   Consider a set of equations

$$a_{11}x_1 + a_{12}x_2 + a_{13}x_3 = y_1$$
$$a_{21}x_1 + a_{22}x_2 + a_{23}x_3 = y_2 \tag{1}$$
$$a_{31}x_1 + a_{32}x_2 + a_{33}x_3 = y_3$$

which can be expressed in the matrix form

$$AX = Y \tag{2}$$

Pre-multiplying by the inverse A^{-1}, we obtain the solution

$$X = A^{-1}Y \tag{3}$$

We can identify the term A^{-1} by Cramer's rule as follows. The solution for x_1 is

$$x_1 = \frac{1}{|A|}\begin{vmatrix} y_1 & a_{12} & a_{13} \\ y_2 & a_{22} & a_{23} \\ y_3 & a_{32} & a_{33} \end{vmatrix}$$

$$= \frac{1}{|A|}\left\{ y_1\begin{vmatrix} a_{22} & a_{23} \\ a_{32} & a_{33} \end{vmatrix} - y_2\begin{vmatrix} a_{12} & a_{13} \\ a_{32} & a_{33} \end{vmatrix} + y_3\begin{vmatrix} a_{12} & a_{13} \\ a_{22} & a_{23} \end{vmatrix}\right\}$$

$$= \frac{1}{|A|}\{y_1 C_{11} + y_2 C_{21} + y_3 C_{31}\}$$

where A is the determinant of the coefficient matrix A, and C_{11}, C_{21}, and C_{31} are the cofactors of A corresponding to elements 11, 21, and 31. We can also write similar expressions for x_2 and x_3 by replacing the second and third columns by the y column respectively. Thus, the complete solution can be written in the matrix form

$$\begin{Bmatrix} x_1 \\ x_2 \\ x_3 \end{Bmatrix} = \frac{1}{|A|}\begin{bmatrix} C_{11} & C_{21} & C_{31} \\ C_{12} & C_{22} & C_{32} \\ C_{13} & C_{23} & C_{33} \end{bmatrix}\begin{Bmatrix} y_1 \\ y_2 \\ y_3 \end{Bmatrix} \tag{4}$$

or

$$\{x\} = \frac{1}{|A|}[C_{ji}]\{y\} = \frac{1}{|A|}[adj\ A]\{y\}$$

Thus, by comparison with Eq. (3) we arrive at the result

$$A^{-1} = \frac{1}{|A|}adj\ A \tag{5}$$

EXAMPLE

Find the inverse of the matrix

$$A = \begin{bmatrix} 1 & 1 & 1 \\ 1 & 2 & 2 \\ 1 & 0 & 3 \end{bmatrix}$$

(a) The determinant of A is $|A| = 3$
(b) The minors of A are

$$M_{11} = \begin{vmatrix} 2 & 2 \\ 0 & 3 \end{vmatrix} = 6,\ M_{12} = \begin{vmatrix} 1 & 2 \\ 1 & 3 \end{vmatrix} = 1,\ \text{etc.}$$

(c) Supply the signs $(-1)^{i+j}$ to the minors to form the cofactors

$$[C_{ij}] = \begin{vmatrix} 6 & -1 & -2 \\ -3 & 2 & 1 \\ 0 & -1 & 1 \end{vmatrix}$$

(d) The adjoint matrix is the transpose of the cofactor matrix, or $[C_{ij}]' = [C_{ji}]$. Thus, the inverse A^{-1} is found to be

$$A^{-1} = \frac{1}{|A|} adj\ A = \frac{1}{3} \begin{bmatrix} 6 & -3 & 0 \\ -1 & 2 & -1 \\ -2 & 1 & 1 \end{bmatrix}$$

(e) The result can be checked as follows

$$A^{-1}A = \frac{1}{3} \begin{bmatrix} 6 & -3 & 0 \\ -1 & 2 & -1 \\ -2 & 1 & 1 \end{bmatrix} \begin{bmatrix} 1 & 1 & 1 \\ 1 & 2 & 2 \\ 1 & 0 & 3 \end{bmatrix}$$

$$= \frac{1}{3} \begin{bmatrix} 3 & 0 & 0 \\ 0 & 3 & 0 \\ 0 & 0 & 3 \end{bmatrix} = \begin{bmatrix} 1 & 0 & 0 \\ 0 & 1 & 0 \\ 0 & 0 & 1 \end{bmatrix}$$

It should be noted that for an inverse to exist, the determinant $|A|$ must not be zero.

Equation (5) for the inverse offers another means of evaluating a determinant. Premultiply Eq. (5) by A

$$AA^{-1} = \frac{A}{|A|} adj\ A = I$$

Thus we obtain the expression

$$|A|I = A\ adj\ A \tag{6}$$

Transpose of a Product

The following operations are given without proof.

$$(AB)' = B'A'$$
$$(A + B)' = A' + B' \tag{7}$$

Orthogonal Transformation. A matrix P is orthogonal if

$$P^{-1} = P'$$

The determinant of an orthogonal matrix is equal to ± 1. If $A =$ symmetric matrix, then

$$P^{-1}AP = D = P'AP \text{ a diagonal matrix} \tag{8}$$

If A = symmetric matrix, then

$$P'A = AP$$

$$\{x\}'A = A\{x\}$$

(9)

Partitioned Matrices

A matrix may be partitioned into submatrices by horizontal and vertical lines as shown by the example below

$$\begin{bmatrix} 2 & 4 & \vdots & -1 \\ 0 & -3 & \vdots & 4 \\ 1 & 2 & \vdots & 2 \\ \cdots & \cdots & \vdots & \cdots \\ 3 & -1 & \vdots & -5 \end{bmatrix} = \begin{bmatrix} [A] & \vdots & [B] \\ \cdots & \vdots & \cdots \\ [C] & \vdots & [D] \end{bmatrix}$$

where the submatrices are

$$A = \begin{bmatrix} 2 & 4 \\ 0 & -3 \\ 1 & 2 \end{bmatrix} \qquad B = \left\{ \begin{array}{c} -1 \\ 4 \\ 2 \end{array} \right\}$$

$$C = \begin{bmatrix} 3 & -1 \end{bmatrix} \qquad D = \begin{bmatrix} -5 \end{bmatrix}$$

Partitioned matrices obey the normal rules of matrix algebra and can be added, subtracted, and multiplied as though the submatrices were ordinary matrix elements. Thus

$$\begin{bmatrix} A & \vdots & B \\ \cdots & \vdots & \cdots \\ C & \vdots & D \end{bmatrix} \left\{ \frac{x}{y} \right\} = \begin{bmatrix} A\{x\} + B\{y\} \\ C\{x\} + D\{y\} \end{bmatrix}$$

$$\begin{bmatrix} A & \vdots & B \\ \cdots & \vdots & \cdots \\ C & \vdots & D \end{bmatrix} \begin{bmatrix} E & \vdots & F \\ \cdots & \vdots & \cdots \\ G & \vdots & H \end{bmatrix} = \begin{bmatrix} AE + BG & \vdots & AF + BH \\ \cdots & \cdots & \vdots & \cdots \cdots \\ CE + DG & \vdots & CF + DH \end{bmatrix}$$

IV   DETERMINATION OF EIGENVECTORS

The eigenvector X_i corresponding to the eigenvalue λ_i can be found from the cofactors of any row of the characteristic equation.

Let $[A - \lambda_i I]X_i = 0$ be written out for a third order system as

$$\begin{bmatrix} (a_{11} - \lambda_i) & a_{12} & a_{13} \\ a_{21} & (a_{22} - \lambda_i) & a_{23} \\ a_{31} & a_{32} & (a_{33} - \lambda_i) \end{bmatrix} \left\{ \begin{array}{c} x_1 \\ x_2 \\ x_3 \end{array} \right\}_i = 0$$

(1)

Its characteristic equation $|A - \lambda_i I| = 0$ written out in determinant form is

$$\begin{vmatrix} (a_{11} - \lambda_i) & a_{12} & a_{13} \\ a_{21} & (a_{22} - \lambda_i) & a_{23} \\ a_{31} & a_{32} & (a_{33} - \lambda_i) \end{vmatrix} = 0 \tag{2}$$

The determinant expanded in terms of the cofactors of the first row is

$$(a_{11} - \lambda_i)C_{11} + a_{12}C_{12} + a_{13}C_{13} = 0 \tag{3}$$

Next replace the first row of the determinant by the second row, leaving the other two rows unchanged. The value of the determinant is still zero because of two identical rows

$$\begin{vmatrix} a_{21} & (a_{22} - \lambda_i) & a_{23} \\ a_{21} & (a_{22} - \lambda_i) & a_{23} \\ a_{31} & a_{32} & (a_{33} - \lambda_i) \end{vmatrix} = 0 \tag{4}$$

Again expand in terms of the cofactors of the first row, which are identical to the cofactors of the previous determinant.

$$a_{21}C_{11} + (a_{22} - \lambda_i)C_{12} + a_{23}C_{13} = 0 \tag{5}$$

Finally replace the first row by the third row and expand in terms of the first row of the new determinant.

$$\begin{vmatrix} a_{31} & a_{32} & (a_{33} - \lambda_i) \\ a_{21} & (a_{22} - \lambda_i) & a_{23} \\ a_{31} & a_{32} & (a_{33} - \lambda_i) \end{vmatrix} = 0 \tag{6}$$

$$a_{31}C_{11} + a_{32}C_{12} + (a_{33} - \lambda_i)C_{13} = 0 \tag{7}$$

Equations (3) (5) and (7) can now be assembled in a single matrix equation

$$\begin{bmatrix} (a_{11} - \lambda_i) & a_{12} & a_{13} \\ a_{21} & (a_{22} - \lambda_i) & a_{23} \\ a_{31} & a_{32} & (a_{33} - \lambda_i) \end{bmatrix} \begin{Bmatrix} C_{11} \\ C_{12} \\ C_{13} \end{Bmatrix} = 0 \tag{8}$$

Comparison of Eqs. (1) and (8) indicates that the eigenvector X_i may be determined from the cofactors of the characteristic equation with $\lambda = \lambda_i$. Since the eigenvectors are relative to a normalized coordinate, the column of cofactors may differ by a multiplying factor.

$$\begin{Bmatrix} x_1 \\ x_2 \\ x_3 \end{Bmatrix}_i = \alpha \begin{Bmatrix} C_{11} \\ C_{12} \\ C_{13} \end{Bmatrix}$$

Instead of the first row, any other row may have been used for the determination of the cofactors.

V CHOLESKY'S METHOD OF SOLUTION*

The matrix equation

$$[A]\{X\} = \{C\} \tag{1}$$

may be solved for $\{X\}$ by premultiplying the equation by the inverse of $[A]$

$$\{X\} = [A]^{-1}\{C\}$$

Cholesky's method avoids the necessity of inverting the matrix $[A]$, the elements of $\{X\}$ being available by successive algebraic steps.

Cholesky's method depends on converting the original equation, Eq. (1), to the form

$$[T]\{X\} = \{k\} \tag{2}$$

where (for a 3 × 3 matrix)

$$[T] = \begin{bmatrix} 1 & t_{12} & t_{13} \\ 0 & 1 & t_{23} \\ 0 & 0 & 1 \end{bmatrix} \tag{3}$$

is an upper triangular matrix with unit diagonal elements. For example, consider a 3 × 3 matrix

$$\begin{bmatrix} 1 & t_{12} & t_{13} \\ 0 & 1 & t_{23} \\ 0 & 0 & 1 \end{bmatrix} \begin{Bmatrix} x_1 \\ x_2 \\ x_3 \end{Bmatrix} = \begin{Bmatrix} k_1 \\ k_2 \\ k_3 \end{Bmatrix} .$$

The elements of $\{X\}$ from the above equation are simply found by a backward substitution as follows

$$x_3 = k_3 \quad \therefore \quad x_3 = k_3$$

$$x_2 + t_{23}x_3 = k_2 \quad \therefore \quad x_2 = k_2 - t_{23}x_3$$

$$x_1 + t_{12}x_2 + t_{13}x_3 = k_3 \quad \therefore \quad x_1 = k_3 - t_{12}x_3 - t_{13}x_3$$

Thus if $[T]$ and $\{k\}$ are known, the solution for $\{X\}$ in Eq. (1) is available.

To determine $[T]$ and $\{k\}$ multiply Eq. (2) by a lower triangular matrix

$$[L] = \begin{bmatrix} l_{11} & 0 & 0 \\ l_{21} & l_{22} & 0 \\ l_{31} & l_{32} & l_{33} \end{bmatrix} \tag{4}$$

*Salvadori and Baron, *Numerical Methods in Engineering.* Englewood Cliffs, N.J.: Prentice-Hall, Inc., 1952, pp. 23–28.

as follows

$$[L][T]\{X\} = [L]\{k\} \tag{5}$$

For this equation to equal the original equation, the following relationships must exist

$$[A] = [L][T] \tag{6}$$

$$\{C\} = [L]\{k\} \tag{7}$$

Writing out the above equations in terms of their elements, we have

$$\begin{bmatrix} a_{11} & a_{12} & a_{13} \\ a_{21} & a_{22} & a_{23} \\ a_{31} & a_{32} & a_{33} \end{bmatrix} = \begin{bmatrix} l_{11} & l_{11}t_{12} & l_{11}t_{13} \\ l_{21} & (l_{21}t_{12} + l_{22}) & (l_{21}t_{13} + l_{22}t_{23}) \\ l_{31} & (l_{31}t_{12} + l_{32}) & (l_{31}t_{13} + l_{32}t_{23} + l_{33}) \end{bmatrix}$$

$$\begin{Bmatrix} c_1 \\ c_2 \\ c_3 \end{Bmatrix} = \begin{Bmatrix} l_{11}k_1 \\ l_{21}k_1 + l_{22}k_2 \\ l_{31}k_1 + l_{32}k_2 + l_{33}k_3 \end{Bmatrix}$$

By equating the elements in these equations we have

$$a_{11} = l_{11}$$
$$a_{21} = l_{21}$$
$$a_{31} = l_{31}$$

$$a_{12} = l_{11}t_{12} \qquad \therefore \quad t_{12} = \frac{a_{12}}{l_{11}}$$

$$a_{13} = l_{11}t_{13} \qquad \therefore \quad t_{13} = \frac{a_{13}}{l_{11}}$$

$$a_{22} = (l_{21}t_{12} + l_{22}) \qquad \therefore \quad l_{22} = a_{22} - l_{21}t_{12}$$

$$a_{23} = (l_{21}t_{13} + l_{22}t_{23}) \qquad \therefore \quad t_{23} = \frac{1}{l_{22}}(a_{23} - l_{21}t_{13})$$

$$a_{32} = (l_{31}t_{12} + l_{32}) \qquad \therefore \quad l_{32} = a_{32} - l_{31}t_{12}$$

$$a_{33} = (l_{31}t_{13} + l_{32}t_{23} + l_{33}) \qquad \therefore \quad l_{33} = a_{33} - l_{31}t_{13} - l_{32}t_{23}$$

$$c_1 = l_{11}k_1 \qquad \therefore \quad k_1 = \frac{c_1}{l_{11}}$$

$$c_2 = l_{21}k_1 + l_{22}k_2 \qquad \therefore \quad k_2 = \frac{1}{l_{22}}(c_2 - l_{21}k_1)$$

$$c_3 = l_{31}k_1 + l_{32}k_2 + l_{33}k_3 \qquad \therefore \quad k_3 = \frac{1}{l_{33}}(c_3 - l_{31}k_1 - l_{32}k_2)$$

Thus the elements of the matrices $[L]$, $[T]$ and $\{k\}$ are now available in terms of the known elements of $[A]$ and $[C]$ and Eq. (1) may be solved without inverting the matrix $[A]$.

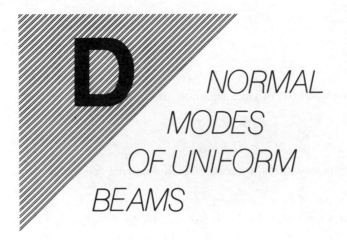

D NORMAL MODES OF UNIFORM BEAMS

We assume the free vibrations of a uniform beam to be governed by Euler's differential equation.

$$EI\frac{\partial^4 y}{\partial x^4} + m\frac{\partial^2 y}{\partial t^2} = 0 \tag{1}$$

To determine the normal modes of vibration, the solution in the form

$$y(x, t) = \phi_n(x)e^{i\omega_n t} \tag{2}$$

is substituted into Eq. (1) to obtain the equation

$$\frac{d^4\phi_n(x)}{dx^4} - \beta_n^4\phi_n(x) = 0 \tag{3}$$

where:

$\phi_n(x)$ = characteristic function describing the deflection of the nth mode

m = mass density per unit length

$\beta_n^4 = m\omega_n^2 / EI$

$\omega_n = (\beta_n l)^2 \sqrt{EI/ml^4}$ = natural frequency of the nth mode.

The characteristic functions $\phi_n(x)$ and the normal-mode frequencies ω_n depend on the boundary conditions, and have been tabulated by Young and Felgar. An abbreviated summary taken from this work is presented here.

REFERENCE

1. Young, D. and R. P. Felgar Jr., *Tables of Characteristic Functions Representing Normal Modes of Vibration of a Beam*. The University of Texas Publication No. 4913, July 1, 1949.

TABLE 1.

CHARACTERISTIC FUNCTIONS AND DERIVATIVES
CLAMPED-CLAMPED BEAM
FIRST MODE

$\dfrac{x}{l}$	ϕ_1	$\phi_1' = \dfrac{1}{\beta_1}\dfrac{d\phi_1}{dx}$	$\phi_1'' = \dfrac{1}{\beta_1^2}\dfrac{d^2\phi_1}{dx^2}$	$\phi_1''' = \dfrac{1}{\beta_1^3}\dfrac{d^3\phi_1}{dx^3}$
0.00	0.00000	0.00000	2.00000	−1.96500
0.04	0.03358	0.34324	1.62832	−1.96285
0.08	0.12545	0.61624	1.25802	−1.94862
0.12	0.26237	0.81956	0.89234	−1.91254
0.16	0.43126	0.95451	0.53615	−1.84732
0.20	0.61939	1.02342	0.19545	−1.74814
0.24	0.81459	1.02986	−0.12305	−1.61250
0.28	1.00546	0.97870	−0.41240	−1.44017
0.32	1.18168	0.87608	−0.66581	−1.23296
0.36	1.33419	0.72992	−0.87699	−0.99452
0.40	1.45545	0.54723	−1.04050	−0.73007
0.44	1.53962	0.33897	−1.15202	−0.44611
0.48	1.58271	0.11478	−1.20854	−0.15007
0.52	1.58271	−0.11478	−1.20854	0.15007
0.56	1.53962	−0.33897	−1.15202	0.44611
0.60	1.45545	−0.54723	−1.04050	0.73007
0.64	1.33419	−0.72992	−0.87699	0.99452
0.68	1.18168	−0.87608	−0.66581	1.23296
0.72	1.00546	−0.97870	−0.41240	1.44017
0.76	0.81459	−1.02986	−0.12305	1.61250
0.80	0.61939	−1.02342	0.19545	1.74814
0.84	0.43126	−0.95451	0.53615	1.84732
0.88	0.26237	−0.81956	0.89234	1.91254
0.92	0.12545	−0.61624	1.25802	1.94862
0.96	0.03358	−0.34324	1.62832	1.96285
1.00	0.00000	0.00000	2.00000	1.96500

1 CLAMPED-CLAMPED BEAM

n	$\beta_n l$	$(\beta_n l)^2$	ω_n/ω_1
1	4.7300	22.3733	1.0000
2	7.8532	61.6728	2.7565
3	10.9956	120.9034	5.4039

2 FREE-FREE BEAM

The natural frequencies of the free-free beam are equal to those of the clamped-clamped beam. The characteristic functions of the free-free beam are related to those of the clamped-clamped beam as follows.

free-free		clamped-clamped
ϕ_n	$=$	ϕ_n''
ϕ_n'	$=$	ϕ_n'''
ϕ_n''	$=$	ϕ_n
ϕ_n'''	$=$	ϕ_n'

3 CLAMPED-FREE BEAM

n	$\beta_n l$	$(\beta_n l)^2$	ω_n/ω_1
1	1.8751	3.5160	1.0000
2	4.6941	22.0345	6.2669
3	7.8548	61.6972	17.5475

4 CLAMPED-PINNED BEAM

n	$\beta_n l$	$(\beta_n l)^2$	ω_n/ω_1
1	3.9266	15.4182	1.0000
2	7.0686	49.9645	3.2406
3	10.2101	104.2477	6.7613

5 FREE-PINNED BEAM

The natural frequencies of the free-pinned beam are equal to those of the clamped-pinned beam. The characteristic functions of the free-pinned beam are related to those of the clamped-pinned beam as follows.

free-pinned		clamped-pinned
ϕ_n	$=$	ϕ_n''
ϕ_n'	$=$	ϕ_n'''
ϕ_n''	$=$	ϕ_n
ϕ_n'''	$=$	ϕ_n'

TABLE 1.
CHARACTERISTIC FUNCTIONS AND DERIVATIVES
CLAMPED-CLAMPED BEAM
SECOND MODE

$\dfrac{x}{l}$	ϕ_2	$\phi_2' = \dfrac{1}{\beta_2}\dfrac{d\phi_2}{dx}$	$\phi_2'' = \dfrac{1}{\beta_2^2}\dfrac{d^2\phi_2}{dx^2}$	$\phi_2''' = \dfrac{1}{\beta_2^3}\dfrac{d^3\phi_2}{dx^3}$
0.00	0.00000	0.00000	2.00000	−2.00155
0.04	0.08834	0.52955	1.37202	−1.99205
0.08	0.31214	0.86296	0.75386	−1.93186
0.12	0.61058	1.00644	0.16713	−1.78813
0.16	0.92602	0.97427	−0.35923	−1.54652
0.20	1.20674	0.79030	−0.79450	−1.21002
0.24	1.41005	0.48755	−1.11133	−0.79651
0.28	1.50485	0.10660	−1.28991	−0.33555
0.32	1.47357	−0.30736	−1.32106	0.13566
0.36	1.31314	−0.70819	−1.20786	0.57665
0.40	1.03457	−1.05271	−0.96605	0.94823
0.44	0.66150	−1.30448	−0.62296	1.21670
0.48	0.22751	−1.43728	−0.21508	1.35744
0.52	−0.22751	−1.43728	0.21508	1.35744
0.56	−0.66150	−1.30448	0.62296	1.21670
0.60	−1.03457	−1.05271	0.96605	0.94823
0.64	−1.31314	−0.70819	1.20786	0.57665
0.68	−1.47357	−0.30736	1.32106	0.13566
0.72	−1.50485	0.10660	1.28991	−0.33555
0.76	−1.41005	0.48755	1.11133	−0.79651
0.80	−1.20674	0.70930	0.79450	−1.21002
0.84	−0.92602	0.97427	0.35923	−1.54652
0.88	−0.61058	1.00644	−0.16713	−1.78813
0.92	−0.31214	0.86296	−0.75386	−1.93186
0.96	−0.08834	0.52955	−1.37202	−1.99205
1.00	0.00000	0.00000	−2.00000	−2.00155

TABLE 1.
CHARACTERISTIC FUNCTIONS AND DERIVATIVES
CLAMPED-CLAMPED BEAM
THIRD MODE

$\dfrac{x}{l}$	ϕ_3	$\phi_3' = \dfrac{1}{\beta_3}\dfrac{d\phi_3}{dx}$	$\phi_3'' = \dfrac{1}{\beta_3^2}\dfrac{d^2\phi_3}{dx^2}$	$\phi_3''' = \dfrac{1}{\beta_3^3}\dfrac{d^3\phi_3}{dx^3}$
0.00	0.00000	0.00000	2.00000	− 1.99993
0.04	0.16510	0.68646	1.12323	− 1.97469
0.08	0.54804	0.99303	0.28189	− 1.82280
0.12	0.98720	0.95006	− 0.45252	− 1.48447
0.16	1.34190	0.62285	− 0.99738	− 0.96698
0.20	1.50782	0.11050	− 1.28572	− 0.33199
0.24	1.42971	− 0.46573	− 1.28637	0.32333
0.28	1.10719	− 0.98087	− 1.01443	0.88956
0.32	0.59186	− 1.32694	− 0.53145	1.26880
0.36	− 0.02445	− 1.43171	0.06438	1.39529
0.40	− 0.62837	− 1.27099	0.65569	1.24912
0.44	− 1.10739	− 0.87257	1.12747	0.86096
0.48	− 1.37174	− 0.31031	1.38852	0.30669
0.52	− 1.37174	0.31031	1.38852	− 0.30669
0.56	− 1.10739	0.87257	1.12747	− 0.86096
0.60	− 0.62837	1.27099	0.65569	− 1.24912
0.64	− 0.02445	1.43171	0.06438	− 1.39529
0.68	0.59186	1.32694	− 0.53145	− 1.26880
0.72	1.10719	0.98087	− 1.01443	− 0.88956
0.76	1.42971	0.46573	− 1.28637	− 0.32333
0.80	1.50782	− 0.11050	− 1.28572	0.33199
0.84	1.34190	− 0.62285	− 0.99738	0.96698
0.88	0.98720	− 0.95006	− 0.45252	1.48447
0.92	0.54804	− 0.99303	0.28189	1.82280
0.96	0.16510	− 0.68646	1.12323	1.97469
1.00	0.00000	0.00000	2.00000	1.99993

TABLE 2.

CHARACTERISTIC FUNCTIONS AND DERIVATIVES
CLAMPED-FREE BEAM
FIRST MODE

$\dfrac{x}{l}$	ϕ_1	$\phi_1' = \dfrac{1}{\beta_1} \dfrac{d\phi_1}{dx}$	$\phi_1'' = \dfrac{1}{\beta_1^2} \dfrac{d^2\phi_1}{dx^2}$	$\phi_1''' = \dfrac{1}{\beta_1^3} \dfrac{d^3\phi_1}{dx^3}$
0.00	0.00000	0.00000	2.00000	-1.46819
0.04	0.00552	0.14588	1.88988	-1.46805
0.08	0.02168	0.28350	1.77980	-1.46710
0.12	0.04784	0.41286	1.66985	-1.46455
0.16	0.08340	0.53400	1.56016	-1.45968
0.20	0.12774	0.64692	1.45096	-1.45182
0.24	0.18024	0.75167	1.34247	-1.44032
0.28	0.24030	0.84832	1.23500	-1.42459
0.32	0.30730	0.93696	1.23889	-1.40410
0.36	0.38065	1.01771	1.02451	-1.37834
0.40	0.45977	1.09070	0.92227	-1.34685
0.44	0.54408	1.15612	0.82262	-1.30924
0.48	0.63301	1.21418	0.72603	-1.26512
0.52	0.72603	1.26512	0.63301	-1.21418
0.56	0.82262	1.30924	0.54408	-1.15612
0.60	0.92227	1.34685	0.45977	-1.09070
0.64	1.02451	1.37834	0.38065	-1.01771
0.68	1.12889	1.40410	0.30730	-0.93696
0.72	1.23500	1.42459	0.24030	-0.84832
0.76	1.34247	1.44032	0.18024	-0.75167
0.80	1.45096	1.45182	0.12774	-0.64692
0.84	1.56016	1.45968	0.08340	-0.53400
0.88	1.66985	1.46455	0.04784	-0.41286
0.92	1.77980	1.46710	0.02168	-0.28350
0.96	1.88988	1.46805	0.00552	-0.14588
1.00	2.00000	1.46819	0.00000	0.00000

$\dfrac{x}{l}$	ϕ_2	$\phi_2' = \dfrac{1}{\beta_2}\dfrac{d\phi_2}{dx}$	$\phi_2'' = \dfrac{1}{\beta_2^2}\dfrac{d^2\phi_2}{dx^2}$	$\phi_2''' = \dfrac{1}{\beta_2^3}\dfrac{d^3\phi_2}{dx^3}$
0.00	0.00000	0.00000	2.00000	−2.03693
0.04	0.03301	0.33962	1.61764	−2.03483
0.08	0.12305	0.60754	1.23660	−2.02097
0.12	0.25670	0.80728	0.86004	−1.98590
0.16	0.42070	0.93108	0.49261	−1.92267
0.20	0.60211	0.99020	0.14007	−1.82682
0.24	0.78852	0.98502	−0.19123	−1.69625
0.28	0.96827	0.92013	−0.49475	−1.53113
0.32	1.13068	0.80136	−0.76419	−1.33373
0.36	1.26626	0.63565	−0.99384	−1.10821
0.40	1.36694	0.43094	−1.17895	−0.86040
0.44	1.42619	0.19593	−1.31600	−0.59748
0.48	1.43920	−0.06012	−1.40289	−0.32772
0.52	1.40289	−0.32772	−1.43920	−0.06012
0.56	1.31600	−0.59748	−1.42619	0.19593
0.60	1.17895	−0.86040	−1.36694	0.43094
0.64	0.99384	−1.10821	−1.26626	0.63565
0.68	0.76419	−1.33373	−1.13068	0.80136
0.72	0.49475	−1.53113	−0.96827	0.92013
0.76	0.19123	−1.69625	−0.78852	0.98502
0.80	−0.14007	−1.82682	−0.60211	0.99020
0.84	−0.49261	−1.92267	−0.42070	0.93108
0.88	−0.86004	−1.98590	−0.25670	0.80428
0.92	−1.23660	−2.02097	−0.12305	0.60754
0.96	−1.61764	−2.03483	−0.03301	0.33962
1.00	−2.00000	−2.03693	0.00000	0.00000

TABLE 2.
CHARACTERISTIC FUNCTIONS AND DERIVATIVES
CLAMPED-FREE BEAM
THIRD MODE

$\dfrac{x}{l}$	ϕ_3	$\phi_3' = \dfrac{1}{\beta_3}\dfrac{d\phi_3}{dx}$	$\phi_3'' = \dfrac{1}{\beta_3^2}\dfrac{d^2\phi_3}{dx^2}$	$\phi_3''' = \dfrac{1}{\beta_3^3}\dfrac{d^3\phi_3}{dx^3}$
0.00	0.00000	0.00000	2.00000	-1.99845
0.04	0.08839	0.52979	1.37287	-1.98892
0.08	0.31238	0.86367	0.75558	-1.92871
0.12	0.61120	1.00785	0.16974	-1.78480
0.16	0.92728	0.97665	-0.35563	-1.54286
0.20	1.20901	0.79394	-0.78975	-1.20575
0.24	1.41376	0.49285	-1.10515	-0.79124
0.28	1.51056	0.11405	-1.28189	-0.32872
0.32	1.48203	-0.29711	-1.31055	0.14479
0.36	1.32534	-0.69422	-1.19398	0.58908
0.40	1.05185	-1.03374	-0.94753	0.96533
0.44	0.68568	-1.27881	-0.59802	1.24030
0.48	0.26103	-1.40247	-0.18130	1.39004
0.52	-0.18130	-1.39004	0.26103	1.40247
0.56	-0.59802	-1.24030	0.68568	1.27881
0.60	-0.94753	-0.96533	1.05185	1.03374
0.64	-1.19398	-0.58908	1.32534	0.69422
0.68	-1.31055	-0.14479	1.48203	0.29711
0.72	-1.28189	0.32872	1.51056	-0.11405
0.76	-1.10515	0.79124	1.41376	-0.49285
0.80	-0.78975	1.20575	1.20901	-0.79394
0.84	-0.35563	1.54236	0.92728	-0.97665
0.88	0.16974	1.78480	0.61120	-1.00785
0.92	0.75558	1.92871	0.31238	-0.86367
0.96	1.37287	1.98892	0.08829	-0.52979
1.00	2.00000	1.99845	0.00000	0.00000

TABLE 3.
CHARACTERISTIC FUNCTIONS AND DERIVATIVES
CLAMPED-PINNED BEAM
FIRST MODE

$\dfrac{x}{l}$	ϕ_1	$\phi_1' = \dfrac{1}{\beta_1}\dfrac{d\phi_1}{dx}$	$\phi_1'' = \dfrac{1}{\beta_1^2}\dfrac{d^2\phi_1}{dx^2}$	$\phi_1''' = \dfrac{1}{\beta_1^3}\dfrac{d^3\phi_1}{dx^3}$
0.00	0.00000	0.00000	2.00000	-2.00155
0.04	0.02338	0.28944	1.68568	-2.00031
0.08	0.08834	0.52955	1.37202	-1.99203
0.12	0.18715	0.72055	1.06060	-1.97079
0.16	0.31214	0.86296	0.75386	-1.93187
0.20	0.45574	0.95776	0.45486	-1.87177
0.24	0.61058	1.00643	0.16712	-1.78812
0.28	0.76958	1.01105	-0.10554	-1.67975
0.32	0.92601	0.97427	-0.35923	-1.54652
0.36	1.07363	0.89940	-0.59009	-1.38932
0.40	1.20675	0.79029	-0.79450	-1.21002
0.44	1.32032	0.65138	-0.96918	-1.01128
0.48	1.41006	0.48755	-1.11133	-0.79652
0.52	1.47245	0.30410	-1.21875	-0.56977
0.56	1.50485	0.10661	-1.28992	-0.33555
0.60	1.50550	-0.09916	-1.32402	-0.09872
0.64	1.47357	-0.30736	-1.32106	0.13566
0.68	1.40913	-0.51224	-1.28180	0.36247
0.72	1.31313	-0.70820	-1.20786	0.57666
0.76	1.18741	-0.88996	-1.10157	0.77340
0.80	1.03457	-1.05270	-0.96606	0.94823
0.84	0.85795	-1.19210	-0.80507	1.09714
0.88	0.66151	-1.30448	-0.62295	1.21670
0.92	0.44974	-1.38693	-0.42455	1.30414
0.96	0.22752	-1.43727	-0.21507	1.35743
1.00	0.00000	-1.45420	0.00000	1.37533

TABLE 3.
CHARACTERISTIC FUNCTIONS AND DERIVATIVES
CLAMPED-PINNED BEAM
SECOND MODE

$\dfrac{x}{l}$	ϕ_2	$\phi_2' = \dfrac{1}{\beta_2}\dfrac{d\phi_2}{dx}$	$\phi_2'' = \dfrac{1}{\beta_2^2}\dfrac{d^2\phi_2}{dx^2}$	$\phi_2''' = \dfrac{1}{\beta_2^3}\dfrac{d^3\phi_2}{dx^3}$
0.00	0.00000	0.00000	2.00000	-2.00000
0.04	0.07241	0.48557	1.43502	-1.99300
0.08	0.25958	0.81207	0.87658	-1.94824
0.12	0.51697	0.98325	0.33937	-1.83960
0.16	0.80176	1.00789	-0.15633	-1.65333
0.20	1.07449	0.90088	-0.58802	-1.38736
0.24	1.30078	0.68345	-0.93412	-1.05012
0.28	1.45308	0.38242	-1.17673	-0.65879
0.32	1.51208	0.02894	-1.30380	-0.23724
0.36	1.46765	-0.34350	-1.31068	0.18649
0.40	1.31923	-0.70122	-1.20092	0.58286
0.44	1.07550	-1.01270	-0.98634	0.92349
0.48	0.75348	-1.25090	-0.68631	1.18364
0.52	0.37700	-1.39515	-0.32640	1.34442
0.56	-0.02536	-1.43265	0.06348	1.39438
0.60	-0.42268	-1.35944	0.45136	1.33056
0.64	-0.78413	-1.18058	0.80569	1.15876
0.68	-1.08158	-0.90972	1.09776	0.89319
0.72	-1.29186	-0.56793	1.30395	0.55537
0.76	-1.39858	-0.18205	1.40755	0.17245
0.80	-1.39351	0.21752	1.40010	-0.22494
0.84	-1.27726	0.59923	1.28198	-0.60506
0.88	-1.05919	0.93288	1.06244	-0.93759
0.92	-0.75676	1.19208	0.75879	-1.19604
0.96	-0.39406	1.35629	0.39504	-1.35983
1.00	0.00000	1.41251	0.00000	-1.41592

TABLE 3.
CHARACTERISTIC FUNCTIONS AND DERIVATIVES
CLAMPED-PINNED BEAM
THIRD MODE

$\dfrac{x}{l}$	ϕ_3	$\phi_3' = \dfrac{1}{\beta_e}\dfrac{d\phi_3}{dx}$	$\phi_3'' = \dfrac{1}{\beta_3^2}\dfrac{d^2\phi_3}{dx^2}$	$\phi_3''' = \dfrac{1}{\beta_3^3}\dfrac{d^3\phi_3}{dx^3}$
0.00	0.00000	0.00000	2.00000	-2.00000
0.04	0.14410	0.65020	1.18532	-1.97961
0.08	0.48626	0.97168	0.39742	-1.85535
0.12	0.89584	0.98593	-0.30845	-1.57331
0.16	1.25604	0.74002	-0.86560	-1.13046
0.20	1.47476	0.30725	-1.21523	-0.56678
0.24	1.49419	-0.21934	-1.32168	0.04683
0.28	1.29662	-0.73864	-1.18195	0.62397
0.32	0.90489	-1.15556	-0.82867	1.07934
0.36	0.37703	-1.39512	-0.32637	1.34445
0.40	-0.20439	-1.41364	0.23807	1.37996
0.44	-0.74658	-1.20525	0.76897	1.18287
0.48	-1.16223	-0.80234	1.17711	0.78746
0.52	-1.38422	-0.26994	1.39411	0.26005
0.56	-1.37687	0.30522	1.38344	-0.31179
0.60	-1.14194	0.82907	1.14631	-0.83344
0.64	-0.71844	1.21582	0.72134	-1.21873
0.68	-0.17628	1.40210	0.17821	-1.40403
0.72	0.39519	1.35742	-0.39391	-1.35870
0.76	0.90188	2.08924	-0.90103	-1.09010
0.80	1.26035	0.64175	-1.25980	-0.64233
0.84	1.41160	0.08860	-1.41124	-0.08900
0.88	1.33072	-0.47918	-1.33049	0.47891
0.92	1.03098	-0.96820	-1.03085	0.96800
0.96	0.56168	-1.29798	-0.56162	1.29782
1.00	0.00000	-1.41429	0.00000	1.41414

ANSWERS TO SELECTED PROBLEMS

CHAPTER 1

1-1 $\dot{x}_{max} = 8.38$ cm/s; $\quad \ddot{x}_{max} = 350.9$ cm/s^2

1-3 $x_{max} = 7.27$ cm, $\quad \tau = 0.10$ s, $\quad \ddot{x}_{max} = 278.1$ m/s^2

1-5 $z = 5e^{0.6435i}$

1-8 $R = 8.697,$ $\quad \theta = 13.29°$

1-9 $x(t) = \dfrac{4}{\pi}(\sin \omega_1 t + \dfrac{1}{3}\sin 3\omega_1 t + \dfrac{1}{5}\sin 5\omega_1 t + \cdots)$

1-11 $x(t) = \dfrac{1}{2} + \dfrac{4}{\pi^2}(\cos \omega_1 t + \dfrac{1}{3^2}\cos 3\omega_1 t + \dfrac{1}{5^2}\cos 5\omega_1 t + \cdots)$

1-13 $\sqrt{\overline{x^2}} = A/2$

1-14 $\overline{x^2} = 1/3$

1-16 $a_0 = 1/3,$ $\quad b_n = \dfrac{1}{n\pi}(1 - \cos\dfrac{2\pi n}{3}),$ $\quad a_n = \dfrac{1}{n\pi}\sin\dfrac{2\pi n}{3}$

1-18 RMS $= 0.3162A$

1-20 Error $= \pm 0.148$ mm

1-22 $x_{peak}/x_{1000} = 39.8$

CHAPTER 2

2-1 5.62 Hz

2-3 0.159 s

2-5 $x(t) = \dfrac{m_2 g}{k}(1 - \cos \omega t) + \dfrac{m_2 \sqrt{2gh}}{\sqrt{k(m_1 + m_2)}} \sin \omega t$

2-7 $J_0 = 9.30$ lb in s^2

2-9 $\kappa = 0.4507$ m

2-11 $\omega = \sqrt{\dfrac{k}{m + J_0/r^2}}$

2-13 $\tau = 1.97$ s

2-15 $\tau = 2\pi \sqrt{\dfrac{J}{Wh}}$

2-17 $\tau = 2\pi \dfrac{L}{a} \sqrt{\dfrac{h}{3g}}$

2-19 $f = \dfrac{1}{2\pi} \sqrt{\dfrac{gab}{h\kappa^2}}$

2-21 $\tau = 2\pi \sqrt{\dfrac{l}{2g}}$

2-23 $m_{\text{eff}} = \dfrac{3}{8} ml$ for each column, $ml = $ mass of column

2-25 $m_{\text{eff}} = \dfrac{33}{140} ml$

2-27 $K_{\text{eff}} = \dfrac{K_1 K_2}{K_1 + K_2} + K_2$

2-29 $J_{\text{eff}} = J_1 + J_2 \left(\dfrac{r_1}{r_2}\right)^2$

2-31 $M = 0.0289$ kg

2-33 $\zeta = 1.45$

2-35 $\omega_n = 27.78$, $\delta = 0.0202$, $\zeta = 0.003215$, $c = 0.405$

2-39 $\omega_d = \sqrt{\dfrac{k}{m}\left(\dfrac{b}{a}\right)^2 - \left(\dfrac{c}{2m}\right)^2}$, $c_{\text{cr.}} = \dfrac{2b}{a} \sqrt{km}$

2-41 $\omega_d = \dfrac{a}{l} \sqrt{\dfrac{3k}{m}} \sqrt{1 - \dfrac{3}{4km}\left(\dfrac{cl}{a}\right)^2}$, $c_{\text{cr}} = \dfrac{2}{3} \dfrac{a}{l} \sqrt{3km}$

2-43 $\dot{x}_{\text{max}} = 92.66$ ft/s, $t = 0.214$ s

2-45 $\zeta_1 = 0.59$ $x_{\text{overshoot}} = 0.379$

2-48   Flexibility $= \dfrac{4}{243} \dfrac{l^3}{EI}$

2-53 $(0.854ml + .5625M)\ddot{x} + 0.5625kx + 2/3c\dot{x} = 0$

CHAPTER 3

3-1 $c = 61.3 \text{ NS/m}$

3-3 $X = 0.797 \text{ cm}, \qquad \phi = 51.43°$

3-5 $\zeta = 0.1847$

3-9 $f_m = 15 \text{ Hz}, \qquad \zeta = 0.0118, \qquad X = 0.149 \text{ cm}, \qquad \phi = 177.68°$

3-12 $f = 1028 \text{ rpm}$

3-14 $F = 1273 \text{ N}, \qquad F = 241.1 \text{ N for } d = 1.905 \text{ cm}$

3-16 $V = \dfrac{L}{2\pi}\sqrt{\dfrac{k}{m}}$

3-21 $k = 18.8 \text{ lb/in each spring}$

3-24 $X = 0.01105 \text{ cm}, \qquad F_T = 42.0 \text{ N}$

3-25 $\omega^2 X = 2.312 \text{ cm/s}^2$

3-34 $c_{eq} = 4D/\pi\omega x$

3-38 $|x_\phi/x_1| = \dfrac{1}{(2p-1)^2}\sqrt{\dfrac{\left[1-\left(\dfrac{\omega_1}{\omega_n}\right)^2\right]^2 + \left[2\zeta\dfrac{\omega_1}{\omega_n}\right]^2}{\left[1-\left\{\dfrac{(2p-1)\omega_1}{\omega_n}\right\}^2\right]^2 + \left[\dfrac{2\zeta(2p-1)\omega_1}{\omega_n}\right]^2}}$

3-44   (a) 624.5 mv,   (b) 3.123 mv

3-47 $E = 25.7 \text{ mv/g}$

CHAPTER 4

4-5 $x = \dfrac{F_0}{k}[\cos \omega_n(t - t_0) - \cos \omega_n t] \qquad t > t_0$

4-10 $z = \dfrac{100}{\omega_n^2}(1 - \cos \omega_n t) - \dfrac{20}{\omega_n}\sin \omega_n t$

$z_{max} = \dfrac{100}{\omega_n^2}\left[1 - \dfrac{5}{\sqrt{25 + \omega_n^2}}\right] - \dfrac{20}{\omega_n}\dfrac{\omega_n}{\sqrt{25 + \omega_n^2}}$

4-13 $\tan \omega_n t = \dfrac{\sqrt{2mgs/k}}{s - mg/2k}$

4-14 $x_{max} = 12.08", \qquad t = 0.392 \text{ s}$

4-20 $x_{max} = 2.34"$

4-29 $x(t) = \dfrac{F_0}{c\omega_n}\left\{\dfrac{e^{-\zeta\omega_n t}}{\sqrt{1 - \zeta^2}}\sin\left(\sqrt{1 - \zeta^2}\,\omega_n t + \sin^{-1}\sqrt{1 - \zeta^2}\right) - \cos \omega_n t\right\}$

CHAPTER 5

5-2 $\omega_1^2 = k/m$ $(X_1/X_2)_1 = 1$
   $\omega_2^2 = 3k/m$ $(X_1/X_2)_2 = -1$

5-4 $\omega_1^2 = 0.570k/m$ $(X_1/X_2)_1 = 3.43$
   $\omega_2^2 = 4.096k/m$ $(X_1/X_2)_2 = -0.096$

5-8 $\omega_n = 15.72$ rad/s

5-10 $\ddot{\theta}_1 + 2g/l\,\theta_1 - g/l\,\theta_2 = 0$
   $\ddot{\theta}_1 + \ddot{\theta}_2 + g/l\,\theta_2 = 0$

5-13 $\omega_1 = 0.796\sqrt{T/ml}$ $(Y_1/Y_2)_1 = 1.365$
   $\omega_2 = 1.538\sqrt{T/ml}$ $(Y_1/Y_2)_2 = -0.366$

5-15 $\omega = \sqrt{\dfrac{g}{l} + \dfrac{k}{ml^2}(1 \pm 1)}$, beat period = 53.02 s

5-20 $\begin{bmatrix} m & 0 \\ 0 & J \end{bmatrix}\begin{Bmatrix} \ddot{x} \\ \ddot{\theta} \end{Bmatrix} + \begin{bmatrix} 2k & kl/4 \\ kl/4 & 5kl/16 \end{bmatrix}\begin{Bmatrix} x \\ \theta \end{Bmatrix} = \{0\}$ x down
   θ clockwise

5-22 both static and dynamic coupling present

5-24 $f_1 = 0.963$ Hz   node 10.9 ft forward of *cg*
   $f_2 = 1.33$ Hz   node 1.48 ft aft of *cg*

5-26 $\omega_1 = 31.6$ rad/s $(X_1/X_2)_1 = 0.50$
   $\omega_2 = 63.4$ rad/s $(X_1/X_2)_2 = -1.00$

5-29 $x_1 = \dfrac{8}{9}\cos\omega_1 t + \dfrac{1}{9}\cos\omega_2 t$; $x_2 = \dfrac{4}{9}\cos\omega_1 t - \dfrac{1}{9}\cos\omega_2 t$

5-30 shear ratio 1st/2nd story = 2.0

5-34 $(\omega/\omega_h)_2 = 2.73$, $(Y_1/Y_0)_2 = -0.74$

5-36 $V_1 = 43.3$ ft/s, $V_2 = 60.3$ ft/s

5-41 $d_2 = 1/2''$

5-43 $w = 11.4$ lb, $k = 17.9$ lb/in

5-45 $\zeta_0 = 0.105$, $\omega/\omega_n = 0.943$

5-48 $y = \left(\dfrac{4}{3\times 81}\dfrac{l^3}{EI}\right)F_0 + \left(\dfrac{4}{9\times 18}\dfrac{l^2}{EI}\right)M_0$ $F_0 = m\omega^2 y$

   $\theta = \left(\dfrac{2}{81}\dfrac{l^2}{EI}\right)F_0 + \left(\dfrac{l}{9EI}\right)M_0$ $M_0 = (J_p - J_d)\omega_1\omega\theta$

5-52 $\begin{bmatrix} J_1 & 0 \\ 0 & J_2 \end{bmatrix}\begin{Bmatrix} \ddot{\theta}_1 \\ \ddot{\theta}_2 \end{Bmatrix} + l^2\begin{bmatrix} \dfrac{9}{16}k_1 & -\dfrac{9}{16}k_1 \\ -\dfrac{9}{16}k_1 & \left(\dfrac{9}{16}k_1 + \dfrac{1}{4}k_2\right) \end{bmatrix}\begin{Bmatrix} \theta_1 \\ \theta_2 \end{Bmatrix} = \{0\}$, $J_1 = m_1\dfrac{l^2}{3}$
   $J_2 = \dfrac{7}{48}m_2 l^2$

CHAPTER 6

6-1 $a_{11} = \dfrac{k_2 + k_3}{\Sigma k_i k_j}$, $a_{21} = a_{12} = \dfrac{k_2}{\Sigma k_i k_j}$, $a_{22} = \dfrac{k_1 + k_2}{\Sigma k_i k_j}$

6-3 $a_{11} = 0.0114 \; l^3/EI$, $a_{21} = a_{12} = 0.0130 \; l^3/EI$, $a_{22} = 0.0192 \; l^3/EI$

6-6 $[K] = \begin{bmatrix} (K_1 + K_2) & -K_2 & 0 \\ -K_2 & (K_2 + K_3) & -K_3 \\ 0 & -K_3 & K_3 \end{bmatrix}$,

$[a] = \begin{bmatrix} 1/K_1 & 1/K_1 & 1/K_1 \\ 1/K_1 & (1/K_1 + 1/K_2) & (1/K_1 + 1/K_2) \\ 1/K_1 & (1/K_1 + 1/K_2) & (1/K_1 + 1/K_2 + 1/K_3) \end{bmatrix}$

6-7 $[a] = \dfrac{l^3}{EI} \begin{bmatrix} 7/96 & 1/8 \\ 1/8 & 5/6 \end{bmatrix}$

6-8 $[a] = \dfrac{l^3}{12EI} \begin{bmatrix} 1 & 1 & 1 & 1 \\ 1 & 2 & 2 & 2 \\ 1 & 2 & 3 & 3 \\ 1 & 2 & 3 & 4 \end{bmatrix}$

6-11 $\begin{Bmatrix} F_1 \\ M_2 \\ M_3 \end{Bmatrix} = \begin{bmatrix} 24EI_1/l_1^3 & -6EI_1/l_1^2 & -6EI_1/l_1^2 \\ -6EI_1/l_1^2 & (4EI_1/l_1 + 4EI_2/l_2) & 2EI_2/l_2 \\ -6EI_1/l_1^2 & 2EI_2/l_2 & (4EI_1/l_1 + 4EI_2/l_2) \end{bmatrix} \begin{Bmatrix} u_1 \\ \theta_1 \\ \theta_2 \end{Bmatrix}$

6-17 $P = \begin{bmatrix} 1.44/l & -8.40/l \\ 1.00 & 1.00 \end{bmatrix}$

6-20 $\tilde{P} = \begin{bmatrix} 0.207 & -1.208 \\ 0.293 & 1.707 \end{bmatrix}$

6-22 $\begin{bmatrix} (m_1 + m_2) & 0 & 0 \\ 0 & J_1 & 0 \\ 0 & 0 & J_2 \end{bmatrix} \begin{Bmatrix} \ddot{x} \\ -\ddot{\theta}_1 \\ \ddot{\theta}_2 \end{Bmatrix} + \dfrac{EI}{l} \begin{bmatrix} 6/l^2 & 3/l & -3/l \\ 3/l & 7 & 2 \\ -3/l & 2 & 7 \end{bmatrix} \begin{Bmatrix} x \\ -\theta_1 \\ \theta_2 \end{Bmatrix} = \{0\}$

$\qquad\qquad\qquad\qquad\quad F\rightarrow \qquad\quad M_1\curvearrowleft \qquad M_2\curvearrowright$

6-24 $[K] = k \begin{bmatrix} 2 & -1 \\ -1 & 2 \end{bmatrix}$, $[C] = c \begin{bmatrix} 2 & -1 \\ -1 & 1 \end{bmatrix}$ $\therefore$ not proportional

6-31 $\ddot{q}_2 + 0.8902 \, \zeta_2 \sqrt{\dfrac{k}{m}} \; \dot{q}_2 + 0.1981 \dfrac{k}{m} q_2 = 0.4068 \, \ddot{u}_0(t)$

$\ddot{q}_3 + 1.4614 \, \zeta_3 \sqrt{\dfrac{k}{m}} \; \dot{q}_3 + 0.5339 \dfrac{k}{m} q_3 = -0.3268 \, \ddot{u}_0(t)$

6-32 $\left| \dfrac{kx(10)}{ma_0} \right| = 1.90 + \sqrt{.610^2 + .369^2} = 2.61$

6-37 $\alpha = \dfrac{2\omega_1\omega_2(\zeta_1\omega_2 - \zeta_2\omega_1)}{\omega_2^2 - \omega_1^2}$, $\beta = \dfrac{2(\zeta_2\omega_2 - \zeta_1\omega_1)}{\omega_2^2 - \omega_1^2}$,

$\zeta_i = \dfrac{\alpha + \beta\omega_i^2}{2\omega_i}$, $\zeta_3 = 0.1867$

6-39 $C_1 = 0.8985$, $C_2 = -0.1477$, $C_3 = 0.3886$

CHAPTER 7

7-2 $f = \dfrac{n}{2l}\sqrt{\dfrac{T}{l}}$ $n = 1, 2, 3, \cdots$

7-3 $\tan\dfrac{\omega l}{c} = -\left(\dfrac{T}{kl}\right)\dfrac{\left(\dfrac{\omega l}{c}\right)}{1 - \left(\dfrac{\omega}{\omega_n}\right)^2}$, $\omega_n = \sqrt{\dfrac{k}{m}}$

7-5 4.792×10^3 m/s

7-15 $\omega_n = (2n - 1)\dfrac{\pi}{l}\sqrt{\dfrac{G}{\rho}}$, $n = 1, 2, 3, \cdots$

7-16 $\tan\dfrac{\omega l}{c} = \dfrac{2\left(\dfrac{J_0}{J_s}\dfrac{\omega l}{c}\right)}{\left(\dfrac{J_0}{J_s}\dfrac{\omega l}{c}\right)^2 - 1}$

7-20 $E = 3.48 \times 10^6$ lb/in^2

7-23 $\omega = \beta^2\sqrt{EI/\rho}$ where β is determined from

$(1 + \cosh\beta l \cdot \cos\beta l) = \beta l\dfrac{W_0}{W_b}(\sinh\beta l \cdot \cos\beta l - \cosh\beta l \cdot \sin\beta l)$

7-32 $\dfrac{EI}{h^2}(y_3 - 2y_2 + y_1) = R_1 h - \dfrac{wh^2}{2}$, $R_1 = $ left reaction

7-34 $\dfrac{EI}{h^2}(y_{10} - 2y_9 + y_8) = R_1(8h) - \dfrac{w}{2}(8h)^2 + R_6(3h)$

$\dfrac{EI}{h^2}(y_{11} - 2y_{10} + y_9) = 0$

CHAPTER 8

8-1 $\tan\theta = \left(\dfrac{l_1}{l_2}\right)^2$

8-3 $\tan\theta = \left(\dfrac{m_2 - m_1}{m_1 + m_2}\right)\dfrac{l}{\sqrt{(2R)^2 + l^2}}$

8-5 $\tan \theta = \dfrac{1}{\mu}$

8-7 $\sin \theta = \dfrac{3}{4}\dfrac{h}{l} - \dfrac{mg}{2kl}$

8-9 $\ddot{\theta}_{\sim} + \dfrac{3}{2}g\left(\dfrac{l_2^2 \cos \theta_0 + l_1^2 \sin \theta_0}{l_1^3 + l_2^3}\right)\theta_{\sim} = 0$

where $\tan \theta_0 = (l_1/l_2)^2$

8-10 $(m_1 + m_2)R^2\,\ddot{\theta}_{\sim} + g\Big[(m_1 + m_2)\sqrt{R^2 + (l/2)^2}\,\cos\theta_0$

$$+ (m_2 - m_1)\dfrac{l}{2}\sin\theta_0\Big]\theta_{\sim} = 0$$

8-13 $m_0(\ddot{r} - r\dot{\theta}^2) + k(r - r_0) = m_0 g \cos \theta$

$m_0 r(r\ddot{\theta} + 2\dot{r}\,\dot{\theta}) + m_{\text{rod}}\dfrac{l^2}{3}\ddot{\theta} + m_0 g(r - r_0)\sin\theta + m_{\text{rod}}g\dfrac{l}{2}\sin\theta = 0$

8-17 $[J_1 + (m_1 + m_2)4l^2]\ddot{q}_1 + [K + l^2(k_1 + 4k_2)]q_1 + 4l^2 k_2 q_2 = 0$

$J_2\ddot{q}_2 + 4l^2 k_2(q_1 + q_2) = 0$

8-19 $[k] = \dfrac{EI}{l^3}\begin{bmatrix} 20.43 & -5.25\,l \\ -5.25\,l & 7.0\,l^2 \end{bmatrix}$

8-21 $R = \dfrac{P(l_1^3/3 + l_1^2 l_2/2) + M(l_1^2/2 + l_1 l_2)}{(l_1^3/3 + 2l_2^3/3 + l_1^2 l_2 + l_1 l_2^2)}$

$M_1 = R(l_1 + l_2) - Pl_1 - M$

CHAPTER 9

9-2 $\omega_1 = 4.63\sqrt{\dfrac{EI}{Ml^3}}$

9-3 $\omega_1 = 1.62\sqrt{\dfrac{EI}{Ml^3}}$

9-9 $\omega_1^2 = \dfrac{3EI/l^3 + 2k}{\dfrac{33}{140}ml + \dfrac{2}{3}m_0}$

9-11 $\omega_1 = 9.96\sqrt{\dfrac{EI}{(0.2188 m_0 l)l^3}}$ where $0.2188 m_0 l$ = total mass

9-12 $\omega_1^2 = \left(\dfrac{3EI}{l^3}\right)\dfrac{1}{27m_1 + 8m_2 + m_3}$

9-15 $f_{11} = 495.2$ cps

9-16 $\left(\dfrac{1}{6} - \dfrac{1}{\pi^2}\right)(ml\omega^2)^2 - \left(\dfrac{\pi^4}{6}\dfrac{EI}{l^3} + \dfrac{k}{2}\right)(ml\omega^2) + \dfrac{\pi^4}{2}\dfrac{EI}{l^3}k = 0$

9-18 $\left(\dfrac{ml}{2}\omega^2\right)^2 - \left[5lEA\left(\dfrac{\pi}{2l}\right)^2 + 2k_0\right]\left(\dfrac{ml}{2}\omega^2\right) + \left[\dfrac{lEA}{2}\left(\dfrac{\pi}{2l}\right)^2 + k_0\right] \cdot$

$\left[\dfrac{lEA}{2}\left(\dfrac{3\pi}{2l}\right)^2 + k_0\right] - k_0^2 = 0$

9-23 $\omega_1 = 0.584\sqrt{\dfrac{k}{m}}$, $\omega_2 = 1.200\sqrt{\dfrac{k}{m}}$, $\omega_3 = 1.642\sqrt{\dfrac{k}{m}}$

CHAPTER 10

10-2 $\omega_1 = 0.629\sqrt{\dfrac{K}{J}}$ $\left\{\begin{matrix}\theta_1\\\theta_2\\\theta_3\end{matrix}\right\}_1 = \left\{\begin{matrix}1.000\\0.604\\0.287\end{matrix}\right\}$

10-4 $\omega_1 = 0.445\sqrt{\dfrac{k}{m}}$ $\left\{\begin{matrix}x_1\\x_2\\x_3\end{matrix}\right\}_1 = \left\{\begin{matrix}1.000\\0.802\\0.445\end{matrix}\right\}$

$\omega_2 = 1.247\sqrt{\dfrac{k}{m}}$ $\left\{\begin{matrix}x_1\\x_2\\x_3\end{matrix}\right\}_2 = \left\{\begin{matrix}-1.000\\0.555\\1.247\end{matrix}\right\}$

$\omega_3 = 1.802\sqrt{\dfrac{k}{m}}$ $\left\{\begin{matrix}x_1\\x_2\\x_3\end{matrix}\right\}_3 = \left\{\begin{matrix}1.000\\-2.247\\1.802\end{matrix}\right\}$

10-10 $\omega_1 = \sqrt{\dfrac{6EI}{Ml^3}\left(1 + \dfrac{n}{2}\right)}$ $y_1/y_2 = -n/2$

10-11 $\omega_1 = 1.651\sqrt{\dfrac{EI}{ml^3}}$ $\left\{\begin{matrix}y_1\\y_2\end{matrix}\right\}_1 = \left\{\begin{matrix}0.320\\1.000\end{matrix}\right\}$

10-13 $u_{43} - \dfrac{u_{41}u_{23}}{u_{21}} = 0$

10-16 $u_{44} - \dfrac{u_{43}u_{24}}{u_{23}} = 0$

10-22 $\omega = 22.7$

10-23 $\omega_1 = 22.5$ $\omega_2 = 52.3$

10-26 $\omega_1 = 101.2$ $\omega_2 = 1836$

10-27 $\omega_1 = 0.5375\sqrt{\dfrac{K}{J}}$ $\omega_2 = 1.805\sqrt{\dfrac{K}{J}}$

$\left\{\begin{matrix}\theta_1\\\theta_2\\\theta_3\\\theta_4\end{matrix}\right\}_1 = \left\{\begin{matrix}1.000\\0.714\\0.239\\-0.326\end{matrix}\right\}$ $\left\{\begin{matrix}\theta_1\\\theta_2\\\theta_3\\\theta_4\end{matrix}\right\}\left\{\begin{matrix}1.000\\-2.270\\1.870\\-0.101\end{matrix}\right\}$

10-34 $\begin{vmatrix} u_{21} & u_{23} \\ u_{41} & u_{43} \end{vmatrix} = 0$

10-36 $\begin{vmatrix} u_{31} & u_{33} \\ u_{41} & u_{43} \end{vmatrix} = 0$

10-42 $\omega_k = 2\sqrt{\dfrac{K}{J}} \ \sin \dfrac{(2k-1)\pi}{2(2N+1)}$

10-44 $\omega_n = 2\sqrt{\dfrac{k}{m}} \ \sin \dfrac{n\pi}{2(N+1)}$

10-47 $-2\cos\beta\left(N + \dfrac{1}{2}\right) \cdot \sin\beta/2 = \dfrac{K_N}{k}\sin\beta N$

CHAPTER 11

11-3 $\Gamma_i = \dfrac{p_0}{l}\displaystyle\int_0^l \phi_i(x)\,dx$

11-8 $y(x,t) = \dfrac{4p_0 l}{\pi M \omega_2^2} \sin\dfrac{2x\pi}{l}(1 - \cos\omega_2 t)$

11-10 Modes absent are 2nd, 5th, 8th, etc.

11-11 $\Gamma = \sqrt{2}\cos(2n-1)\pi/6,\ D_n = (1 - \cos\omega_n t)$

$$u = \frac{2F_0 l}{AE}\left[\frac{\cos(\pi/6)\cos(\pi/2)(x/l)}{(\pi/2)^2}D_1 + \frac{\cos(5\pi/6)\cos(5\pi/2)(x/l)}{(5\pi/2)^2}D_2 + \cdots \right]$$

11-14 $\Gamma_1 = \dfrac{1}{l}\displaystyle\int_0^l \phi_1(x)\,dx = 0.784$

$\Gamma_2 = \dfrac{1}{l}\displaystyle\int_0^l \phi_2(x)\,dx = 0.434$

$\Gamma_3 = \dfrac{1}{l}\displaystyle\int_0^l \phi_3(x)\,dx = 0.254$

11-19 $\left\{ 1 + \dfrac{K\varphi_2'^2(0)}{M\omega_1^2\left[1 - (\omega/\omega_1)^2\right]} \right\}\left\{ 1 + \dfrac{K\varphi_2'^2(0)}{M\omega_2^2\left[1 - (\omega/\omega_2)^2\right]} \right\}$

$$= \left\{ \frac{K\varphi_1'(0)\varphi_2'(0)}{M\omega_1^2\left[1 - (\omega/\omega_1)^2\right]} \right\}\left\{ \frac{K\varphi_1'(0)\varphi_2'(0)}{M\omega_2^2\left[1 - (\omega/\omega_2)^2\right]} \right\}$$

$\varphi_1 = \sqrt{2}\sin\dfrac{\pi x}{l}, \qquad \varphi_1' = \dfrac{\pi}{l}\sqrt{2}\cos\dfrac{\pi x}{l}$, etc.

One mode approximation gives

$$\left(\frac{\omega}{\omega_1}\right)^2 = 1 + \frac{2K}{M\omega_1^2}\left(\frac{\pi}{l}\right)^2, \qquad \omega_1 = \pi^2\sqrt{\frac{EI}{Ml^3}}$$

11-20 One mode approximation

$$\left(\frac{\omega}{\omega_1}\right)^2 = 1 + \frac{4K}{M\omega_1^2}\left(\frac{\pi}{l}\right)^2$$

11-21   Using one free-free mode and translation mode of M_0

$$\left(\frac{\omega}{\omega_1}\right)^2 = \frac{M_1}{M_1 + M_0 \varphi_1^2(0) - \left[M_0^2 \varphi_1^2(0)/(M_0 + 2ml) \right]}$$

where $M_1 = \int \varphi_1^2(x) m \, dx = 2ml$, $\omega_1 = 22.4 \sqrt{\dfrac{EI}{m(2l)^4}}$

CHAPTER 12

12-2 $m\ddot{x} + \dfrac{2T_0}{l_0} x \left[1 + \dfrac{1}{2} \left(\dfrac{EA}{T_0} - 1 \right) \left(\dfrac{x}{l} \right)^2 \right] = 0$

12-3 $m\ddot{x} + \dfrac{\pi r_0^2 \rho}{3(h - x_0)^2} [(h - x)^3 - (h - x_0)^3]$

r_0 = radius of circle at water line

$\rho = wt/\text{vol of water}$

12-8 $V = \sqrt{y^2 + \omega_n^4 x^2}$ $x = y = 0$ for equilibrium

12-11   Shift origin of phase plane to π in Fig. 11.5-2

12-13 $\lambda_{1,2} = 3, 4$ $\left\{ \begin{matrix} x \\ y \end{matrix} \right\}_1 = \left\{ \begin{matrix} .50 \\ 1.00 \end{matrix} \right\}_1$, $\left\{ \begin{matrix} x \\ y \end{matrix} \right\}_2 = \left\{ \begin{matrix} 1 \\ 1 \end{matrix} \right\}$

12-14 $\left\{ \begin{matrix} x \\ y \end{matrix} \right\} = \left[\begin{matrix} 0.5 & 1 \\ 1 & 1 \end{matrix} \right] \left\{ \begin{matrix} \xi \\ \eta \end{matrix} \right\}$

12-25 $y = \dfrac{-x(\omega_n^2 + \mu \, x^2)}{c}$ $c = \dfrac{dy}{dx}$

12-27 $\dfrac{dy}{dx} = \dfrac{-(x + \delta)}{y}$ where $\delta = \left(\dfrac{c}{m} y + \dfrac{\mu}{\omega_n^2 m} x^3 \right)$

$$\omega_n^2 = k/m$$

$$\tau = \omega_n t, \qquad y = \dfrac{dx}{d\tau}$$

12-30 $\tau = 4 \sqrt{\dfrac{l}{g}} \displaystyle\int_0^{60°} \dfrac{d\theta}{\sqrt{2(\cos\theta - \cos\theta_0)}} = 4 \sqrt{\dfrac{l}{g}} \displaystyle\int_0^{\pi/2} \dfrac{d\theta}{\sqrt{1 - k^2 \sin^2 \phi}}$

where $k = \sin \dfrac{\theta_0}{2}$, $\sin \dfrac{\theta}{2} = k \sin \phi$

12-34

$$\tau \cong \sqrt{\dfrac{l}{g}} \left(1 + \dfrac{1}{16} \theta_0^2 \right)$$

12-35

12-36 $m\ddot{x}_1 + \left(\dfrac{2T_0}{l_0} \right) x_1 + \left(\dfrac{2K}{l_0^2} \right) x_1^3 = 0$, $T = T_0 + K \dfrac{x^2}{l_0}$

$m\ddot{x}_2 + \left(\dfrac{2T_0}{l_0} \right) x_2 + 3\alpha \, x_1^2 x_2 = 0$, $\alpha = \dfrac{2K}{l_0^2}$

$m\ddot{x}_2 + \left[\left(\dfrac{2T_0}{l_0} + \dfrac{3}{2} \alpha \, A^2 \right) + \dfrac{3}{2} \alpha \, A^2 \cos 2\omega_n t \right] x_2 = 0$

CHAPTER 13

13-5 $\bar{x} = 0.50, \qquad \overline{x^2} = 0.333$

13-6 $\bar{x} = A_0, \qquad \overline{x^2} = A_0^2 + \dfrac{1}{2}A_1^2$

13-14 A triangle of twice the base, symmetric about $t = 0$

13-15 $R(\tau) = 5$ at $\tau = 0$ and linearly decrease to $R(1) = 1$.

13-18 RMS $= 53.85g = 528.3 \ m/s^2$

13-21 RMS $= 1.99$ in, $\qquad \sigma = 0.9798$

13-24 $f(t) = \dfrac{2}{\pi}\left[\sin \omega_1 t - \dfrac{1}{2}\sin 2\omega_1 t + \dfrac{1}{3}\sin 3\omega_1 t - \cdots \right]$

$s(\omega) = \sum_n \dfrac{1}{2}C_n^2 = \dfrac{2}{\pi^2}\sum_n \dfrac{1}{n^2}$

13-26 $x(t) = \dfrac{2A}{\pi} \sum\limits_{n=-\infty}^{\infty} \dfrac{1}{in}e^{in\omega_1 t} \qquad n - \text{odd}$

$\qquad = \dfrac{4A}{\pi}\left[\sin \omega_1 t + \dfrac{1}{3}\sin 3\omega_1 t + \dfrac{1}{5}\sin 5\omega_1 t + \cdots \right]$

$S(\omega_n) = \dfrac{C_n C_n^*}{2} = \dfrac{4A^2}{n^2\pi^2}$

13-27 $f_{1,2} = f_n\left(1 \mp \dfrac{1}{2Q} \right)$

13-31 $53.2g = 2\sigma, \qquad P[x > 2\sigma] = \ 4.6\%$
$\qquad\qquad\qquad\qquad P[X > 2\sigma] = 13.5\%$

13-32 $F(t) = 10^6\left[\dfrac{1}{4} + \sum\limits_{n=1}^{\infty}\dfrac{2}{n\pi}\sin\dfrac{n\pi}{4}\cdot\cos n\dfrac{2\pi}{T}t \right]$

$S_F(\omega_n) = \dfrac{1}{2}\times 10^{12}\left[\left(\dfrac{1}{16}\right)_{n=0} + \sum\limits_{n=1}^{\infty}\left(\dfrac{2}{n\pi}\right)^2 \sin^2\dfrac{n\pi}{4} \right]$

$\overline{y^2} = \sum \dfrac{1}{k^2}\dfrac{S_F(\omega_n)}{\left[1 - \left(\dfrac{n\omega_1}{2\pi}T\right)^2\right]^2}, \qquad k = \left(\dfrac{2\pi}{T}\right)^2 m$

13-35 $\overline{y^2} = \sigma^2 \cong \dfrac{S_0}{k^2}\dfrac{f_n\pi}{4\zeta} = .00438, \qquad \sigma = .0662m$

$P[|y| > 0.132 = 2\sigma] = 4.6\%$

13-36 $\sigma = .0039m \qquad P[|y| > 0.012] = 0.3\%$

INDEX

A

Absorber, vibration, 149, 151
Accelerometer, 79, 81
Accelerometer error, 82
Adjoint matrix, 184, 456
Amplitude:
  complex, 5
  of forced vibration, 50
  of normal modes, 134, 199
  relative, 63
  resonant, 51
Analog simulation, 390
Aperiodic motion, 29
Arbitrary excitation, 94
Argand diagram, 4
Aurocorrelation, analyzer, 410, 412
  of random function, 411
  of sine wave, 412
Autonomous system, 318
Average Delta method, 380
Average value, 8, 402

B

Balancing:
  disks, 55
  dynamic, 55
  long rotor, 57
  static, 55
Barton, M.V., 435
Base excitation, 95
Bathe, K.S., 123
Beam vibration, 218, 270
  centrifugal effect, 310
  coupled-flexure torsion, 311
  on elastic foundation, 228
  flexure formula, 270
  influence coefficients, 177
  lumped mass, 275
  mode summation, 341
  natural frequency table, 220
  orthogonality, 345
  Rayleigh method, 21, 268
  rotary inertia and shear, 221, 346

Beat phenomena, 137
Bellman, R., 394
Bendat, J.S., 418, 435
Bifilar suspension, 170
Bilinear hysteresis, 390
Blackman, R.B., 435
Branched torsional system, 320
Brock, J.E., 394
Building vibration, 163, 168, 255, 198, 280
Butenin, N.V., 394

C

Caughey, T.K., 390
Centrifugal pendulum, 151
Characteristic equation, 133
Cholesky's inversion, 462
Circular frequency, 3
Clarkson, B.L., 435
Complex algebra, 5
Complex stiffness, 75
Component mode synthesis, 355
Computer flow diagrams, 112, 117, 146, 301, 307
Computer program, beam vibration, 308
Computer program, torsional system, 300
Concentrated mass, frequency effect, 280, 349
Conjugate complex quantities, 5, 405
Conservative nonlinear system, 369
Conservation of energy, 18
Consistent mass, 258
Consistent stiffness, 267
Constrained structures, 347
Constraint equations, 239, 243
Continuous spectrum, 415
Convolution integral, 95
Coordinate coupling, 139
Coordinate transformation, 195
Correlation, 410
Coulomb damping, 34, 74
Coulomb friction, 74, 388

Coupled pendulum, 136
Coupling:
 dynamic, static, 141
Cramer's rule, 182
Crandall, S.H., 123, 435
Crank mechanism, 12
Crede, C.E., 66
Critical damping, 26, 29
Critical speed, 61
Cross correlation, 413
Cross spectral density, 426
Cumulative probability, 405
Cunningham, W.J., 394

D

D'Alembert's principle, 246
Damped vibration, 25
Damper, vibration:
 Lanchester, torsional, 153
 untuned, 154
Damping, 25
 Coulomb, 34
 critical, 26, 29
 energy dissipated by, 69
 equivalent viscous, 72
 ratio, factor, 27
 Rayleigh, 197
 solid, structural, 74
 viscous, 25
Davis, H.T., 394
Decay of oscillation, 31
Decibel, 9
Decoupling of equations, 195
Decrement, logarithmic, 30
Degrees of freedom, 2
Delta function, 93
Delta phase plane method, 377
Den Hartog, J.P., 154
Determinant, 452, 453
Diagonalization of matrices, 194
Difference equation, 328
Digital computer programs:
 beam vibration, 308
 damped systems, 116

Digital computer programs (*Contd.*)
  finite difference, 110
  initial conditions, 114
  Runge-Kutta, 119, 229
  two DOF system, 147
Diode limiting circuit, 390
Discrete spectrum, 415
Drop test, 102
Duffing's equation, 383
Dunkerley's equation, 276
Dynamic absorber, 149, 151
Dynamic coupling, 141
Dynamic load factor, 342
Dynamic unbalance, 56

E

Effective mass:
  of beams, 24
  of levers, 23
  of springs, 22
Eigenvalues, eigenvectors, 183, 185
Elastic energy, 248
Energy dissipated by damping, 68
Energy method, 18
Ensemble of random functions, 401
Equilibrium state, 370, 372
Equivalent viscous damping, 72
  of viscoelastic system, 326
Ergodic process, 401
Euler beam equation, 218
Excitation:
  arbitrary, 94
  impulsive, 93
  step, 96
Expected value, 402

F

Felgar, R.P., 465
Field matrix, 314
Finite difference:
  beams, table, 225, 226
  second order equations, 111

Flexibility matrix, 175
Flexure formula for beams, 270
Flexure-torsion vibration, 311
Flow diagrams, 112, 117, 146, 301, 307
Force:
  rotational, 53
  transmitted, 64
Forced harmonic motion, 143
  1 DOF, 49
  2 DOF, 145
  multi DOF, 196, 197
Forced normal modes, 197
Forced vibration, 48
  matrix notation, 144
  peak resonant amplitude, 52
  vectors, phase, 49
Fortran program, 118, 147, 300, 308
Fourier series, 6, 402
Fourier spectrum, 8
Fourier transforms, 421
Framed structure vibration, 256
Free vibration:
  damped, 25
  undamped, 13
Frequency:
  damped oscillation, 28
  higher modes, 287
  natural, 14
  peak amplitude, 85
  resonant, 52
  response function, 427
  spectrum, 423

G

Gaussian distribution, 407
Geared system, 319
Generalized:
  coordinates, 238
  force, 249
  mass, stiffness, 248
Gyroscopic effect, 158

H

Half power points, 76
Harmonic analysis, 6
Harmonic motion, 2
Hayashi, C., 394
Higher mode matrix iteration, 287
Holonomic constraint, 239
Holzer computer program, 300
Holzer method, 296
Houdaille damper, 155
Hurty, W.C., 356
Hysteresis damping, 70, 390

I

Impulse, 92
Inertia unbalance, 57
Influence coefficients, 174
Initial conditions, 15, 28, 93, 189
Instruments, vibration measuring, 78
Integrating method for beams, 272
Inversion, Laplace transform, 447, 448
Inversion of matrices, 184, 195, 457
Isoclines, 375
Isolation of vibration, 64
Iteration, matrix method, 383

J

Jacobsen, L.S., 377
Jump phenomena, 385

K

Kimball, A.L., 74
Kinetic energy of vibration, 18, 247

L

Lagrange's equation, 253
Lagrangian, 253
Lanchester damper, 153
Laplace transform, 100, 446–51

Lazan, B.S., 74
Leckie, F.A., 313
Levers, 23
Limit cycle, 376, 392
Linear systems, definition, 1
Logarithmic decrement, 30
Longitudinal vibration:
  of missiles, 350
  of rods, 212
  of triangular plates, 283
Loss coefficient, 70
Lumped mass beams, 304

M

Malkin, I.G., 394
Mass addition, natural frequency
    effect, 278, 280, 349
Mathieu equation, 382
Matrices, 454–56
Matrix iteration, higher modes, 285, 287
Mean square value, 9
Mean value, 8
Membrane, 223, 236
Mindlin, R.D., 102
Minorsky, N., 394
Modal damping, 196
Modal matrix, 192
Modal matrix, weighted, 194
Mode:
  acceleration method, 353
  orthogonality of, 189
  participation factor, 342
  summation method, 340
Myklestad, N.O.:
  coupled flexure-torsion, 311
  method for beams, 304
  rotating beams, 310

N

Narrow-band spectral density, 417
Natural frequency:
  of beams, table, 220

Natural frequency (*Contd.*)
  membranes, 223, 236
  rods, 214, 216
  strings, 212
Nishikawa, Y., 394
Node position, 136, 142, 299, 309
Nonlinear differential equation, 367
Normal coordinates, 195
Normal modes, 132
  of beams, 220
  of constrained structures, 347
  of coupled systems, 135, 139
  summation of, 195
  of torsional systems, 299
Nth power of a matrix, 327

O

Octave, 10
Orthogonality, 188
  with rotary inertia and shear, 345
Orthogonal matrix, 456

P

Parseval's theorem, 424
Partial fractions, 144
Partitioned matrices, 156, 260, 336,
      337, 358
Peak value, 8, 199
Pendulum:
  absorber, 151
  bifilar, 170
  coupled, 136
  nonlinear oscillation, 375, 395,
      396, 399
  torsional, 17
Periodic motion, response, 5, 77
Period of vibration, 3
  of nonlinear system, 381
Perturbation method, 380
Pestel, E.C., 313
Phase, 4, 7
  distortion, 83

Phase (*Contd.*)
  of harmonic motion, 4, 49
  plane, 368
Piezoelectric instruments, 83
Point matrix, 314
Popov, E.P., 176
Potential energy:
  of beams, 270
  of nonlinear system, 370
Power, 71
Power spectral density, analyzer,
      414, 418
Principal coordinates, 195
Probability:
  cumulative density, 405
  density, 405
  distribution, 404
  of instantaneous value, 406
  of peak values, 408
Proportional damping, 196
Pseudo response spectrum, 107

Q

Q-sharpness of resonance, 76

R

Ralston, A., 123
Random time function, 401
Rauscher, M., 394
Rayleigh:
  damping, 197
  distribution, 406
  method, 21, 268
  –Ritz method, 281
Reciprocity, 182
Relative amplitude, 63
Repeated:
  impulse, 98
  roots, 190
  structures, 325, 328
Resonance, 2
Response spectrum, 103, 109

Rice, S.O., 409, 435
Robson, J.D., 435
Rod:
    longitudinal vibration, 213
    torsional vibration, 215
Root locus damping, 27
Root mean square, 9
Rotary inertia, 221
Rotating beam, 310
Rotating shaft, 58, 161
Rotating unbalance, 58
Rotational motion, 17
Rotor balancing, 55
Runge-Kutta computation:
    for beams, 229
    for nonlinear equations, 391
    for second order equations, 119

S

Salvadori, M.G., 123
Seismic instruments, 79
Seismometer, 79, 80
Self-excited oscillation, 388
Sensitivity of instruments, 83
Separatrices, 371
Shaft vibration, 58, 158
Sharpness of resonance, 76
Shear deformation of beams, 221
Shock response spectra:
    drop test, 103
    rectangular pulse, 105
    sine pulse, 105
    step-ramp, 105
    triangular pulse, 128
Side bands, 76
Singular points, 368
Solid damping, 74
Specific damping, 70
Spectral density, analyzer, 414, 418
Spring constraint, 348
Springs, table of stiffness, 35
Stability of equilibrium, 372

Stability of oscillation, 370
Standard deviation, 402
State vector, 313
Static balance, 55
    coupling, 139
    deflection, natural frequency, 14
Stationary process, 401
Step function, 96
Stiffness:
    matrix, for frames, 175, 179
    table for beam elements, 176
Stoker, J.J., 394
String vibration, 210
Structural damping, 74
Successive approximation, 383
Superfluous coordinates, 239
Superposition integral, 95
Support (base) motion, 62, 95
Sweeping matrix, 288
Synchronous whirl, 59, 161
System transfer function, 428

T

Timoshenko equation, 223
Thomson, W.T., 435
Torsional damper, 153
Torsional vibration:
    with damping, 316
    Holzer's method, 296
Trace of a matrix, 276, 454
Trajectory of phase plane, 368
Transducer, seismic, 80
Transfer matrices:
    of beams, 323
    with damping, 317
    flexure-torsion, 324
    lumped spring mass, 314
    repeated structures, 325
    torsional system, 316
Transient time function, 92
Transmissibility, 65
Transpose matrix, 455

Traveling waves, 210
Triangular pulse, digital solution, 115
Tukey, J.W., 435

U

Unbalance, vibration, 55
Unit step function, 96
Untuned viscous damper, 155

V

Van der Pol equation, 376
Variance, 402
Vectors, steady state vibration, 5, 49
Vehicle suspension, 87, 142
Velocity excitation of base, 97, 130
Vibration absorber, 149, 151
    bounds, 444

Vibration absorber (*Contd.*)
    damper, 153
    isolation, 64
    testing, mass effect, 277
Virtual work, 244, 246
Viscoelastic damping, 326
Viscous damping force, 25

W

Wave equation, 210
Wave velocity, 210
Whirling of shafts, 58, 161
Wide-band spectral density, 417
Wiener-Khintchine, 426
Work due to damping, 69

Y

Young, D., 465